Arteriogenesis, Angiogenesis and Vascular Remodeling

Arteriogenesis, Angiogenesis and Vascular Remodeling

Elisabeth Deindl
Paul Quax

Basel • Beijing • Wuhan • Barcelona • Belgrade • Novi Sad • Cluj • Manchester

Elisabeth Deindl
Walter-Brendel-Centre of Experimental Medicine
Ludwig-Maximilians-Universität Munich
Germany

Paul Quax
Einthoven Laboratory for Experimental Vascular Medicine
Leiden University
Leiden
Netherlands

Editorial Office
MDPI AG
Grosspeteranlage 5
4052 Basel, Switzerland

This is a reprint of articles from the Special Issue published online in the open access journal *International Journal of Molecular Sciences* (ISSN 1422-0067) (available at: www.mdpi.com/journal/ijms/special_issues/Angiogenesis_Remodeling).

For citation purposes, cite each article independently as indicated on the article page online and using the guide below:

Lastname, A.A.; Lastname, B.B. Article Title. *Journal Name* **Year**, *Volume Number*, Page Range.

ISBN 978-3-7258-2052-8 (Hbk)
ISBN 978-3-7258-2051-1 (PDF)
https://doi.org/10.3390/books978-3-7258-2051-1

Cover image courtesy of Larissa Deindl

© 2024 by the authors. Articles in this book are Open Access and distributed under the Creative Commons Attribution (CC BY) license. The book as a whole is distributed by MDPI under the terms and conditions of the Creative Commons Attribution-NonCommercial-NoDerivs (CC BY-NC-ND) license (https://creativecommons.org/licenses/by-nc-nd/4.0/).

Contents

About the Editors vii

Paul H. A. Quax and Elisabeth Deindl
The Intriguing World of Vascular Remodeling, Angiogenesis, and Arteriogenesis
Reprinted from: *Int. J. Mol. Sci.* **2024**, *25*, 6376, doi:10.3390/ijms25126376 1

Nathan K. P. Wong, Emma L. Solly, Richard Le, Victoria A. Nankivell, Jocelyne Mulangala and Peter J. Psaltis et al.
TRIM2 Selectively Regulates Inflammation-Driven Pathological Angiogenesis without Affecting Physiological Hypoxia-Mediated Angiogenesis
Reprinted from: *Int. J. Mol. Sci.* **2024**, *25*, 3343, doi:10.3390/ijms25063343 6

Judith A. H. M. Peeters, Hendrika A. B. Peters, Anique J. Videler, Jaap F. Hamming, Abbey Schepers and Paul H. A. Quax
Exploring the Effects of Human Bone Marrow-Derived Mononuclear Cells on Angiogenesis In Vitro
Reprinted from: *Int. J. Mol. Sci.* **2023**, *24*, 13822, doi:10.3390/ijms241813822 22

Malaika K. Motlana and Malebogo N. Ngoepe
Computational Fluid Dynamics (CFD) Model for Analysing the Role of Shear Stress in Angiogenesis in Rheumatoid Arthritis
Reprinted from: *Int. J. Mol. Sci.* **2023**, *24*, 7886, doi:10.3390/ijms24097886 35

Suzanne L. Laboyrie, Margreet R. de Vries, Roel Bijkerk and Joris I. Rotmans
Building a Scaffold for Arteriovenous Fistula Maturation: Unravelling the Role of the Extracellular Matrix
Reprinted from: *Int. J. Mol. Sci.* **2023**, *24*, 10825, doi:10.3390/ijms241310825 51

Dilara Z. Gatina, Ilnaz M. Gazizov, Margarita N. Zhuravleva, Svetlana S. Arkhipova, Maria A. Golubenko and Marina O. Gomzikova et al.
Induction of Angiogenesis by Genetically Modified Human Umbilical Cord Blood Mononuclear Cells
Reprinted from: *Int. J. Mol. Sci.* **2023**, *24*, 4396, doi:10.3390/ijms24054396 71

Mariam Anis, Janae Gonzales, Rachel Halstrom, Noman Baig, Cat Humpal and Regaina Demeritte et al.
Non-Muscle MLCK Contributes to Endothelial Cell Hyper-Proliferation through the ERK Pathway as a Mechanism for Vascular Remodeling in Pulmonary Hypertension
Reprinted from: *Int. J. Mol. Sci.* **2022**, *23*, 13641, doi:10.3390/ijms232113641 90

Philipp Götz, Sharon O. Azubuike-Osu, Anna Braumandl, Christoph Arnholdt, Matthias Kübler and Lisa Richter et al.
Cobra Venom Factor Boosts Arteriogenesis in Mice
Reprinted from: *Int. J. Mol. Sci.* **2022**, *23*, 8454, doi:10.3390/ijms23158454 104

Ritesh Urade, Yan-Hui Chiu, Chien-Chih Chiu and Chang-Yi Wu
Small GTPases and Their Regulators: A Leading Road toward Blood Vessel Development in Zebrafish
Reprinted from: *Int. J. Mol. Sci.* **2022**, *23*, 4991, doi:10.3390/ijms23094991 119

Ferdinand le Noble and Christian Kupatt
Interdependence of Angiogenesis and Arteriogenesis in Development and Disease
Reprinted from: *Int. J. Mol. Sci.* **2022**, *23*, 3879, doi:10.3390/ijms23073879 134

Tao Wang, Liang Yang, Mingjie Yuan, Charles R. Farber, Rosanne Spolski and Warren J. Leonard et al.
MicroRNA-30b Is Both Necessary and Sufficient for Interleukin-21 Receptor-Mediated Angiogenesis in Experimental Peripheral Arterial Disease
Reprinted from: *Int. J. Mol. Sci.* **2021**, *23*, 271, doi:10.3390/ijms23010271 . . . . **147**

About the Editors

Elisabeth Deindl

Elisabeth Deindl graduated at the ZMBH in Heidelberg, Germany, where she worked on hepatitis B viruses. Thereafter, she joined the lab of Wolfgang Schaper at the Max-Planck-Institute in Bad Nauheim, where she started to decipher the molecular mechanisms of arteriogenesis. After a short detour on stem cells, she focused again on arteriogenesis becoming a leading expert in the field. By using a peripheral model of arteriogenesis, she demonstrated that collateral artery growth is a matter of innate immunity and presents a blueprint of sterile inflammation, which is locally triggered by extracellular RNA.

Paul Quax

Paul Quax obtained his PhD at the University of Leiden, the Netherlands, on the role of plasminogen activators in tissue remodeling. He kept on working on this topic in relation to vascular remodeling, first at the Gaubius Laboratory TNO and later at the Leiden University Medical Center as professor in experimental vascular medicine. His interest in arteriogenesis was driven by the lack of therapeutic options for patients with peripheral arterial disease. Therapeutic arteriogenesis and angiogenesis induced by gene therapy, growth factors, modulation of inflammatory and immune response but also by modulation of microRNAs and other noncoding RNAs in small animal model are topics of his research.

Editorial

The Intriguing World of Vascular Remodeling, Angiogenesis, and Arteriogenesis

Paul H. A. Quax [1,*] and Elisabeth Deindl [2,3,*]

1   Einthoven Laboratory for Experimental Vascular Medicine, Department of Surgery, Leiden University Medical Center, 2300 RC Leiden, The Netherlands
2   Walter-Brendel-Centre of Experimental Medicine, University Hospital, Ludwig-Maximilians-Universität, 81377 Munich, Germany
3   Biomedical Center, Institute of Cardiovascular Physiology and Pathophysiology, Ludwig-Maximilians-Universität München, Planegg-Martinsried, 82152 Munich, Germany
*   Correspondence: p.h.a.quax@lumc.nl (P.H.A.Q.); elisabeth.deindl@med.uni-muenchen.de (E.D.); Tel.: +31-71-5261584 (P.H.A.Q.)

Citation: Quax, P.H.A.; Deindl, E. The Intriguing World of Vascular Remodeling, Angiogenesis, and Arteriogenesis. *Int. J. Mol. Sci.* **2024**, 25, 6376. https://doi.org/10.3390/ijms25126376

Received: 17 April 2024
Revised: 5 June 2024
Accepted: 7 June 2024
Published: 9 June 2024

Copyright: © 2024 by the authors. Licensee MDPI, Basel, Switzerland. This article is an open access article distributed under the terms and conditions of the Creative Commons Attribution (CC BY) license (https://creativecommons.org/licenses/by/4.0/).

Vascular remodeling is a very general feature related to angiogenesis and arteriogenesis, which are involved in neovascularization processes. Vascular remodeling is of course a crucial element in development [1,2], but also in pathological vascular remodeling processes such as atherosclerosis [3] and aneurysm formation.

In this Special Issue, many aspects of vascular remodeling are addressed. Zebrafish are an excellent model for studying blood vessel development [4,5]. The role of small GTPases in vascular development in zebrafish is addressed by Urade et al. [6], who describe the role of the small GTPase subfamilies in the development of the vascular system in zebrafish. Critical processes in blood vessel development in zebrafish relate to vascular endothelial growth factor (VEGF) signaling, Notch signaling, and bone morphogenetic protein (BMP) signaling [7]. Impaired small GTPases can contribute to vascular remodeling [8]. Despite this importance, their regulatory role in vascular development is unclear. This review discusses the role of the small GTPases along with their regulators in blood vessel development in zebrafish [6].

Another group of small regulators of vascular remodeling are microRNAs. microRNAs are regulators for multifactorial processes such as angiogenesis, arteriogenesis, and vascular remodeling [9,10]. Evidence on the regulation of vascular remodeling by various microRNAs is still growing. Here, Wang et al. [11] studied the role of microRNA-30b in the regulation angiogenesis. They show that miR-30b appears both necessary and sufficient for IL21/IL-21R-mediated angiogenesis and may present a new therapeutic option to treat peripheral arterial disease [12] if IL21R is not available for activation.

The stimulation of angiogenesis and neovascularization by different cells with a pro-angiogenic character remains very attractive from a therapeutic point of view [13,14]. Gatina et al. focused on umbilical-cord-derived mononuclear cells and genetically modified them using adenoviral constructs, adVEGF, adFGF2, and SDF1, and utilized adGFP as a control. They showed that these genetically modified UBC-MC were able to induce neovascularization in in vivo Matrigel plug tubule formations in mice [15].

Shear stress can generally be recognized as a crucial trigger in vascular remodeling processes [16]. In this Special Issue, Motlana et al. [17] describe how a computational fluid dynamics model can be used for analyzing the role of shear stress in angiogenesis in rheumatoid arthritis, as it represents an important factor [18]; they show that the magnitude of wall shear stress relates to the degree and extent of new blood vessel development.

Next to angiogenesis, arteriogenesis plays a crucial role in neovascularization [19,20]. However, the processes of angiogenesis and arteriogenesis, although different at regulatory levels [21,22], cannot be regarded as non-related processes. Le Noble and Kupatt show, very elegantly and convincingly, that these two processes, arteriogenesis and angiogenesis, are two interdependent processes in development and disease [23]. They describe the mechanisms and signals that contribute to the synchronized growth of micro- and macrovascular structures after ischemic challenges, as well as during development. They conclude that a long-term successful revascularization strategy should aim to both remove obstructions in the proximal part of the arterial tree and restore "bottom-up" vascular communication based on striking similarities between micro- and macrovascular structures.

The induction of arteriogenesis is a process that attracts a lot of attention from the clinical point of view in relation to peripheral arterial disease (PAD) patients [14,24], as good therapeutic options are still lacking. The link with the immune system [25–27] and the innate immune system, in particular [27–30], has raised a lot of attention with regard to new therapeutic options. In this Special Issue, Götz et al. provide some compelling data how cobra venom factor, known to be a very potent inhibitor of Complement Factor C3 [31], is able to boost arteriogenesis in mice after femoral artery ligation, demonstrating the role of the innate immune system in arteriogenesis [32]. Furthermore, the importance of the adaptive immune system in arteriogenesis is shown in the study by Kumaraswami et al. [33], who found an impaired arteriogenic response in mice deficient in Rag1, i.e., lacking both T cells and B cells. This is in line with previous work supporting the role of T and B cells [25,34,35]. In this Special Issue, Wong et al. study the role of tripartite-motif-containing protein 2 (TRIM2) in vascular remodeling and did not observe any effects on blood flow recovery in a hind limb ischemia model of TRIM2-deficient mice, despite the fact that TRIM2 knockdown in endothelial cells in vitro attenuated the inflammation-driven induction of critical angiogenic mediators.

Despite cell therapy being regarded as a promising approach for inducing neovascularization in PAD patients, many clinical trials using autologous bone marrow transplantation to induce neovascularization have failed or yielded ambiguous results [36,37]. In an attempt to unravel the effect of human-bone-marrow-derived mononuclear cells on angiogenesis, Peeters et al. studied the effect of human mononuclear cells on angiogenic responses of human cells in a large set of in vitro angiogenesis models. Despite their extensive studies, in which proliferation, migration, tube formation and aortic ring assays were studied, none of the conditions tested showed a pro-angiogenic effect of human-bone-marrow-derived cells [38].

Another very important aspect of vascular remodeling is the formation and maturation of arteriovenous fistulas (AVFs), required for vascular access in dialysis patients. It is well known that these AVFs fail very frequently [39,40]. In this Special Issue, Laboyrie et al. review the role of the extracellular matrix in AVF maturation. Moreover, they examine the effect of chronic kidney failure on vasculature remodeling and ECM changes post AVF surgery, and describe current ECM interventions to improve AVF maturation. Furthermore, they discuss the suitability of ECM interventions as a therapeutic target for AVF maturation [41].

Lastly, this Special Issue contains a paper by Anis et al. [42] that describes the vascular remodeling processes in pulmonary hypertension; more specifically, how non-muscle MLCK contributes to endothelial cell hyperproliferation in vascular remodeling in these pathological processes.

Conclusions

In conclusion, in this Special Issue we hope to have demonstrated the broad role of vascular remodeling in pathological and physiological processes, and we have illustrated some of the complex regulatory mechanisms involved. Moreover, we discuss several potential therapeutic targets for intervening in vascular remodeling.

Author Contributions: Conceptualization P.H.A.Q. and E.D.; writing—original draft preparation, P.H.A.Q. and E.D.; writing—review and editing, P.H.A.Q. and E.D. All authors have read and agreed to the published version of the manuscript.

Conflicts of Interest: The authors declare no conflicts of interest.

List of Contributions:

A. Urade, R.; Chiu, Y.-H.; Chiu, C.-C.; Wu, C.-Y. Small GTPases and Their Regulators: A Leading Road toward Blood Vessel Development in Zebrafish. *Int. J. Mol. Sci.* **2022**, *23*, 4991. https://doi.org/10.3390/ijms23094991

B. Wang, T.; Yang, L.; Yuan, M.; Farber, C.R.; Spolski, R.; Leonard, W.J.; Ganta, V.C.; Annex, B.H. MicroRNA-30b Is Both Necessary and Sufficient for Interleukin-21 Receptor-Mediated Angiogenesis in Experimental Peripheral Arterial Disease. *Int. J. Mol. Sci.* **2022**, *23*, 271. https://doi.org/10.3390/ijms23010271

C. Gatina, D.Z.; Gazizov, I.M.; Zhuravleva, M.N.; Arkhipova, S.S.; Golubenko, M.A.; Gomzikova, M.O.; Garanina, E.E.; Islamov, R.R.; Rizvanov, A.A.; Salafutdinov, I.I. Induction of Angiogenesis by Genetically Modified Human Umbilical Cord Blood Mono-nuclear Cells. *Int. J. Mol. Sci.* **2023**, *24*, 4396. https://doi.org/10.3390/ijms24054396

D. Motlana, M.K.; Ngoepe, M.N. Computational Fluid Dynamics (CFD) Model for Analysing the Role of Shear Stress in Angiogenesis in Rheumatoid Arthritis. *Int. J. Mol. Sci.* **2023**, *24*, 7886. https://doi.org/10.3390/ijms24097886

E. le Noble, F.; Kupatt, C. Interdependence of Angiogenesis and Arteriogenesis in Development and Disease. *Int. J. Mol. Sci.* **2022**, *23*, 3879. https://doi.org/10.3390/ijms23073879

F. Götz, P.; Azubuike-Osu, S.O.; Braumandl, A.; Arnholdt, C.; Kübler, M.; Richter, L.; Lasch, M.; Bobrowski, L.; Preissner, K.T.; Deindl, E. Cobra Venom Factor Boosts Arteriogenesis in Mice. *Int. J. Mol. Sci.* **2022**, *23*, 8454. https://doi.org/10.3390/ijms23158454

G. Wong, N.K.P.; Solly, E.L.; Le, R.; Nankivell, V.A.; Mulangala, J.; Psaltis, P.J.; Nicholls, S.J.; Ng, M.K.C.; Bursill, C.A.; Tan, J.T.M. TRIM2 Selectively Regulates Inflammation-Driven Pathological Angiogenesis without Affecting Physiological Hypoxia-Mediated Angiogenesis. *Int. J. Mol. Sci.* **2024**, *25*, 3343. https://doi.org/10.3390/ijms25063343

H. Peeters, J.A.H.M.; Peters, H.A.B.; Videler, A.J.; Hamming, J.F.; Schepers, A.; Quax, P.H.A. Exploring the Effects of Human Bone Marrow-Derived Mononuclear Cells on Angiogenesis In Vitro. *Int. J. Mol. Sci.* **2023**, *24*, 13822. https://doi.org/10.3390/ijms241813822

I. Laboyrie, S.L.; de Vries, M.R.; Bijkerk, R.; Rotmans, J.I. Building a Scaffold for Arteriovenous Fistula Maturation: Unravelling the Role of the Extracellular Matrix. *Int. J. Mol. Sci.* **2023**, *24*, 10825. https://doi.org/10.3390/ijms241310825

J. Anis, M.; Gonzales, J.; Halstrom, R.; Baig, N.; Humpal, C.; Demeritte, R.; Epshtein, Y.; Jacobson, J.R.; Fraidenburg, D.R. Non-Muscle MLCK Contributes to Endothelial Cell Hyper-Proliferation through the ERK Pathway as a Mechanism for Vascular Remodeling in Pulmonary Hypertension. *Int. J. Mol. Sci.* **2022**, *23*, 13641. https://doi.org/10.3390/ijms232113641

References

1. Eichmann, A.; Yuan, L.; Moyon, D.; Lenoble, F.; Pardanaud, L.; Breant, C. Vascular development: From precursor cells to branched arterial and venous networks. *Int. J. Dev. Biol.* **2005**, *49*, 259–267. [CrossRef]
2. Risau, W.; Flamme, I. Vasculogenesis. *Annu. Rev. Cell Dev. Biol.* **1995**, *11*, 73–91. [CrossRef]
3. Parma, L.; Baganha, F.; Quax, P.H.A.; de Vries, M.R. Plaque angiogenesis and intraplaque hemorrhage in atherosclerosis. *Eur. J. Pharmacol.* **2017**, *816*, 107–115. [CrossRef]
4. Schuermann, A.; Helker, C.S.; Herzog, W. Angiogenesis in zebrafish. *Semin. Cell Dev. Biol.* **2014**, *31*, 106–114. [CrossRef] [PubMed]
5. Ny, A.; Autiero, M.; Carmeliet, P. Zebrafish and Xenopus tadpoles: Small animal models to study angiogenesis and lymphangiogenesis. *Exp. Cell Res.* **2006**, *312*, 684–693. [CrossRef]
6. Urade, R.; Chiu, Y.-H.; Chiu, C.-C.; Wu, C.-Y. Small GTPases and Their Regulators: A Leading Road toward Blood Vessel Development in Zebrafish. *Int. J. Mol. Sci.* **2022**, *23*, 4991. [CrossRef] [PubMed]
7. Kim, J.D.; Lee, H.W.; Jin, S.W. Diversity is in my veins: Role of bone morphogenetic protein signaling during venous morphogenesis in zebrafish illustrates the heterogeneity within endothelial cells. *Arterioscler. Thromb. Vasc. Biol.* **2014**, *34*, 1838–1845. [CrossRef] [PubMed]
8. Bryan, B.A.; D'Amore, P.A. What tangled webs they weave: Rho-GTPase control of angiogenesis. *Cell Mol. Life Sci.* **2007**, *64*, 2053–2065. [CrossRef]
9. Welten, S.M.; Goossens, E.A.; Quax, P.H.A.; Nossent, A.Y. The multifactorial nature of microRNAs in vascular remodelling. *Cardiovasc. Res.* **2016**, *110*, 6–22. [CrossRef]
10. Bonauer, A.; Boon, R.A.; Dimmeler, S. Vascular microRNAs. *Curr. Drug Targets* **2010**, *11*, 943–949. [CrossRef]
11. Wang, T.; Yang, L.; Yuan, M.; Farber, C.R.; Spolski, R.; Leonard, W.J.; Ganta, V.C.; Annex, B.H. MicroRNA-30b Is Both Necessary and Sufficient for Interleukin-21 Receptor-Mediated Angiogenesis in Experimental Peripheral Arterial Disease. *Int. J. Mol. Sci.* **2022**, *23*, 271. [CrossRef] [PubMed]

12. Wang, T.; Cunningham, A.; Houston, K.; Sharma, A.M.; Chen, L.; Dokun, A.O.; Lye, R.J.; Spolski, R.; Leonard, W.J.; Annex, B.H. Endothelial interleukin-21 receptor up-regulation in peripheral artery disease. *Vasc. Med.* **2016**, *21*, 99–104. [CrossRef] [PubMed]
13. Silvestre, J.S.; Smadja, D.M.; Lévy, B.I. Postischemic revascularization: From cellular and molecular mechanisms to clinical applications. *Physiol. Rev.* **2013**, *93*, 743–802. [CrossRef]
14. Annex, B.H. Therapeutic angiogenesis for critical limb ischaemia. *Nat. Rev. Cardiol.* **2013**, *10*, 387–396. [CrossRef] [PubMed]
15. Gatina, D.Z.; Gazizov, I.M.; Zhuravleva, M.N.; Arkhipova, S.S.; Golubenko, M.A.; Gomzikova, M.O.; Garanina, E.E.; Islamov, R.R.; Rizvanov, A.A.; Salafutdinov, I.I. Induction of Angiogenesis by Genetically Modified Human Umbilical Cord Blood Mono-nuclear Cells. *Int. J. Mol. Sci.* **2023**, *24*, 4396. [CrossRef] [PubMed]
16. Souilhol, C.; Serbanovic-Canic, J.; Fragiadaki, M.; Chico, T.J.; Ridger, V.; Roddie, H.; Evans, P.C. Endothelial responses to shear stress in atherosclerosis: A novel role for developmental genes. *Nat. Rev. Cardiol.* **2020**, *17*, 52–63. [CrossRef]
17. Motlana, M.K.; Ngoepe, M.N. Computational Fluid Dynamics (CFD) Model for Analysing the Role of Shear Stress in Angiogenesis in Rheumatoid Arthritis. *Int. J. Mol. Sci.* **2023**, *24*, 7886. [CrossRef]
18. Elshabrawy, H.A.; Chen, Z.; Volin, M.V.; Ravella, S.; Virupannavar, S.; Shahrara, S. The pathogenic role of angiogenesis in rheumatoid arthritis. *Angiogenesis* **2015**, *18*, 433–448. [CrossRef]
19. Buschmann, I.; Schaper, W. Arteriogenesis Versus Angiogenesis: Two Mechanisms of Vessel Growth. *Physiology* **1999**, *14*, 121–125. [CrossRef]
20. Deindl, E.; Schaper, W. The Art of Arteriogenesis. *Cell Biochem. Biophys.* **2005**, *43*, 1–15. [CrossRef]
21. Germain, S.; Monnot, C.; Muller, L.; Eichmann, A. Hypoxia-driven angiogenesis: Role of tip cells and extracellular matrix scaffolding. *Curr. Opin. Hematol.* **2010**, *17*, 245–251. [CrossRef] [PubMed]
22. Silvestre, J.S.; Mallat, Z.; Tedgui, A.; Lévy, B.I. Post-ischaemic neovascularization and inflammation. *Cardiovasc. Res.* **2008**, *78*, 242–249. [CrossRef] [PubMed]
23. le Noble, F.; Kupatt, C. Interdependence of Angiogenesis and Arteriogenesis in Development and Disease. *Int. J. Mol. Sci.* **2022**, *23*, 3879. [CrossRef] [PubMed]
24. Cooke, J.P.; Losordo, D.W. Modulating the vascular response to limb ischemia: Angiogenic and cell therapies. *Circ. Res.* **2015**, *116*, 1561–1578. [CrossRef] [PubMed]
25. van Weel, V.; Toes, R.E.; Seghers, L.; Deckers, M.M.; de Vries, M.R.; Eilers, P.H.; Sipkens, J.; Schepers, A.; Eefting, D.; van Hinsbergh, V.W.M.; et al. Natural killer cells and CD4+ T-cells modulate collateral artery development. *Arterioscler. Thromb. Vasc. Biol.* **2007**, *27*, 2310–2318. [CrossRef] [PubMed]
26. Hellingman, A.A.; van der Vlugt, L.E.; Lijkwan, M.A.; Bastiaansen, A.J.; Sparwasser, T.; Smits, H.H.; Hamming, J.F.; Quax, P.H.A. A limited role for regulatory T cells in post-ischemic neovascularization. *J. Cell. Mol. Med.* **2012**, *16*, 328–336. [CrossRef] [PubMed]
27. la Sala, A.; Pontecorvo, L.; Agresta, A.; Rosano, G.; Stabile, E. Regulation of collateral blood vessel development by the innate and adaptive immune system. *Trends. Mol. Med.* **2012**, *18*, 494–501. [CrossRef] [PubMed]
28. Monaco, C. Innate immunity meets arteriogenesis: The versatility of toll-like receptors. *J. Mol. Cell. Cardiol.* **2011**, *50*, 9–12. [CrossRef] [PubMed]
29. Chillo, O.; Kleinert, E.C.; Lautz, T.; Lasch, M.; Pagel, J.I.; Heun, Y.; Troidl, K.; Fischer, S.; Caballero-Martinez, A.; Mauer, A.; et al. Perivascular Mast Cells Govern Shear Stress-Induced Arteriogenesis by Orchestrating Leukocyte Function. *Cell Rep.* **2016**, *16*, 2197–2207. [CrossRef]
30. Bot, I.; Velden, D.V.; Bouwman, M.; Kröner, M.J.; Kuiper, J.; Quax, P.H.A.; de Vries, M.R. Local Mast Cell Activation Promotes Neovascularization. *Cells* **2020**, *9*, 701. [CrossRef]
31. Vogel, C.W.; Fritzinger, D.C. Cobra venom factor: Structure, function, and humanization for therapeutic complement depletion. *Toxicon* **2010**, *56*, 1198–1222. [CrossRef] [PubMed]
32. Götz, P.; Azubuike-Osu, S.O.; Braumandl, A.; Arnholdt, C.; Kübler, M.; Richter, L.; Lasch, M.; Bobrowski, L.; Preissner, K.T.; Deindl, E. Cobra Venom Factor Boosts Arteriogenesis in Mice. *Int. J. Mol. Sci.* **2022**, *23*, 8454. [CrossRef] [PubMed]
33. Kumaraswami, K.; Arnholdt, C.; Deindl, E.; Lasch, M. Rag1 Deficiency Impairs Arteriogenesis in Mice. *Int. J. Mol. Sci.* **2023**, *24*, 12839. [CrossRef] [PubMed]
34. Simons, K.H.; Aref, Z.; Peters, H.A.B.; Welten, S.P.; Nossent, A.Y.; Jukema, J.W.; Hamming, J.F.; Arens, R.; de Vries, M.R.; Quax, P.H.A. The role of CD27-CD70-mediated T cell co-stimulation in vasculogenesis, arteriogenesis and angiogenesis. *Int. J. Cardiol.* **2018**, *260*, 184–190. [CrossRef] [PubMed]
35. Simons, K.H.; de Jong, A.; Jukema, J.W.; de Vries, M.R.; Arens, R.; Quax, P.H.A. T cell co-stimulation and co-inhibition in cardiovascular disease: A double-edged sword. *Nat. Rev. Cardiol.* **2019**, *16*, 325–343. [CrossRef] [PubMed]
36. Rigato, M.; Monami, M.; Fadini, G.P. Autologous Cell Therapy for Peripheral Arterial Disease: Systematic Review and Meta-Analysis of Randomized, Nonrandomized, and Noncontrolled Studies. *Circ. Res.* **2017**, *120*, 1326–1340. [CrossRef]
37. Annex, B.H.; Cooke, J.P. New Directions in Therapeutic Angiogenesis and Arteriogenesis in Peripheral Arterial Disease. *Circ. Res.* **2021**, *128*, 1944–1957. [CrossRef]
38. Peeters, J.A.H.M.; Peters, H.A.B.; Videler, A.J.; Hamming, J.F.; Schepers, A.; Quax, P.H.A. Exploring the Effects of Human Bone Marrow-Derived Mononuclear Cells on Angiogenesis In Vitro. *Int. J. Mol. Sci.* **2023**, *24*, 13822. [CrossRef]
39. Rothuizen, T.C.; Wong, C.; Quax, P.H.A.; van Zonneveld, A.J.; Rabelink, T.J.; Rotmans, J.I. Arteriovenous access failure: More than just intimal hyperplasia? *Nephrol. Dial. Transplant.* **2013**, *28*, 1085–1092. [CrossRef]

40. Shiu, Y.T.; Rotmans, J.I.; Geelhoed, W.J.; Pike, D.B.; Lee, T. Arteriovenous conduits for hemodialysis: How to better modulate thepathophysiological vascular response to optimize vascular access durability. *Am. J. Physiol. Renal. Physiol.* **2019**, *316*, F794–F806. [CrossRef]
41. Laboyrie, S.L.; de Vries, M.R.; Bijkerk, R.; Rotmans, J.I. Building a Scaffold for Arteriovenous Fistula Maturation: Unravelling the Role of the Extracellular Matrix. *Int. J. Mol. Sci.* **2023**, *24*, 10825. [CrossRef] [PubMed]
42. Anis, M.; Gonzales, J.; Halstrom, R.; Baig, N.; Humpal, C.; Demeritte, R.; Epshtein, Y.; Jacobson, J.R.; Fraidenburg, D.R. Non-Muscle MLCK Contributes to Endothelial Cell Hyper-Proliferation through the ERK Pathway as a Mechanism for Vascular Remodeling in Pulmonary Hypertension. *Int. J. Mol. Sci.* **2022**, *23*, 13641. [CrossRef] [PubMed]

Disclaimer/Publisher's Note: The statements, opinions and data contained in all publications are solely those of the individual author(s) and contributor(s) and not of MDPI and/or the editor(s). MDPI and/or the editor(s) disclaim responsibility for any injury to people or property resulting from any ideas, methods, instructions or products referred to in the content.

Article

TRIM2 Selectively Regulates Inflammation-Driven Pathological Angiogenesis without Affecting Physiological Hypoxia-Mediated Angiogenesis

Nathan K. P. Wong [1,2,3], Emma L. Solly [1,4], Richard Le [1,5], Victoria A. Nankivell [1,4], Jocelyne Mulangala [1,6], Peter J. Psaltis [1,4], Stephen J. Nicholls [7], Martin K. C. Ng [2,8], Christina A. Bursill [1,2,4,†] and Joanne T. M. Tan [1,2,4,*,†]

1. Vascular Research Centre, Lifelong Health Theme, South Australian Health and Medical Research Institute, Adelaide, SA 5000, Australia; nwon9940@alumni.sydney.edu.au (N.K.P.W.); emma.solly@sahmri.com (E.L.S.); richard.le@flinders.edu.au (R.L.); victoria.nankivell@sahmri.com (V.A.N.); jocelyne.mulangala@heartfoundation.org.au (J.M.); peter.psaltis@sahmri.com (P.J.P.); christina.bursill@sahmri.com (C.A.B.)
2. Faculty of Medicine and Health, The University of Sydney School of Medicine, Camperdown, NSW 2050, Australia; mkcng@med.usyd.edu.au
3. Department of Cardiology, St. Vincent's Hospital, Darlinghurst, NSW 2010, Australia
4. Adelaide Medical School, Faculty of Health and Medical Sciences, University of Adelaide, Adelaide, SA 5005, Australia
5. College of Medicine and Public Health, Flinders University, Adelaide, SA 5042, Australia
6. Heart Foundation, Brisbane, QLD 4000, Australia
7. Victorian Heart Institute, Monash University, Clayton, VIC 3800, Australia; stephen.nicholls@monash.edu
8. Department of Cardiology, Royal Prince Alfred Hospital, Camperdown, NSW 2050, Australia
* Correspondence: joanne.tan@sahmri.com; Tel.: +61 8 8128 4789
† These authors contributed equally to this work.

Citation: Wong, N.K.P.; Solly, E.L.; Le, R.; Nankivell, V.A.; Mulangala, J.; Psaltis, P.J.; Nicholls, S.J.; Ng, M.K.C.; Bursill, C.A.; Tan, J.T.M. TRIM2 Selectively Regulates Inflammation-Driven Pathological Angiogenesis without Affecting Physiological Hypoxia-Mediated Angiogenesis. *Int. J. Mol. Sci.* **2024**, *25*, 3343. https://doi.org/10.3390/ijms25063343

Academic Editor: Riccardo Alessandro

Received: 13 February 2024
Revised: 5 March 2024
Accepted: 13 March 2024
Published: 15 March 2024

Copyright: © 2024 by the authors. Licensee MDPI, Basel, Switzerland. This article is an open access article distributed under the terms and conditions of the Creative Commons Attribution (CC BY) license (https://creativecommons.org/licenses/by/4.0/).

Abstract: Angiogenesis is a critical physiological response to ischemia but becomes pathological when dysregulated and driven excessively by inflammation. We recently identified a novel angiogenic role for tripartite-motif-containing protein 2 (TRIM2) whereby lentiviral shRNA-mediated TRIM2 knockdown impaired endothelial angiogenic functions in vitro. This study sought to determine whether these effects could be translated in vivo and to determine the molecular mechanisms involved. CRISPR/Cas9-generated $Trim2^{-/-}$ mice that underwent a periarterial collar model of inflammation-induced angiogenesis exhibited significantly less adventitial macrophage infiltration relative to wildtype (WT) littermates, concomitant with decreased mRNA expression of macrophage marker *Cd68* and reduced adventitial proliferating neovessels. Mechanistically, TRIM2 knockdown in endothelial cells in vitro attenuated inflammation-driven induction of critical angiogenic mediators, including nuclear HIF-1α, and curbed the phosphorylation of downstream effector eNOS. Conversely, in a hindlimb ischemia model of hypoxia-mediated angiogenesis, there were no differences in blood flow reperfusion to the ischemic hindlimbs of $Trim2^{-/-}$ and WT mice despite a decrease in proliferating neovessels and arterioles. TRIM2 knockdown in vitro attenuated hypoxia-driven induction of nuclear HIF-1α but had no further downstream effects on other angiogenic proteins. Our study has implications for understanding the role of TRIM2 in the regulation of angiogenesis in both pathophysiological contexts.

Keywords: inflammation; ischemia; neovascularization; HIF-1α; eNOS

1. Introduction

Angiogenesis is the process in which new blood vessels are formed from pre-existing vessels. It is crucial in many physiological contexts, such as in wound healing, and as an adaptive response to hypoxia and ischemia [1]. However, imbalance in angiogenic regulation can cause deleterious effects, as it accelerates inflammation-driven pathologies, as seen in atherosclerosis and cancer [2,3]. New vessels provide additional conduits for the delivery of inflammatory cells and cytokines that promote atherosclerotic plaque development and rupture. They also deliver the oxygen and nutrients necessary to sustain tumor growth and serve as potential routes for metastatic spread [4].

Angiogenesis-associated conditions are highly prevalent globally, with cardiovascular disease (CVD) and cancer among the leading causes of morbidity and mortality worldwide. Current anti-angiogenic agents are limited, as they can interfere with physiological angiogenic processes, while pro-angiogenic therapies can potentially exacerbate chronic inflammation and inadvertently precipitate tumorigenesis [5,6]. Given the critical role of angiogenesis across such diverse pathologies, any agent capable of differentially modulating angiogenesis in a context-specific manner would be of great therapeutic value.

We previously identified a novel angiogenic role for tripartite motif-containing protein 2 (TRIM2) [7]. Lentiviral short hairpin (sh)RNA knockdown of TRIM2 impaired endothelial cell tubule formation in both hypoxia and inflammatory conditions in vitro [7]. We have also shown that TRIM2 knockdown attenuates the ability of human coronary artery endothelial cells (HCAECs) to migrate and proliferate in response to hypoxic and inflammatory stimuli. However, whether these effects are translated in vivo, and what the molecular mechanisms are underlying these, remains unknown.

In this study, we used CRISPR/Cas9-generated homozygous *Trim2* null ($Trim2^{-/-}$) mice to evaluate the functional importance of TRIM2 in two well-validated models of pathological inflammation-driven angiogenesis and physiological hypoxia-mediated angiogenesis, namely the periarterial cuff and hindlimb ischemia models, respectively. In $Trim2^{-/-}$ mice, we report markedly attenuated infiltration of adventitial macrophages in response to femoral artery cuff placement, when compared to wildtype (WT) littermates concomitant with a reduction in mRNA levels of the macrophage marker cluster of differentiation 68 (*Cd68*). Mechanistically, we show that TRIM2 knockdown in human coronary artery endothelial cells (HCAECs) attenuates the induction of key mediators involved in the classical inflammation-driven angiogenic signaling pathway, including nuclear translocation of hypoxia-inducible factor (HIF)-1α and phosphorylation of downstream mediator endothelial nitric oxide synthase (eNOS).

In contrast, we find no significant differences in blood flow reperfusion despite a reduction in proliferating neovessels and arterioles in the ischemic hindlimbs of $Trim2^{-/-}$ and WT mice. In vitro, while the hypoxia-mediated induction of HIF-1α was tempered by TRIM2 knockdown, further downstream activation of angiogenic signaling proteins were unaffected. These findings collectively highlight a novel role for TRIM2 in the regulation of inflammation-driven angiogenesis and delineate the mechanistic basis for these effects. We propose TRIM2 to be a potential therapeutic target for diseases driven by pathological angiogenesis, unlimited by the usual adverse effects associated with inhibiting physiological angiogenesis.

2. Results

2.1. Plasma Glucose and Lipid Concentrations Are Not Affected by TRIM2 Knockdown

Deletion of *Trim2* from the tissues of $Trim2^{-/-}$ mice was confirmed by qPCR, with only $7.0 \pm 4.6\%$ and $1.7 \pm 0.5\%$ of residual *Trim2* expression detected in the gastrocnemius muscle and liver tissues of $Trim2^{-/-}$ mice, respectively (Supplementary Materials Figure S1). All WT and $Trim2^{-/-}$ animals were monitored regularly pre- and post-operatively until the conclusion of each study. This included the measurement of daily weights, which provided an opportunity to observe the mice for any major phenotypic differences that might develop. We observed no obvious differences noted with respect to general neurological or motor function, nor were there any clear differences in cardiovascular health. There were no differences in total body weights at the conclusion of the periarterial cuff or hindlimb ischemia studies or the weights of various individual organs between WT and $Trim2^{-/-}$ mice (Supplementary Materials Figure S2). As part of the phenotypic evaluation of $Trim2^{-/-}$

mice, plasma glucose and lipid levels were determined. These metabolic parameters were invariably lower in $Trim2^{-/-}$ mice relative to WT mice, though no statistically significant differences were observed (Table 1).

Table 1. Plasma glucose and lipids measured in WT and $Trim2^{-/-}$ mice.

Plasma Parameter	WT (N = 23)	$Trim2^{-/-}$ (N = 23)
Glucose (mM)	14.5 ± 0.3	14.0 ± 0.4
Total cholesterol (mg/dL)	256.3 ± 7.4	239.7 ± 7.2
HDL cholesterol (mg/dL)	122.2 ± 4.1	118.2 ± 4.4
LDL cholesterol (mg/dL)	134.1 ± 6.0	121.5 ± 5.7
Triglycerides (mg/dL)	100.4 ± 7.6	97.0 ± 6.2

Data are shown as mean ± SEM.

2.2. Trim2 Deletion Inhibits Infiltration of Adventitial Macrophages and Attenuates Angiogenic Responses to Inflammation In Vivo

In the periarterial cuff study, CD68$^+$ macrophage infiltration into the adventitia of cuffed arteries was markedly reduced in $Trim2^{-/-}$ mice (33.9 ± 10.2% vs. WT: 100.0 ± 27.4%, $p < 0.05$, Figure 1a). Furthermore, $Trim2^{-/-}$ mice had fewer CD31$^+$ vessels in the adventitia of cuffed arteries (73.0 ± 12.8% vs. WT: 100.0 ± 10.7%, Figure 1b), though this difference did not reach statistical significance ($p = 0.1221$). Notably, $Trim2^{-/-}$ mice had significantly fewer Ki67$^+$CD31$^+$ proliferating neovessels (40.5 ± 10.5% vs. WT: 100.0 ± 16.4%, $p < 0.01$, Figure 1c). However, no differences were observed in the presence of CD34$^+$ endothelial tip cells (Figure 1d). The intima-to-media ratio, a measure of inflammation-driven neointima formation, was similar between WT and $Trim2^{-/-}$ mice (Figure 1e).

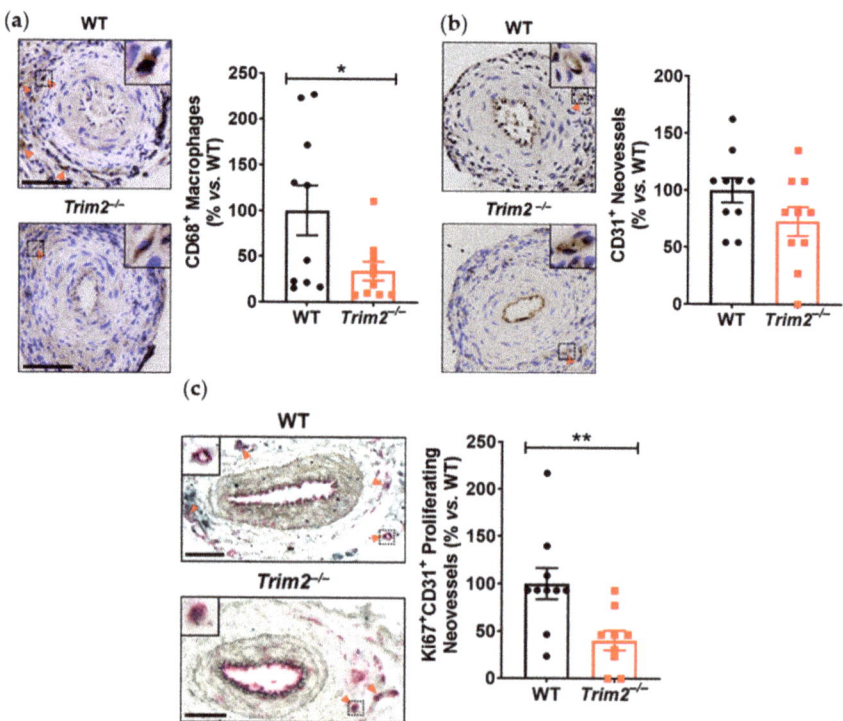

Figure 1. *Cont.*

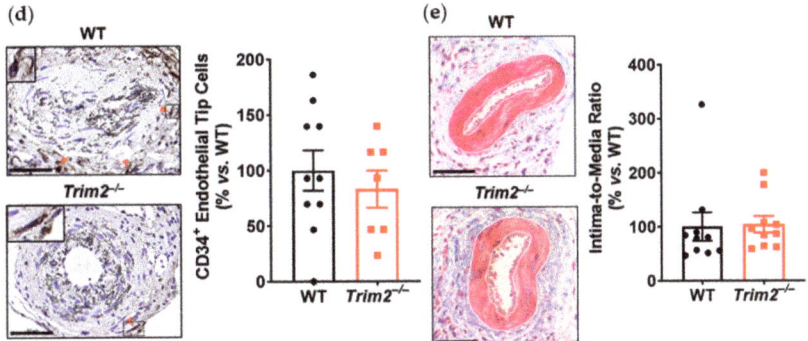

Figure 1. Trim2 deletion inhibits inflammatory-driven adventitial macrophage infiltration in vivo. A non-occlusive polyethylene cuff (2 mm) was placed around the left femoral arteries of wild-type (WT) and $Trim2^{-/-}$ mice (N = 10/group) for 21 days to trigger localized inflammatory responses. Femoral arteries were sectioned for immunohistochemical detection of adventitial (**a**) $CD68^+$ macrophages (brown staining, red arrowheads), (**b**) $CD31^+$ neovessels (brown staining, red arrowheads), (**c**) $Ki67^+CD31^+$ proliferating neovessels (brown and red dual staining, red arrowheads) and (**d**) $CD34^+$ endothelial tip cells (brown staining, red arrowheads). (**e**) Masson's trichrome staining was performed to assess intima-to-media ratio. Representative images of cuffed artery sections from WT and $Trim2^{-/-}$ mice were taken at 40× magnification. Scale bars: 50 µm. Results are mean ± SEM. * $p < 0.05$, ** $p < 0.01$ using unpaired Student's t-test.

Together with reduced macrophage infiltration, there was attenuated induction of key inflammatory markers at 24 h post-surgery in the cuffed arteries of $Trim2^{-/-}$ mice. Firstly, comparison between WT and $Trim2^{-/-}$ mRNA levels of cuffed arteries showed a significant decrease in *Cd68* mRNA expression in the $Trim2^{-/-}$ animals (55.9 ± 13.0% vs. WT: 100.0 ± 12.6%, $p < 0.05$, Figure 2a), consistent with the $CD68^+$ macrophage staining. However, no differences were observed in *Ccl2* (Figure 2b) and *Rela* (Figure 2c) levels. We also assessed if global Trim2 knockout attenuates the extent of cuff-induced inflammatory response. When compared to their respective non-cuffed control arteries, there was a 42-fold increase in *Cd68* mRNA levels in the cuffed arteries of WT mice ($p < 0.0001$, Figure 2d). Cuff placement also induced Cd68 expression in $Trim2^{-/-}$ mice; however, the extent of stimulation was less pronounced (30-fold, $p < 0.001$). We observed a similar pattern with *Ccl2* mRNA levels such that cuff placement induced a 200-fold increase in *Ccl2* in WT mice while there was a 150-fold induction in $Trim2^{-/-}$ mice (Figure 2e). WT and $Trim2^{-/-}$ had similar levels of *Rela* induction (Figure 2f).

2.3. Inflammation-Induced Activation of Angiogenic Signaling Mediators Is Attenuated by TRIM2 Knockdown In Vitro

To understand the mechanistic basis for the effects observed in vivo, we examined the modulation of several key angiogenic signaling mediators in vitro following lentiviral shRNA knockdown of TRIM2 in HCAECs. Comparison of TNFα-stimulated cells alone showed that nuclear NF-κB p65 protein levels was 30% lower in shTRIM2 cells (70.1 ± 29.0%, Figure 3a) when compared to shControl cells (100.0 ± 26.1%), but this did not reach statistical significance ($p = 0.4952$). However, nuclear HIF-1α protein levels were significantly lower in shTRIM2 cells (32.9 ± 8.3% vs. shControl: 100.0 ± 25.4%, $p < 0.05$, Figure 3b). No differences were observed in either PHD3 protein levels (Figure 3c). Interestingly, while no differences were observed in VEGFA protein levels (Figure 3d), TRIM2 knockdown in vitro augmented VEGFR2 activation (142.4 ± 13.4% vs. shControl: 100.0 ± 10.5%, $p < 0.05$, Figure 3e). No differences were observed in p38 MAPK phosphorylation (Figure 3f); however, eNOS phosphorylation was significantly attenuated (shTRIM2: 63.5 ± 9.3% vs. shControl: 100.0 ± 8.5%, $p < 0.05$, Figure 3g).

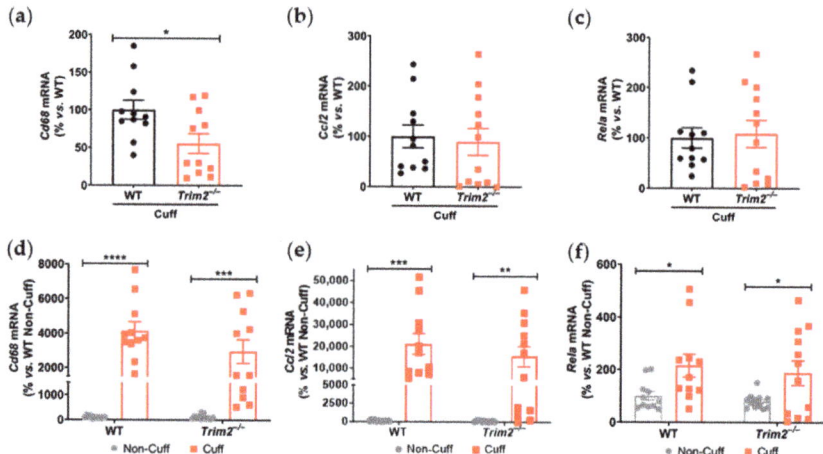

Figure 2. *Trim2* deletion attenuates angiogenic responses to inflammation in vivo. A non-occlusive polyethylene cuff (2 mm) was placed around the left femoral arteries of wildtype (WT) and *Trim2*$^{-/-}$ mice (N = 10/group) for 24 h to trigger localized inflammatory responses. (**a**) *Cd68*, (**b**) *Ccl2* and (**c**) NF-κB p65 (*Rela*) mRNA levels in cuffed arteries of WT and *Trim2*$^{-/-}$ mice, normalized using the $^{\Delta\Delta}$Ct method to *36B4* and WT cuffed arteries. (**d**) *Cd68*, (**e**) *Ccl2* and (**f**) NF-κB p65 (*Rela*) mRNA levels in non-cuffed and cuffed arteries of WT and *Trim2*$^{-/-}$ mice, normalized using the $^{\Delta\Delta}$Ct method to *36B4* and WT non-cuffed arteries. Results are mean ± SEM. * $p < 0.05$, ** $p < 0.01$, *** $p < 0.001$, **** $p < 0.0001$ using unpaired Student's *t*-test or two-way ANOVA with Bonferroni's post hoc analysis.

We also compared the extent of TNFα-induced inflammatory response when compared to respective baseline (No TNFα) controls. In shControl-transduced cells, inflammatory stimulation with TNFα significantly increased nuclear NF-κB p65 ($p < 0.05$, Figure 3h). While nuclear NF-κB p65 levels were also higher in TNFα-stimulated shTRIM2 cells, this did not reach statistical significance ($p = 0.1999$) when compared to its respective non-stimulated control. Strikingly, nuclear HIF-1α levels were significantly reduced in shTRIM2 cells when compared to TNFα-stimulated shControl cells ($p < 0.05$, Figure 3i). TNFα also significantly induced protein levels of PHD3 ($p < 0.01$, Figure 3j) and VEGFA ($p < 0.05$, Figure 3k) in shControl cells. However, these inflammatory-driven inductions were not seen with TRIM2 knockdown. VEGFR2 phosphorylation was reduced in response to TNFα stimulation in both shControl and shTRIM2 cells (Figure 3l). No differences were observed with p38 MAPK phosphorylation irrespective of conditions (Figure 3m). TRIM2 knockdown significantly reduced eNOS activation compared to both unstimulated shTRIM2 cells and TNFα-stimulated shControl cells ($p < 0.05$ for both, Figure 3n).

2.4. Trim2 Deletion Does Not Affect Angiogenic Responses to Ischemia In Vivo

In the hindlimb ischemia study, there were no differences between *Trim2*$^{-/-}$ and WT mice in their capacity for blood flow reperfusion, as monitored by longitudinal laser Doppler imaging (Figure 4a). WT and *Trim2*$^{-/-}$ mice were indistinguishable in their motor functions and the appearance of their distal limbs throughout the study.

Angiogenic responses to ischemia were further assessed histologically in the distal gastrocnemius muscle of the ischemic hindlimbs (Figure 4b). In *Trim2*$^{-/-}$ mice, the density of CD31$^+$ neovessels relative to the number of myocytes was increased (133.5 ± 9.5%, $p < 0.05$) when compared to that of WT mice (Figure 4c). However, the density of α-SMA$^+$ arterioles relative to the number of myocytes was not different between the ischemic hindlimbs of *Trim2*$^{-/-}$ and WT mice (Figure 4d). Interestingly, the number of Ki67$^+$CD31$^+$ proliferating neovessels was significantly decreased in *Trim2*$^{-/-}$ mice (51.6 ± 7.9% vs. WT:

100.0 ± 15.7%, $p < 0.05$, Figure 4e). We also observed a decrease in Ki67$^+$ α-SMA$^+$ arterioles in $Trim2^{-/-}$ mice (56.4 ± 9.3% vs. WT: 100.0 ± 15.4%, $p < 0.05$, Figure 4f).

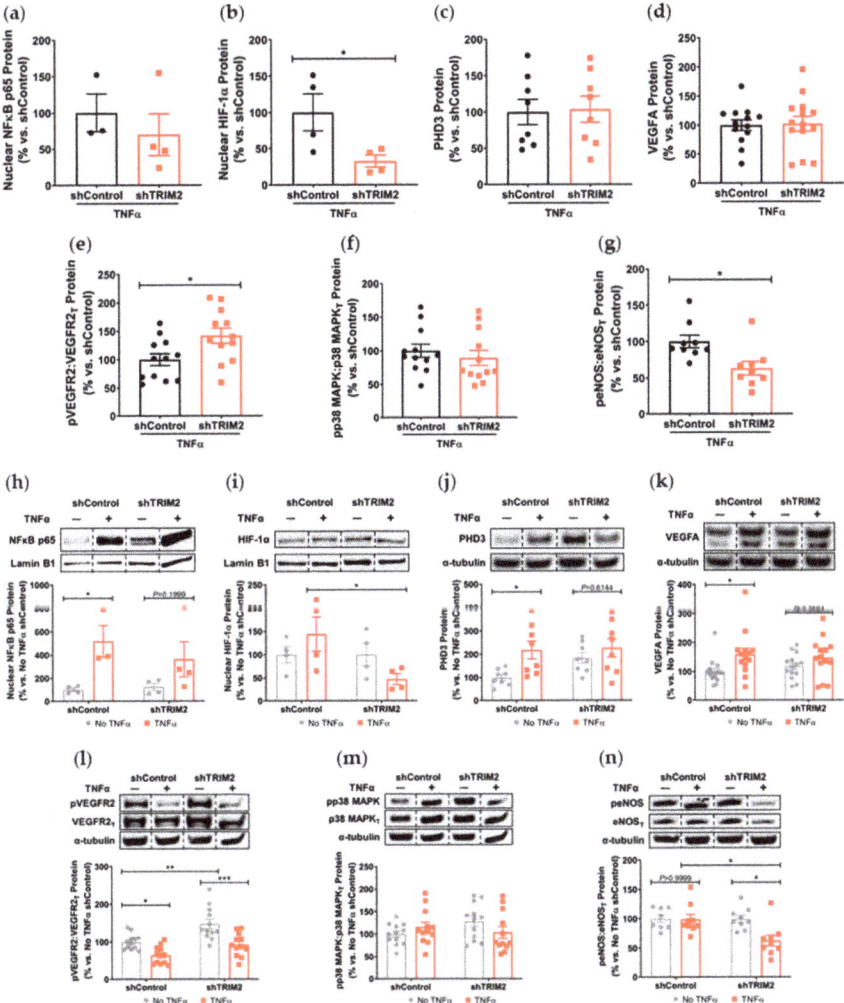

Figure 3. Inflammation-induced activation of angiogenic signaling mediators is attenuated by TRIM2 knockdown in vitro. Lentiviral-transduced shControl and shTRIM2 HCAECs were incubated without and with TNFα (0.6 ng/mL, 4.5 h). Protein levels of (**a**) nuclear NF-κB p65, (**b**) nuclear HIF-1α, (**c**) PHD3, (**d**) VEGFA, (**e**) Phospho:Total VEGFR2, (**f**) Phospho:Total p38 MAPK and (**g**) Phospho:Total eNOS, presented as percent change relative to TNFα-treated shControl cells. Protein levels of (**h**) nuclear NF-κB p65, (**i**) nuclear HIF-1α, (**j**) PHD3, (**k**) VEGFA, (**l**) Phospho:Total VEGFR2, (**m**) Phospho:Total p38 MAPK and (**n**) Phospho:Total eNOS, presented as percent change relative to No TNFα shControl cells. Dotted lines separate noncontiguous lanes from the same gel. The cropped blots are used in the figure, and the full blots are presented in Supplementary Materials Figures S3 and S4. Each experiment was conducted at least three times independently with triplicates for each condition. Results are mean ± SEM. * $p < 0.05$, ** $p < 0.01$ *** $p < 0.001$ using unpaired Student's t-test or two-way ANOVA with Bonferroni's post hoc analysis.

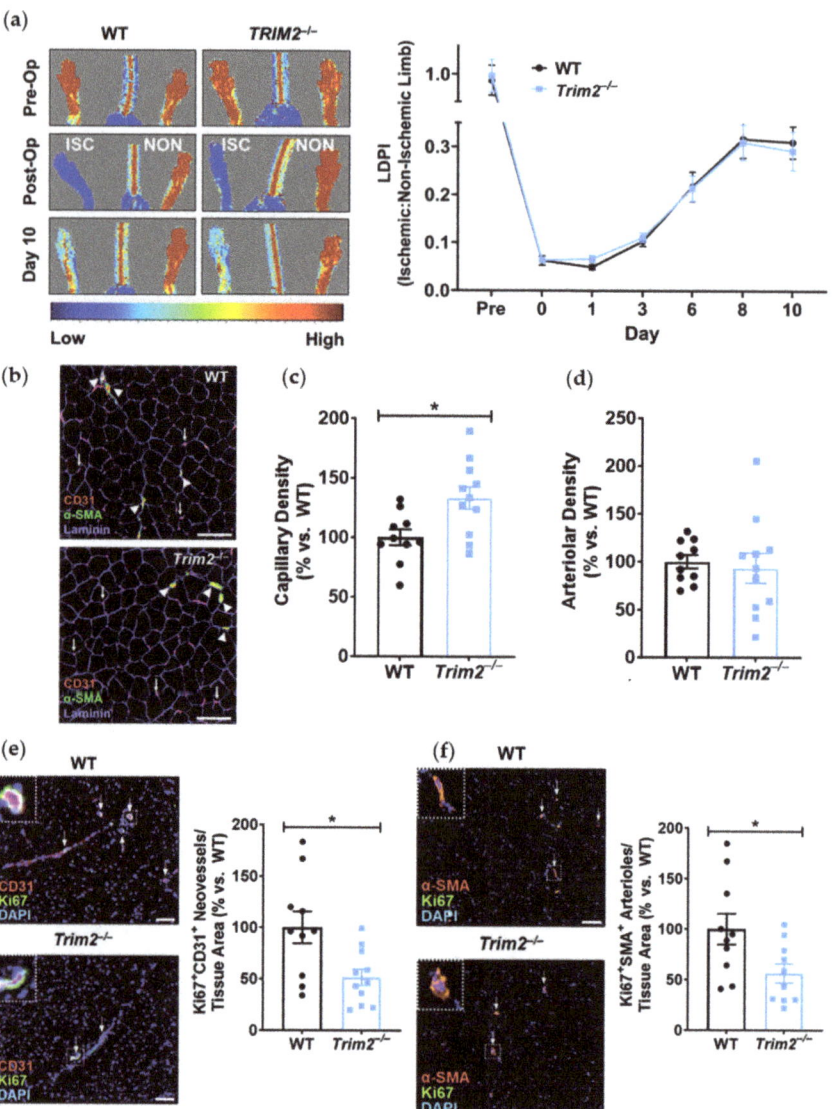

Figure 4. Trim2 does not affect angiogenic responses to ischemia in vivo. Wildtype (WT) and $Trim2^{-/-}$ mice (N = 10/group) underwent ligation and excision of the left femoral artery and vein. (**a**) Blood flow reperfusion was measured by laser Doppler imaging over 10 days. Representative images show high (red) to low (blue) blood flow in the ischemic (ISC) and non-ischemic (NON) hindlimbs. Laser Doppler perfusion index (LDPI) was calculated as the ratio of flow in the ISC:NON hindlimbs. (**b**) Representative sections taken across the medial plane of the ischemic gastrocnemius muscles of WT and $Trim2^{-/-}$ mice taken at 20× magnification showing $CD31^+$ neovessels (red/purple staining, arrows), $\alpha\text{-}SMA^+$ arterioles (green staining, arrowheads) and laminin-stained basement membrane of the muscle fibers (blue staining). (**c**) The density of $CD31^+$ neovessels per myocyte, normalized to WT mice. (**d**) The density of $\alpha\text{-}SMA^+$ arterioles per myocyte, normalized to WT mice. Immunofluorescent co-staining was performed to detect (**e**) $Ki67^+CD31^+$ proliferating neovessels and (**f**) $Ki67^+\alpha\text{-}SMA^+$ proliferating arterioles. Scale bars: 100 μm. Results are mean ± SEM. * $p < 0.05$ using unpaired Student's t-test.

2.5. TRIM2 Knockdown Attenuates Hypoxia-Mediated Induction of Nuclear HIF-1α, PHD3 and VEGFA but Not Downstream Signaling Pathways In Vitro

In vitro, comparison of shTRIM2 and shControl cells exposed to hypoxia showed that TRIM2 knockdown significantly reduced nuclear HIF-1α protein levels (31.8 ± 9.3% vs. shControl: 100.0 ± 25.3%, $p < 0.05$, Figure 5a), with no differences observed in PHD3, VEGFA and activation of VEGFR2, p38 MAPK and eNOS (Figure 5b–f). We then compared the extent of hypoxic induction when compared to respective baseline controls. Under hypoxic conditions, nuclear HIF-1α levels were significantly reduced in shTRIM2 cells (44 ± 13%, $p < 0.05$, Figure 5g) when compared to hypoxia-stimulated shControl cells (138 ± 35%). Hypoxia also significantly increased PHD3 ($p < 0.05$, Figure 5h) and VEGFA ($p < 0.05$, Figure 5i) protein levels in shControl cells relative to their respective normoxia controls. However, these hypoxia-driven inductions were not observed in shTRIM2 cells. Meanwhile, VEGFR2, p38 MAPK and eNOS phosphorylation was unaffected by TRIM2 knockdown in HCAECs in hypoxia (Figure 5j–l).

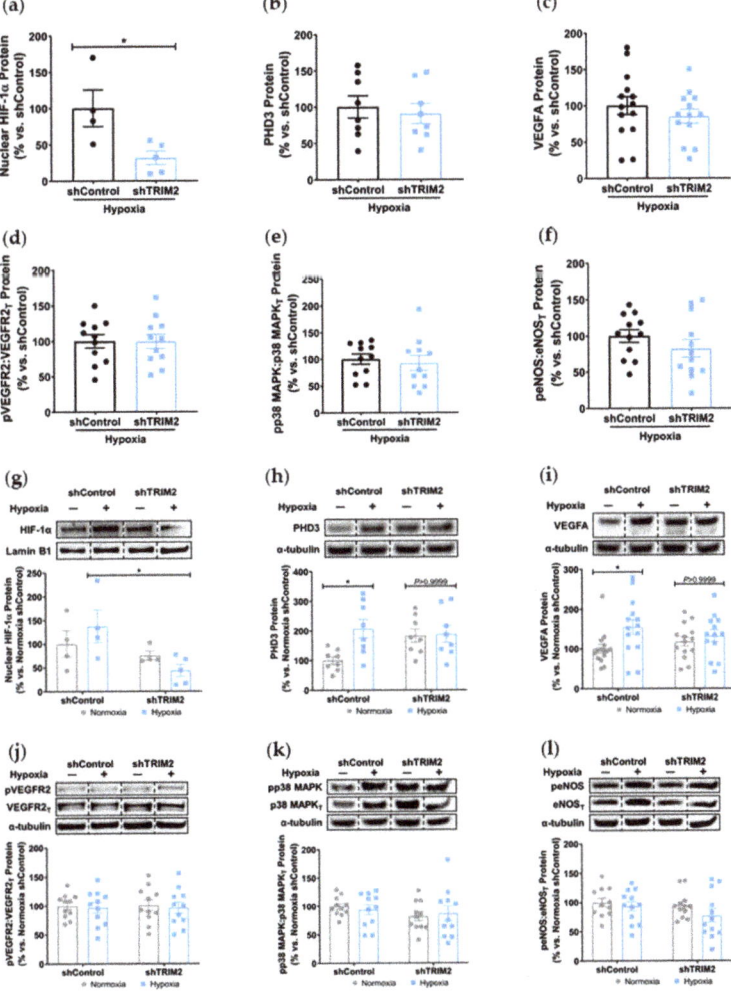

Figure 5. TRIM2 knockdown in vitro attenuates hypoxia-driven induction of nuclear HIF-1α, PHD3 and VEGFA but not downstream signaling pathways. Lentiviral-transduced shControl and shTRIM2

HCAECs were incubated in normoxia or hypoxia (1.2% O_2, 6 h). Protein levels of (**a**) nuclear HIF-1α, (**b**) PHD3, (**c**) VEGFA, (**d**) Phospho:Total VEGFR2, (**e**) Phospho:Total p38 MAPK and (**f**) Phospho:Total eNOS, presented as percent change relative to hypoxia-stimulated shControl cells. Protein levels of (**g**) nuclear HIF-1α, (**h**) PHD3, (**i**) VEGFA, (**j**) Phospho:Total VEGFR2, (**k**) Phospho:Total p38 MAPK and (**l**) Phospho:Total eNOS, presented as percent change relative to normoxia shControl cells. Dotted lines separate noncontiguous lanes from the same gel. The cropped blots are used in the figure, and the full blots are presented in Supplementary Materials Figures S5 and S6. Each experiment was conducted at least three times independently with triplicates for each condition. Results are mean ± SEM. * $p < 0.05$ using unpaired Student's *t*-test or two-way ANOVA with Bonferroni's post hoc analysis.

3. Discussion

Dysregulated angiogenesis crucially underpins a wide range of chronic and debilitating diseases including atherosclerotic CVD and cancer. TRIM2 has emerged as a promising novel target that may differentially modulate both inflammation-driven pathological angiogenesis and hypoxia-stimulated physiological angiogenesis, as our previous studies have demonstrated impaired endothelial tubule formation in HCAECs with TRIM2 knockdown in vitro [7]. Here, we report markedly reduced adventitial macrophage infiltration following *Trim2* deletion in a murine periarterial cuff model of inflammation-driven angiogenesis, concomitant with reduced proliferating adventitial neovessels and attenuated the induction of the inflammatory response. Correspondingly, we find that TRIM2 knockdown in HCAECs suppresses the TNFα-driven induction of several classical angiogenic mediators, particularly nuclear HIF-1α and reduced activation of the eNOS angiogenic signaling pathway. *Trim2* deletion, however, did not alter the capacity for blood flow reperfusion nor the extent of neovascularization in the murine hindlimb ischemia model despite a reduction in proliferating neovessels and arterioles. While TRIM2 knockdown in vitro suppressed the hypoxia-driven stimulation of nuclear HIF-1α, it did not affect downstream expression and activation of pro-angiogenic signaling pathways.

Our findings, particularly from the periarterial cuff model and the mechanistic studies, are consistent with our previous work showing the inhibition of inflammation-induced endothelial tubule formation after TRIM2 knockdown [7]. The reduction in CD68$^+$ macrophage infiltration into the inflamed arteries of $Trim2^{-/-}$ mice suggests that TRIM2 may be involved in broader mechanisms of inflammatory activation, which enhance vessel growth by stimulating a wealth of pro-angiogenic growth factors and mediators [8]. The concomitant reduction in *Cd68* mRNA levels in the cuffed arteries of $Trim2^{-/-}$ mice further support the idea that TRIM2 plays a key role in regulating inflammation-driven pathological angiogenesis, particularly in the early stages of macrophage recruitment to the site of injury. The reduction in proliferating adventitial neovessels in the cuffed arteries of $Trim2^{-/-}$ mice is reflective of this. This was associated with a trend towards reduced total neovessels. Had the study duration been extended beyond 3 weeks, it could be hypothesized that a larger reduction in adventitial neovessels would be observed, as the blunted macrophage response would lead to fewer pro-angiogenic factors being released. No differences were observed in the presence of CD34$^+$ endothelial tip cells. The intima-to-media ratio, though, which assesses the formation of a thickened neointima as an 'outside-in' response to adventitial inflammation, was not affected by *Trim2* deletion, indicating that the reduction in adventitial neovessels was a specific effect on angiogenesis and not a consequence of the development of a smaller neointimal or media. Furthermore, recent clinical studies have identified TRIM2 as a potential oncogene in human cancer cell lines including colorectal carcinoma, epithelial ovarian carcinoma and osteosarcoma [9–11]. These observations fit with a postulated role for TRIM2 in inflammation-driven angiogenesis, a hallmark of cancer development and progression.

Our findings also suggest a potential mechanistic pathway by which TRIM2 may be directing angiogenic responses to inflammation. Direct comparisons of TNFα-stimulated cells in vitro showed a significant reduction in nuclear HIF-1α levels and eNOS activation.

Intriguingly, when compared to their respective baseline unstimulated controls, we found that the extent of inflammatory induction of nuclear NF-κB, PHD3 and VEGFA were less pronounced in shTRIM2 cells. Under stimulation with cytokines like TNFα, it is possible that TRIM2, functioning as a ubiquitin ligase [12], may contribute to the stabilization and nuclear translocation of the transcription factors NF-κB and HIF-1α, likely by promoting proteasomal degradation of their cytosolic inhibitors, such as PHD3 in the latter case [13]. These transcription factors, in turn, promote VEGFA expression, which activates endothelial cell migration and tubule formation through numerous intracellular pathways [14,15], including the phosphorylation of p38 MAPK and eNOS. NF-κB p65 and HIF-1α may also stimulate angiogenesis downstream of TRIM2 by upregulating inflammatory cytokines and chemokines like CCL2, leading to the recruitment and activation of macrophages that help to potentiate the inflammation-driven angiogenic response.

It is not clear whether TRIM2 mediates VEGFA-related effects via VEGFR2, as only a modest decrease in Tyr^{1175} phosphorylation was seen following TRIM2 knockdown. TRIM2 could be targeting alternative VEGFR2 phosphorylation sites. While there was no change in VEGFR2 phosphorylation at Tyr^{1175}, it is possible that other key tyrosine sites like Tyr^{801}, Tyr^{1054} or Tyr^{1059} may be involved, each of which may activate a distinct set of signal transduction mechanisms and cellular responses [14]. Specifically, VEGFR2 phosphorylation at Tyr^{801} also contributes to Akt-dependent eNOS activation and nitric oxide release from endothelial cells [16,17]. Another potential target of TRIM2 is neuropilin-1 (NRP1), one of the key co-receptors for VEGFR2. NRP1 is highly expressed in endothelial cells and neurons and can bind to both VEGFA and the class 3 semaphorins, a family of axonal guidance proteins, thus forming a key link between angiogenesis and neurogenesis [15,18]. Given the association of TRIM2 with axonal outgrowth and development [12,19], and now angiogenesis, it is possible that NRP1 and/or its semaphorin ligands may be involved in regulating angiogenic function by TRIM2. Further studies examining a possible link between NRP1 and TRIM2 would be useful to clarify the mechanistic pathway.

Consistent with our previous work which demonstrated impairment of hypoxia-stimulated tubule formation in vitro with TRIM2 knockdown [7], *Trim2* deletion reduced the number of proliferating neovessels and arterioles in the ischemic tissue in vivo. Interestingly, there was an increase in total CD31$^+$ vessels in the ischemic hindlimbs of *Trim2*$^{-/-}$ mice, suggestive of a potential negative feedback loop. In the late stages of angiogenesis, vessel pruning occurs, whereby capillaries disintegrate, to facilitate mature vessel formation during tissue remodeling [20]. We therefore postulate that the changes seen with increased capillary density yet reduced number of proliferating neovessels may be indicative of vessel pruning at this late stage post-ischemia. Furthermore, revascularization in the hindlimb ischemia model is primarily facilitated through arteriogenesis [21]. No differences were observed in overall arteriolar density, which is consistent with the lack of change seen in blood flow reperfusion. However, while there were significant changes at a cellular/tissue level, this did not seem to impact the recovery of blood flow reperfusion to the ischemic hindlimb. These incongruous findings may reflect inherent angiogenic compensatory mechanisms in vivo that may be activated in response to *Trim2* deletion, perhaps starting early in embryonic development and thereby rendering *Trim2* redundant. Mechanistically, while the hypoxia-driven increase in nuclear HIF-1α was attenuated in TRIM2-deficient HCAECs, downstream angiogenic signaling was not altered. The paucity of effects on these intracellular mediators may explain the lack of differences between WT and *Trim2*$^{-/-}$ mice in their angiogenic responses to hypoxia. This could prove clinically useful, as anti-TRIM2 therapies may be developed to suppress pathological inflammatory angiogenesis, without the adverse effects of impairing hypoxia-driven physiological angiogenesis.

The discordant in vitro and in vivo findings may also reflect the activation of compensatory angiogenic mechanisms in response to *Trim2* deletion in vivo. Future studies could explore inducible and endothelial cell-specific *Trim2* silencing to exclude such effects as angiogenic responses may be countered by *Trim2* deletion in other cell types, like vascular smooth muscle cells, pericytes and immune cells [22]. It is also plausible that TRIM2 may target non-classical pathways downstream of VEGFA to confer its angiogenic effects in hypoxia, or it could be modulating other angiogenic factors like the fibroblast growth factors or angiopoietins. Future studies could evaluate a broader range of signaling targets to better elucidate the mechanistic basis by which TRIM2 may be modulating endothelial responses to hypoxia.

Overall, we have shown, for the first time, that TRIM2 is functionally important in regulating pathological angiogenic responses to inflammation in vivo, via modulation of classical angiogenic mediators HIF-1α, NF-κB p65, and VEGFA and downstream targets of VEGFA. Given that TRIM2 has no effect on physiological ischemia-driven angiogenesis, targeted TRIM2 inhibition could prove therapeutically useful for diseases driven predominantly by pathological angiogenesis including atherosclerosis and cancer, without the adverse effects of inhibiting physiological angiogenesis.

4. Materials and Methods

4.1. Animal Studies

All experimental procedures were conducted with approval from the SAHMRI Animal Ethics Committee (#SAM335) and conformed to the Australian Code for the Care and Use of Animals for Scientific Purposes (National Health and Medical Research Council, Australia). A $Trim2^{-/-}$ mouse line was generated by the South Australian Genome Editing facility using a CRISPR/Cas9 approach. In brief, Cas9 protein was injected into C57BL/6J murine embryos along with two guide RNA sequences. These guide RNAs were designed such that non-homologous end joining of the DNA following CRISPR/Cas9 activity would result in excision of a DNA fragment containing exon 2 of $Trim2$, leading to a frameshift in the coding sequence and an early stop codon in exon 3. The founder male carrying this mutant $Trim2$ allele was back-crossed to wildtype (WT) female C57BL/6J mice, generating identical heterozygous offspring which were subsequently crossed to generate homozygous $Trim2$ knockout ($Trim2^{-/-}$) mice. Male wildtype (WT) and $Trim2^{-/-}$ mouse littermates were housed in a temperature and humidity-controlled environment under a 12 h light/dark cycle with ad libitum access to water and standard mouse chow. They underwent surgery at 8 weeks of age.

4.2. Plasma Glucose and Lipid Analyses

Plasma glucose concentrations were determined using a glucometer (Accu-Chek® Performa, Roche, Basel, Switzerland), while total plasma and HDL cholesterol concentrations were measured enzymatically (439-17501, Wako Diagnostics, Richmond, VA, USA). HDL cholesterol concentrations were determined following polyethylene glycol precipitation of apoB-containing lipoproteins, while LDL cholesterol concentrations were calculated by subtracting HDL from total cholesterol concentrations. Triglyceride concentrations were determined using a colorimetric assay (290-63701, Wako Diagnostics).

4.3. Periarterial Cuff Model

The femoral periarterial cuff model is an established model of inflammation-driven neointima formation and adventitial angiogenesis [23,24], processes which are known to contribute to atherosclerotic plaque development. A non-occlusive 2 mm length of polyethylene cuff was placed around the left femoral artery to trigger a localized inflammatory response, while a sham operation was performed on the right femoral artery as a parallel control. The animals were sacrificed 21 days post-surgery by overdose of isoflurane and intracardiac puncture, followed by perfusion with phosphate-buffered saline (PBS) via the left ventricle. The femoral arteries (complete with cuff) were excised for histochemical analyses.

Excised femoral arteries were fixed in 10% (v/v) formalin for 24 h then embedded in 3% (w/v) agarose prior to tissue processing and paraffin embedding. Angiogenic responses to cuff placement were assessed via immunohistochemistry on 5 μm sections, probing for CD68 (Bio-Rad, Hercules, CA, USA, Cat# MCA1957GA, RRID:AB_324217) to assess macrophage infiltration and CD31 (Abcam, Cambridge, UK, Cat# ab28365, RRID:AB_726365) to detect adventitial vessels. Proliferating neovessels were determined by co-staining tissue sections with proliferation marker Ki-67 (Thermo Fisher Scientific, Waltham, MA, USA, Cat# 14-5698-82, RRID:AB_10854564) and CD31. Endothelial tip cells were determined by staining sections with CD34 (Abcam, Cat# ab8158, RRID:AB_306316). Masson's trichrome staining was performed with a Trichrome Stain Kit (ab150686, Abcam) to assess intima-to-media ratio as a measure of neointimal responses to inflammatory stimulation. All histological sections were photographed with a Zeiss Axio Scan.Z1 Digital Slide Scanner (Carl Zeiss Microscopy, Oberkochen, Baden-Württemberg, Germany), and image analysis was performed using Image-Pro Premier software (v9.0.4, Media Cybernetics, Rockville, MD, USA).

An additional cohort of mice underwent the same procedure and were sacrificed 24 h post-surgery for gene expression analysis. Total RNA was isolated from the femoral arteries with TRI® reagent (Sigma-Aldrich, St. Louis, MO, USA) and quantitated spectrophotometrically. Then, 200 ng of total RNA was reverse transcribed using the iScript cDNA synthesis kit (Bio-Rad). Quantitative real-time PCR was performed for *Cd68* (F: 5′-GGACAGCTTACCTTTGGATTCAA-3′; R: 5′-CTGTGGGAAGGACACATTGTATTC-3′), *Ccl2* (F: 5′-GCTGGAGCATCCACGTGTT-3′; R: 5′-ATCTTGCTGGTGAATGAGTAGCA-3′), NF-κB p65 (*Rela*, forward [F]: 5′-AGTATCCATAGCTTCCAGAACC-3′; reverse [R]: 5′-ACTGC-ATTCAAGTCATAGTCC-3′) and *36B4* (F: 5′-CAACGGCAGCA-TTTATAACCC-3′; R: 5′-CCCATTGATGATGGAGTGTGG-3′). Relative gene expression was calculated using the $^{\Delta\Delta}$Ct method, normalized to *36B4* and WT non-cuffed arteries.

4.4. Hindlimb Ischemia Model

The hindlimb ischemia model is a well-validated model of physiological angiogenesis in response to tissue ischemia [25]. Hindlimb ischemia was induced by ligation and excision of the left superficial and deep femoral arteries, along with the left femoral vein down to the saphenous artery. A sham procedure was performed on the contralateral hindlimb as a parallel control. Hindlimb blood reperfusion was determined by laser Doppler imaging (moorLDI2-IR, Moor Instruments, Devon, UK), performed prior to and immediately following surgery, then at days 1, 3, 6, 8 and 10 post-surgery. Animals were sacrificed 10 days post-surgery by isoflurane overdose and intracardiac puncture, and the gastrocnemius muscles of both hindlimbs were collected for histological analyses.

Gastrocnemius muscles from both ischemic and non-ischemic hindlimbs were OCT-embedded and frozen on dry ice. Sections were taken across the medial plane of the gastrocnemius muscle (anterior distal hindlimb). This region is known to provide the most consistent and uniform responses to ischemic induction [25–27]. To histologically assess angiogenic responses to ischemia, immunofluorescence was performed on 5 µm tissue sections, staining with CD31 (Abcam, Cat# ab28364, RRID:AB_726362) to detect neovessels, α-smooth muscle actin (α-SMA, Sigma-Aldrich, Cat# F3777, RRID:AB_476977) to detect arterioles and laminin (Millipore, Burlington, MA, USA, Cat# MAB1905, RRID:AB_94392) to detect myocytes. Proliferating neovessels and arterioles were determined by co-staining tissue sections with Ki-67 (Thermo Fisher Scientific, Cat# 11-5698-82, RRID:AB_11151330) and either CD31 or α-SMA, respectively. Images were taken using an Eclipse Ni-E fluorescent microscope (Nikon Instruments, Tokyo, Japan). CD31$^+$ neovessels and α-SMA$^+$ arterioles were quantified using CellProfiler software (www.cellprofiler.org, accessed on 9 January 2023, Broad Institute of MIT and Harvard, Boston, MA, USA), while the myocytes were manually quantified using ImageJ (https://imagej.net/ij/, accessed on 9 January 2023, National Institutes of Health, Bethesda, MD, USA).

4.5. Lentiviral shRNA Knockdown of TRIM2 In Vitro

Human coronary artery endothelial cells (HCAECs, Cell Applications, San Diego, CA, USA) were cultured in MesoEndo Cell Growth Medium (212-500, Cell Applications) and used at passages 3–4. HCAECs were seeded at 5×10^4 cells/well in 6-well plates and cultured at 37 °C and 5% CO_2 overnight. The cells were exposed to 1×10^4 infectious units (IFU)/mL of lentiviral particles containing shRNA against TRIM2 (shTRIM2) or a random control sequence (shControl) for 24 h in the presence of polybrene. Transduced HCAECs were trypsinized, counted and seeded at a density of 1.5×10^5 cells/well and 8×10^4 cells/well for the inflammation and hypoxia experiments, respectively. HCAECs were then either incubated for 4.5 h with 0.6 ng/mL TNFα (to mimic inflammation) or for 6 h at 5% CO_2 and 1.2% O_2 balanced with N_2 (to mimic hypoxia). To measure phosphorylated proteins, HCAECs were stimulated with 10 ng/mL recombinant human $VEGF_{165}$ protein (R&D Systems) 15 min prior to harvest. Nuclear proteins were isolated from cell lysates using the NE-PER® Nuclear and Cytoplasmic Extraction kit (Thermo Fisher Scientific). Whole-cell protein lysates were extracted using RIPA buffer [7,28]. Each experiment was performed at least four times independently with triplicates for each condition.

4.6. Protein Expression

Nuclear and whole-cell protein extracts were subjected to Western blot analysis and probed with primary antibodies for NF-κB p65 (Abcam, Cat# ab16502, RRID:AB_443394), HIF-1α (Novus Biologicals, Centennial, CO, USA, Cat# NB100-105, RRID:AB_10001154), PHD3 (Novus, Cat# NB100-303, RRID:AB_10003302), VEGFA (Abcam, Cat# ab46154, RRID:AB_2212642), phosphorylated (Tyr^{1175}) VEGFR2 (Cell Signaling Technology, Danvers, MA, USA, Cat# 2478, RRID:AB_331377), total VEGFR2 (Cell Signaling Technology, Cat# 2479, RRID:AB_2212507), phosphorylated (Thr^{180}/Tyr^{182}) p38 MAPK (Cell Signaling Technology, Cat# 4511, RRID:AB_2139682), total p38 MAPK (Cell Signaling Technology, Cat# 8690, RRID:AB_10999090), phosphorylated (Ser^{1177}) eNOS (BD Biosciences, Franklin Lakes, NJ, USA, Cat# 612393, RRID:AB_399751) and total eNOS (BD Biosciences, Cat# 610297, RRID:AB_397691). Even protein loading was confirmed with lamin B1 (Abcam, Cat# ab16048, RRID:AB_443298) for nuclear fractions or α-tubulin (Abcam, Cat# ab40742, RRID:AB_880625) for whole-cell lysates.

4.7. Statistics

Data are expressed as mean ± SEM. Comparisons were made using unpaired Student's *t*-tests or two-way ANOVA followed by post hoc analysis using Bonferroni's multiple comparison tests. Significance was set at a two-sided $p < 0.05$.

5. Conclusions

In conclusion, we have shown, for the first time, that TRIM2 is functionally important in regulating pathological angiogenic responses to inflammation. We found that $Trim2^{-/-}$ mice that underwent a periarterial collar model of inflammation-induced angiogenesis exhibited significantly less adventitial macrophage infiltration, concomitant with decreased *Cd68* mRNA levels. $Trim2^{-/-}$ mice also had reduced adventitial proliferating neovessels. Mechanistically, our in vitro findings show that TRIM2 knockdown inhibits nuclear HIF-1α translocation and eNOS phosphorylation (Figure 6). Given that TRIM2 appears to have limited bearing on physiological ischemia-driven angiogenesis, TRIM2-directed therapies may represent safer alternatives to current anti-angiogenic strategies for the treatment of atherosclerotic CVD, cancer and chronic rheumatological conditions.

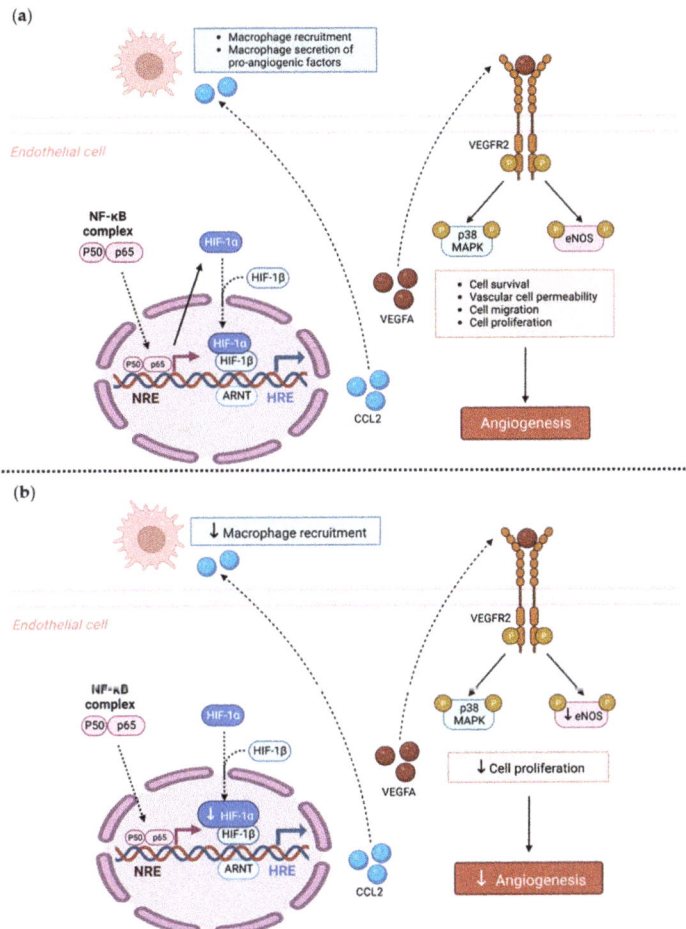

Figure 6. Proposed mechanistic role of TRIM2 in inflammatory-driven angiogenesis. (**a**) In response to an inflammatory stimulus such as TNFα, nuclear translocation of the key inflammatory transcription factor NF-κB occurs. The NF-κB complex comprises the p50/p65 subunits, forming a dimer that translocates into the nucleus. The p50/p65 dimer binds to NF-κB response elements (NREs) and upregulates a range of inflammatory and angiogenic targets including HIF-1α and VEGFA. Nuclear HIF-1α (nHIF-1α) dimerizes with HIF-1β, and together they bind to hypoxia-response elements (HREs), leading to upregulation of VEGFA, among many other pro-angiogenic genes. VEGFA binds to the VEGFR2 receptor, resulting in autophosphorylation of tyrosine residues in the cytoplasmic domain of VEGFR2. This leads to the phosphorylation (denoted by P circles) and activation of downstream signaling mediators including eNOS and p38 MAPK, resulting in angiogenesis. Additionally, NF-κB activates pro-inflammatory cytokines such as CCL2, which facilitate the recruitment of macrophages. Macrophages contribute to inflammatory-driven angiogenesis by secreting pro-inflammatory angiogenic factors. The solid arrows indicate activation, and dashed arrows indicate translocation. (**b**) Our study found that $Trim2^{-/-}$ mice that underwent a periarterial collar model of inflammation-induced angiogenesis exhibited significantly less adventitial macrophage infiltration, concomitant with decreased *Cd68* mRNA levels. $Trim2^{-/-}$ mice also had reduced adventitial proliferating neovessels. Mechanistically, our in vitro findings show that TRIM2 knockdown inhibits nuclear HIF-1α translocation and eNOS phosphorylation. Figure adapted from [28]. Created with BioRender.com (accessed on 9 January 2023).

Supplementary Materials: The supporting information can be downloaded at: https://www.mdpi.com/article/10.3390/ijms25063343/s1.

Author Contributions: Conceptualization, C.A.B. and J.T.M.T.; formal analysis, N.K.P.W. and J.T.M.T.; funding acquisition, C.A.B. and J.T.M.T.; investigation, N.K.P.W., E.L.S., R.L., V.A.N. and J.M.; methodology, N.K.P.W.; project administration, N.K.P.W. and J.T.M.T.; supervision, C.A.B. and J.T.M.T.; visualization, N.K.P.W. and J.T.M.T.; writing—original draft, N.K.P.W. and J.T.M.T.; writing—review and editing, P.J.P., S.J.N. and M.K.C.N. All authors have read and agreed to the published version of the manuscript.

Funding: C.A. Bursill holds a Lin Huddleston Heart Foundation Fellowship. P.J. Psaltis receives research fellowships from the National Heart Foundation of Australia (Future Leader Fellowship FLF102056) and National Health and Medical Research Council of Australia (CDF1161506).

Institutional Review Board Statement: The animal study protocol was approved by the Animal Ethics Committee of the South Australian Health and Medical Research Institute (protocol code: #SAM335; date of approval: 23 May 2018).

Informed Consent Statement: Not applicable.

Data Availability Statement: The original contributions presented in the study are included in the article/Supplementary Materials. Further inquiries can be directed to the corresponding author.

Conflicts of Interest: The authors declare no conflicts of interest.

References

1. Semenza, G.L. Vascular responses to hypoxia and ischemia. *Arterioscler. Thromb. Vasc. Biol.* **2010**, *30*, 648–652. [CrossRef]
2. Carmeliet, P.; Jain, R.K. Angiogenesis in cancer and other diseases. *Nature* **2000**, *407*, 249–257. [CrossRef]
3. Moulton, K.S.; Vakili, K.; Zurakowski, D.; Soliman, M.; Butterfield, C.; Sylvin, E.; Lo, K.M.; Gillies, S.; Javaherian, K.; Folkman, J. Inhibition of plaque neovascularization reduces macrophage accumulation and progression of advanced atherosclerosis. *Proc. Natl. Acad. Sci. USA* **2003**, *100*, 4736–4741. [CrossRef] [PubMed]
4. Carmeliet, P. Angiogenesis in health and disease. *Nat. Med.* **2003**, *9*, 653–660. [CrossRef] [PubMed]
5. Meadows, K.L.; Hurwitz, H.I. Anti-VEGF Therapies in the Clinic. *Cold Spring Harb. Perspect. Med.* **2012**, *2*, a006577. [CrossRef] [PubMed]
6. Ylä-Herttuala, S.; Rissanen, T.T.; Vajanto, I.; Hartikainen, J. Vascular Endothelial Growth Factors. *J. Am. Coll. Cardiol.* **2007**, *49*, 1015–1026. [CrossRef] [PubMed]
7. Wong, N.; Cheung, H.; Solly, E.; Vanags, L.; Ritchie, W.; Nicholls, S.; Ng, M.; Bursill, C.; Tan, J. Exploring the Roles of CREBRF and TRIM2 in the Regulation of Angiogenesis by High-Density Lipoproteins. *Int. J. Mol. Sci.* **2018**, *19*, 1903. [CrossRef]
8. Costa, P.Z.; Soares, R. Neovascularization in diabetes and its complications. Unraveling the angiogenic paradox. *Life Sci.* **2013**, *92*, 1037–1045. [CrossRef]
9. Cao, H.; Fang, Y.; Liang, Q.; Wang, J.; Luo, B.; Zeng, G.; Zhang, T.; Jing, X.; Wang, X. TRIM2 is a novel promoter of human colorectal cancer. *Scand. J. Gastroenterol.* **2019**, *54*, 210–218. [CrossRef]
10. Chen, X.; Dong, C.; Law, P.T.; Chan, M.T.; Su, Z.; Wang, S.; Wu, W.K.; Xu, H. MicroRNA-145 targets TRIM2 and exerts tumor-suppressing functions in epithelial ovarian cancer. *Gynecol. Oncol.* **2015**, *139*, 513–519. [CrossRef]
11. Qin, Y.; Ye, J.; Zhao, F.; Hu, S.; Wang, S. TRIM2 regulates the development and metastasis of tumorous cells of osteosarcoma. *Int. J. Oncol.* **2018**, *53*, 1643–1656. [CrossRef]
12. Balastik, M.; Ferraguti, F.; Pires-da Silva, A.; Lee, T.H.; Alvarez-Bolado, G.; Lu, K.P.; Gruss, P. Deficiency in ubiquitin ligase TRIM2 causes accumulation of neurofilament light chain and neurodegeneration. *Proc. Natl. Acad. Sci. USA* **2008**, *105*, 12016–12021. [CrossRef] [PubMed]
13. Fong, G.H.; Takeda, K. Role and regulation of prolyl hydroxylase domain proteins. *Cell Death Differ.* **2008**, *15*, 635–641. [CrossRef] [PubMed]
14. Abhinand, C.S.; Raju, R.; Soumya, S.J.; Arya, P.S.; Sudhakaran, P.R. VEGF-A/VEGFR2 signaling network in endothelial cells relevant to angiogenesis. *J. Cell Commun. Signal.* **2016**, *10*, 347–354. [CrossRef] [PubMed]
15. Simons, M.; Gordon, E.; Claesson-Welsh, L. Mechanisms and regulation of endothelial VEGF receptor signalling. *Nat. Rev. Mol. Cell Biol.* **2016**, *17*, 611–625. [CrossRef] [PubMed]
16. Blanes, M.G.; Oubaha, M.; Rautureau, Y.; Gratton, J.P. Phosphorylation of tyrosine 801 of vascular endothelial growth factor receptor-2 is necessary for Akt-dependent endothelial nitric-oxide synthase activation and nitric oxide release from endothelial cells. *J. Biol. Chem.* **2007**, *282*, 10660–10669. [CrossRef] [PubMed]
17. Wang, X.; Bove, A.M.; Simone, G.; Ma, B. Molecular Bases of VEGFR-2-Mediated Physiological Function and Pathological Role. *Front. Cell Dev. Biol.* **2020**, *8*, 599281. [CrossRef] [PubMed]
18. Gu, C.; Rodriguez, E.R.; Reimert, D.V.; Shu, T.; Fritzsch, B.; Richards, L.J.; Kolodkin, A.L.; Ginty, D.D. Neuropilin-1 Conveys Semaphorin and VEGF Signaling during Neural and Cardiovascular Development. *Dev. Cell* **2003**, *5*, 45–57. [CrossRef]

19. Khazaei, M.R.; Bunk, E.C.; Hillje, A.L.; Jahn, H.M.; Riegler, E.M.; Knoblich, J.A.; Young, P.; Schwamborn, J.C. The E3-ubiquitin ligase TRIM2 regulates neuronal polarization. *J. Neurochem.* **2011**, *117*, 29–37. [CrossRef]
20. Korn, C.; Augustin, H.G. Mechanisms of Vessel Pruning and Regression. *Dev. Cell* **2015**, *34*, 5–17. [CrossRef]
21. Moraes, F.; Paye, J.; Mac Gabhann, F.; Zhuang, Z.W.; Zhang, J.; Lanahan, A.A.; Simons, M. Endothelial cell-dependent regulation of arteriogenesis. *Circ. Res.* **2013**, *113*, 1076–1086. [CrossRef] [PubMed]
22. Potente, M.; Gerhardt, H.; Carmeliet, P. Basic and therapeutic aspects of angiogenesis. *Cell* **2011**, *146*, 873–887. [CrossRef] [PubMed]
23. Prosser, H.C.; Tan, J.T.; Dunn, L.L.; Patel, S.; Vanags, L.Z.; Bao, S.; Ng, M.K.; Bursill, C.A. Multifunctional regulation of angiogenesis by high-density lipoproteins. *Cardiovasc. Res.* **2014**, *101*, 145–154. [CrossRef] [PubMed]
24. Ridiandries, A.; Tan, J.T.; Ravindran, D.; Williams, H.; Medbury, H.J.; Lindsay, L.; Hawkins, C.; Prosser, H.C.; Bursill, C.A. CC-chemokine class inhibition attenuates pathological angiogenesis while preserving physiological angiogenesis. *FASEB J.* **2017**, *31*, 1179–1192. [CrossRef] [PubMed]
25. Limbourg, A.; Korff, T.; Napp, L.C.; Schaper, W.; Drexler, H.; Limbourg, F.P. Evaluation of postnatal arteriogenesis and angiogenesis in a mouse model of hind-limb ischemia. *Nat. Protoc.* **2009**, *4*, 1737–1746. [CrossRef]
26. Bonauer, A.; Carmona, G.; Iwasaki, M.; Mione, M.; Koyanagi, M.; Fischer, A.; Burchfield, J.; Fox, H.; Doebele, C.; Ohtani, K.; et al. MicroRNA-92a controls angiogenesis and functional recovery of ischemic tissues in mice. *Science* **2009**, *324*, 1710–1713. [CrossRef]
27. Lee, J.J.; Arpino, J.M.; Yin, H.; Nong, Z.; Szpakowski, A.; Hashi, A.A.; Chevalier, J.; O'Neil, C.; Pickering, J.G. Systematic Interrogation of Angiogenesis in the Ischemic Mouse Hind Limb: Vulnerabilities and Quality Assurance. *Arterioscler. Thromb. Vasc. Biol.* **2020**, *40*, 2454–2467. [CrossRef]
28. Cannizzo, C.M.; Adonopulos, A.A.; Solly, E.L.; Ridiandries, A.; Vanags, L.Z.; Mulangala, J.; Yuen, S.C.G.; Tsatralis, T.; Henriquez, R.; Robertson, S.; et al. VEGFR2 is activated by high-density lipoproteins and plays a key role in the proangiogenic action of HDL in ischemia. *FASEB J.* **2018**, *32*, 2911–2922. [CrossRef]

Disclaimer/Publisher's Note: The statements, opinions and data contained in all publications are solely those of the individual author(s) and contributor(s) and not of MDPI and/or the editor(s). MDPI and/or the editor(s) disclaim responsibility for any injury to people or property resulting from any ideas, methods, instructions or products referred to in the content.

Article

Exploring the Effects of Human Bone Marrow-Derived Mononuclear Cells on Angiogenesis In Vitro

Judith A. H. M. Peeters [1,2], Hendrika A. B. Peters [1,2], Anique J. Videler [1,2], Jaap F. Hamming [1], Abbey Schepers [1] and Paul H. A. Quax [1,2,*]

[1] Department of Surgery, Leiden University Medical Center, 2300 RC Leiden, The Netherlands; a.h.m.peeters@lumc.nl (J.A.H.M.P.); h.a.b.peters@lumc.nl (H.A.B.P.); anique.videler@hotmail.com (A.J.V.); j.f.hamming@lumc.nl (J.F.H.); a.schepers@lumc.nl (A.S.)
[2] Einthoven Laboratory for Experimental Vascular Medicine, Leiden University Medical Center, 2300 RC Leiden, The Netherlands
* Correspondence: p.h.a.quax@lumc.nl; Tel.: +31-71-5261584; Fax: +31-71-5266750

Abstract: Cell therapies involving the administration of bone marrow-derived mononuclear cells (BM-MNCs) for patients with chronic limb-threatening ischemia (CLTI) have shown promise; however, their overall effectiveness lacks evidence, and the exact mechanism of action remains unclear. In this study, we examined the angiogenic effects of well-controlled human bone marrow cell isolates on endothelial cells. The responses of endothelial cell proliferation, migration, tube formation, and aortic ring sprouting were analyzed in vitro, considering both the direct and paracrine effects of BM cell isolates. Furthermore, we conducted these investigations under both normoxic and hypoxic conditions to simulate the ischemic environment. Interestingly, no significant effect on the angiogenic response of human umbilical vein endothelial cells (HUVECs) following treatment with BM-MNCs was observed. This study fails to provide significant evidence for angiogenic effects from human bone marrow cell isolates on human endothelial cells. These in vitro experiments suggest that the potential benefits of BM-MNC therapy for CLTI patients may not involve endothelial cell angiogenesis.

Keywords: peripheral artery disease; chronic limb-threatening ischemia; angiogenesis; bone marrow-derived mononuclear cells; arteriogenesis; cell therapy

Citation: Peeters, J.A.H.M.; Peters, H.A.B.; Videler, A.J.; Hamming, J.F.; Schepers, A.; Quax, P.H.A. Exploring the Effects of Human Bone Marrow-Derived Mononuclear Cells on Angiogenesis In Vitro. *Int. J. Mol. Sci.* **2023**, *24*, 13822. https://doi.org/10.3390/ijms241813822

Academic Editor: Abdelkrim Hmadcha

Received: 9 August 2023
Revised: 29 August 2023
Accepted: 1 September 2023
Published: 7 September 2023

Copyright: © 2023 by the authors. Licensee MDPI, Basel, Switzerland. This article is an open access article distributed under the terms and conditions of the Creative Commons Attribution (CC BY) license (https://creativecommons.org/licenses/by/4.0/).

1. Introduction

Peripheral arterial disease (PAD) is a chronic condition where peripheral blood flow is restricted due to stenosis or blockage of the arteries. In an advanced state, PAD can lead to chronic limb-threatening ischemia (CLTI), resulting in patients suffering from rest pain and/or ischemic ulcers or gangrene. The current treatment for CLTI is directed at restoring the blood flow to the limb with endovascular or surgical interventions, in addition to standard drug therapy and cardiovascular risk management. Unfortunately, the success and patency rates of these interventions are around 60%. Due to the severity of the disease and shortcomings of current therapies, there is a need for new effective therapies.

In the last decades, the interest in the field of cell therapy, including stem cells, is rising, since this could be a promising alternative to conventional therapy. Cell therapy came to light early this century in 2002, when the first clinical study reported that bone marrow-derived mononuclear cells (BM-MNCs) could be safe and effectively used to treat CLTI [1]. Due to their potential ability to promote angiogenesis, BM-MNCs have been used in various clinical trials, showing beneficial effects for ulcer healing and limb salvage [2–6]. However, detailed analyses of various randomized controlled trials have failed to show clinically relevant beneficial effects [7]. Mononuclear cells are a mixture of different types of hematological cells, including lymphocytes, monocytes, and hematopoietic stem cells. They have been shown to have regenerative properties and the ability to promote angiogenesis [8,9]. However, the composition of cell therapies is largely variable, with various preparation methods and different routes of administration.

One of the promising new approaches is based on the use of REX-001, a highly standardized autologous bone marrow-derived mononuclear cell product that has shown significant blood flow recovery by increasing vascular density and functional neovascularization, which correlated with clinical benefits [10]. Due to these promising results, currently a phase III clinical trial is being conducted (ClinicalTrials.gov Identifier: NCT03174522). However, the exact mechanism of action of REX-001 is still unknown.

The neovascularization processes that lead to restoring the blood flow comprise both arteriogenesis and angiogenesis. Arteriogenesis is the recruitment of collaterals from pre-existing arterioles, and is mainly inflammatory driven. Angiogenesis is the formation of new capillary blood vessels, and plays a crucial role in various physiological and pathological processes involving the sprouting and remodeling of blood vessels from the pre-existing vasculature. The cell types that contribute to neovascularization are endothelial cells, circulating monocytes, smooth muscle cells, and pericytes [11,12]. Endothelial cells, which are important elements of blood vessels, play a pivotal role in angiogenesis. The proliferation and migration of endothelial cells are crucial events contributing to the formation of new vessels and the formation of a functional vascular network, and are driven by multiple growth factors and cytokines including vascular endothelial growth factor (VEGF), platelet-derived growth factor, insulin-like growth factor 1, interleukin 1, interleukin 6 (IL-6), and interleukin 8 (IL-8) [13,14]. In addition to proliferation, endothelial cell migration allows endothelial cells to navigate through the extracellular matrix to form new blood vessels. Endothelial cell migration is regulated by various signaling molecules, including VEGF and angiopoietins. Activated endothelial cells can release chemoattractants such as monocyte chemoattractant protein-1 (MCP-1), initiating the recruitment of monocytes to the angiogenic site [15–17].

Bone marrow-derived mononuclear cells (BM-MNCs) consist of a variety of cell types including lymphocytes, granulocytes, monocytes, and progenitor cells. It is hypothesized that BM-MNCs induce neovascularization, i.e., arteriogenesis and angiogenesis. In the current study the effect of BM-MNCs on angiogenesis was explored by studying endothelial cell proliferation, cell migration, angiogenic tube formation, and sprouting in different set-ups.

2. Results

2.1. Isolation and Quality Control of Bone Marrow-Derived Cells

The bone marrow mononuclear cell isolates used in this study were obtained from healthy volunteers (Hemacare, Charles River, Wilmington, MA, USA) and isolated according to a strict protocol that met strict specifications, as defined by Rojas-Torres et al. [18]. The first step was to isolate the BM-MNC cells according this protocol. As shown schematically in Figure 1, the BM-MNCs were isolated from heparinized bone marrow via Ficoll gradient separation. The characteristics of the product are defined in Table 1.

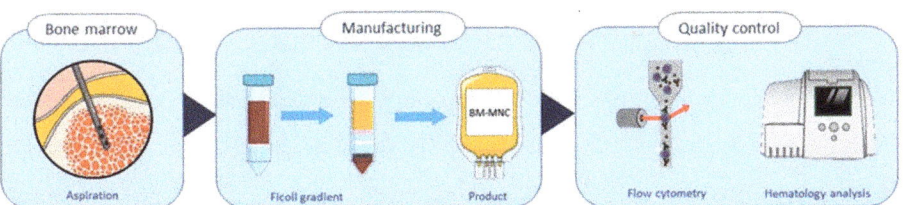

Figure 1. Illustration of the manufacturing process of BM-MNC isolates, starting with bone marrow aspiration, followed by manufacturing the product via Ficoll gradient cell separation, and finally a quality assessment was performed using flow cytometry and hematology analysis.

Table 1. The ranges of process performance indicators and cell populations from six times manufacturing the BM-MNC isolate.

Process Performance Indicator	BM-MNCs Relative to Bone Marrow (%)	Acceptance Criteria (%)
Leukocyte recovery	13.05 *–22.28	>15%
Erythrocyte depletion	99.91–100	>96%
Thrombocyte depletion	95.79–100	>60%
Cell Population	**BM-MNCs (%)**	
Viability	99.50–99.90	>80%
Leukocytes (CD45+) [1]	63.63–86.87	-
B lymphocytes (CD19+)	1.08–4.63	-
T lymphocytes (CD3+)	4.84–9.74	-
CD4+ T lymphocytes	2.52–4.68	-
CD8+ T lymphocytes	1.36–3.42	-
Granulocytes (CD16+, CD14−)	68.23–84.40	>30%
Monocytes (CD14+)	2.76–10.50	-
CD34+ leukocytes	1.34–3.59	>0.1%

* This cell isolate was not used in experiments. [1] of single viable cells.

The quality of the isolated BM-MNCs was analyzed using both hematology analysis and flow cytometry to demonstrate that the manufactured cell isolate met the quality acceptance criteria of the bone marrow cells isolates, as in the REX-001 clinical trial. The cell isolates in this study were produced according to the REX-001 manufacturing protocol.

Not all of the BM-MNC samples met the criteria of >15% leukocyte recovery; one sample only had 13.05% leukocyte recovery, and was not used in experiments (Table 1). All of the samples had >96% erythrocyte depletion and >60% thrombocyte depletion, meeting the quality criteria. Furthermore, all of the BM-MNC isolates met the following criteria: viability above 80%, containing >30% granulocytes, and the presence of CD34+/CD45+ cells (>0.1%).

In addition to the required quality assessment, a more extensive flow cytometry panel was used to characterize the cell composition of the BM-MNC isolates in more detail. The CD45+ cell fraction was analyzed further to determine the percentages of B lymphocytes and T lymphocytes. Subsequently, the T lymphocytes were further characterized to CD4+ and CD8+ T cells. In addition, the percentages of monocytes in the BM-MNC isolates were determined. The flow cytometry gating strategy is shown in Supplementary Figure S1.

2.2. BM-MNCs Have No Effect on Endothelial Cell Proliferation

To determine whether BM-MNCs have an effect on endothelial cell proliferation, directly or indirectly, human umbilical vein endothelial cells (HUVECs) were incubated either with increasing numbers of freshly isolated BM-MNCs or increasing concentrations of BM-MNC-conditioned medium, and the proliferation was analyzed with MTT assays. Based on previous experiments, the endpoints of both assays were set at 24 h after treatment with BM-MNCs.

To explore a direct effect of BM-MNCs on endothelial cell proliferation, BM-MNCs were added directly to the HUVEC cultures. None of the doses of BM-MNCs tested (625, 1250, 2500, and 5000 cells) resulted in a change in HUVEC proliferation in the MTT assay as compared to the negative control, i.e., EBM2 medium with 0.2% serum. The proliferation was significantly lower than in the positive control group that was exposed to the EMB2 medium supplemented with growth factors. If any effect could be observed, this would be that with the higher BM-MNC dose, slightly less endothelial cell proliferation occurred (Figure 2A and Supplementary Figure S2A). The data shown in Figure 2A are from one representative experiment. All of the experiments with BM isolates for different donors showed a similar pattern, with no effects on HUVEC proliferation (Figure S2).

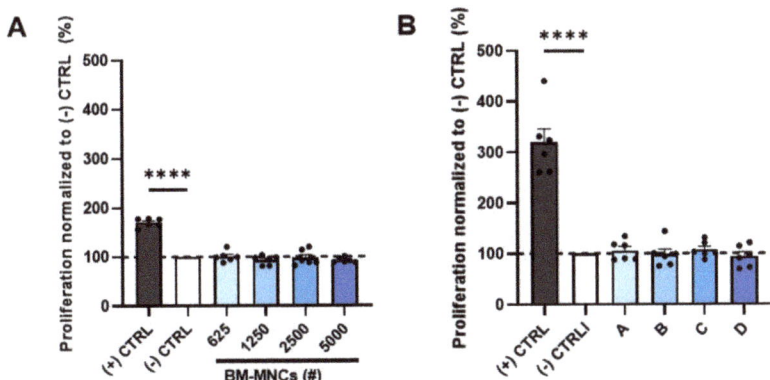

Figure 2. Quantification of HUVEC proliferation after treatment with either (**A**) BM-MNCs (625, 1250, 2500, or 5000 cells added indicated as BM-MNCs (#)) or with (**B**) BM-MNC-conditioned media A, B, C, and D, respectively, representative for 2500, 5000, 10,000, or 20,000 BM-MNCs. Graph 2A is representative for 6 experiments performed with BM-MNCs manufactured from 6 different bone marrow samples. Graph 2B is representative for 2 experiments performed with BM-MNC-conditioned media from 2 BM-MNC products. The positive control is EBM2 medium containing 2% serum, and the negative control is EBM2 medium containing 0.2% serum. Data are presented as mean ± SEM with datapoints (indicated as (•)) in sextuplicate. **** $p \leq 0.0001$ via one-way ANOVA.

In addition to the direct effects on HUVEC proliferation by BM-MNC isolates, we studied whether proliferation could be induced by paracrine factors present in the isolate. For this investigation, we incubated HUVECs with increasing concentrations of BM-MNC-conditioned media. The concentration is defined as the equivalent of BM-MNC cells secreting their paracrine factors into the conditioned medium, representative for 2500, 5000, 10,000, or 20,000 BM-MNCs (Figure 2B).

To evaluate if the BM-MNCs would have an indirect effect on HUVEC proliferation in other conditions, conditioned medium was also prepared in media with less or more serum added. To evoke a potential effect, the assays were also executed in hypoxic conditions, since hypoxia induces vascular endothelial growth factor (VEGF), which is an angiogenic factor (Figure S3). Nevertheless, adding BM-MNCs in hypoxic conditions did not increase HUVEC proliferation. To evaluate whether the kind of culture medium led to different BM-MNC-conditioned medium with different effects on HUVEC proliferation, these experiments were also performed using immune cell-suitable culture media to optimize the culturing conditions for the BM-MNCs, OptiMEM, and AIMV, in order to prepare BM-MNC-conditioned medium (Figure S4). HUVEC proliferation after adding BM-MNC-conditioned medium in OptiMEM did not show any differences, whereas HUVEC proliferation after adding BM-MNC-conditioned medium in AIMV showed a decrease in HUVEC proliferation under hypoxic circumstances. Since in none of these conditions was any difference in HUVEC proliferation observed compared to the negative control, this suggests that BM-MNCs exert no paracrine effects on endothelial cell proliferation.

2.3. BM-MNCs Do Not Affect Endothelial Cell Migration

Next to endothelial cell proliferation, endothelial cell migration is a key process involved in the formation of new vessels that might be stimulated by bone marrow cell isolates. To evaluate the effect of BM-MNCs on HUVEC migration, wound healing assays were performed. After culturing HUVEC for 24 h, a scratch wound was introduced into monolayers of HUVECs using the Incucyte Woundmaker Tool. Subsequently, these wounded cultures were treated with different doses of BM-MNCs. The plates were incubated in the IncuCyte S3, and pictures were taken after 12 h. The percentage of scratch-wound closure after 12 h was calculated. Figure 3A clearly shows the increasing concentration of BM-MNCs that was added at t = 0, visualized as cells over the wounded area.

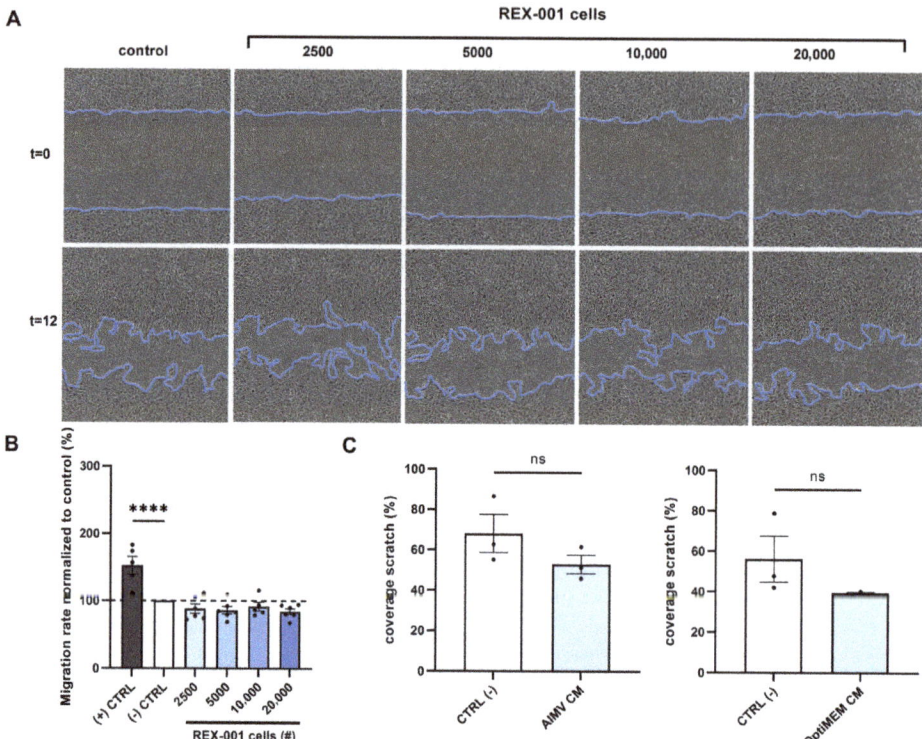

Figure 3. Representative pictures (**A**) and quantification of HUVEC scratch wound healing after (**B**) treatment with BM-MNCs (2500, 5000, 10,000, or 20,000 cells added) or (**C**) after treatment with BM-MNC-conditioned medium (AIMV or OptiMEM incubated with 1.33×10^6 BM-MNCs/mL). Graphs (**B**) and (**C**) are representative for 6 and 2 experiments performed, respectively, with BM-MNC isolates manufactured from different bone marrow samples. Data points represent technical replicates, six and two, respectively, and are presented as mean ± SEM. ns = non-significant, **** $p \leq 0.0001$ via one-way ANOVA.

Quantification of the scratch wound closure rate of HUVECs treated with BM-MNCs was performed after 12 h. The experiments were performed in six-fold with different BM-MNC isolates, and the results did not show an increase in migration rate (Figure 3B and Supplementary Figure S5A). Figure 3B shows a non-significant decreased migration rate of BM-MNCs in all dosages, whereas some graphs shown in Figure S5A also show a non-significant increase in scratch wound coverage. However, a clear induction of endothelial cells migration after adding BM-MNC isolates cannot be observed. Moreover, in one of the six experiments, a significantly lower migration rate was observed when adding 20,000 BM-MNCs.

In this scratch wound set-up, we also studied the potential paracrine effects; conditioned medium was prepared in immune cell-suitable culture media AIMV or OptiMEM mixed 1:1 with endothelial cell culture medium EBM2 containing 0.2% serum, and added to the wounded HUVEC cultures. After 12 h, the scratch wound cultures showed no significant difference in migration rate (Figure 3C). The conditioned media in other batches of BM-MNC isolates also did not lead to any changes in the migration rate (Figure S5B).

2.4. No Effect of BM-MNCs on Endothelial Cell Tube Formation

The angiogenic capacity of endothelial cells in general can be studied using a Matrigel tube formation assay. Therefore, we also studied the effect of BM-MNCs on the capacity of HUVECs to form tubes in a Matrigel tube formation assay. Here, we determined the total length of the tubes formed after 12 h of incubating HUVECs with different doses of BM-MNC isolates (Figure 4 and Supplementary Figure S6). The photos clearly show the increasing BM-MNC doses that were added at t = 0, visualizable as more cells adhering to the tubular structures. Quantification of the length, however, showed no differences between the different numbers of cells added.

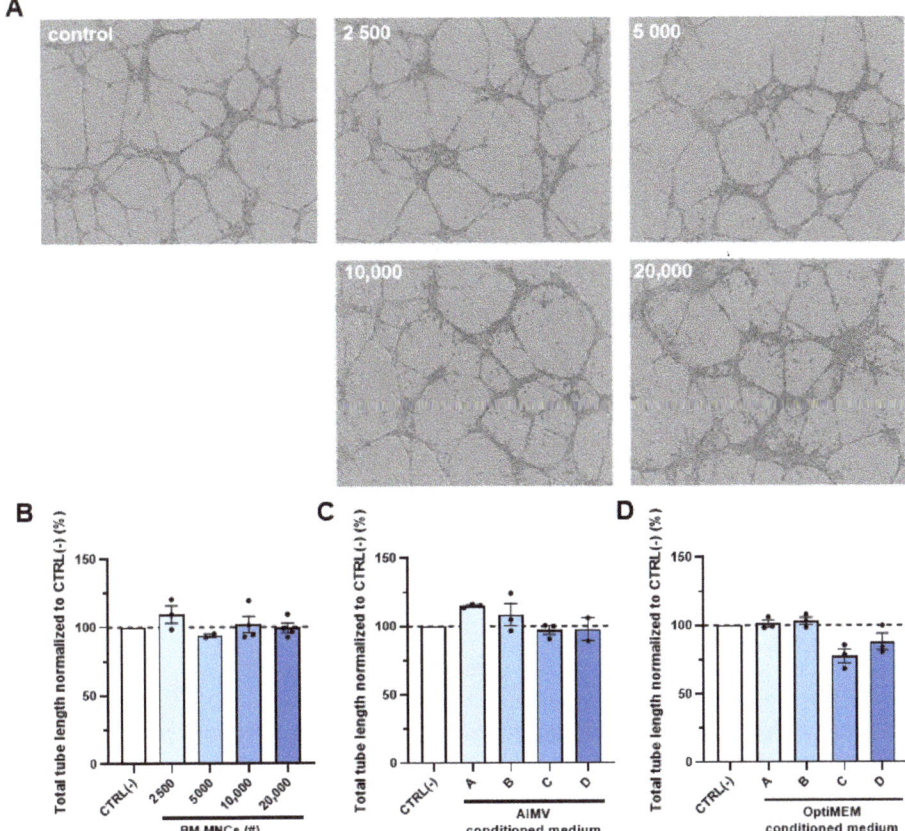

Figure 4. Representative microscopy photos (10×) of HUVEC tube formation with the presence of BM-MNCs (**A**), quantification of HUVEC tube formation length after treatment with (**B**) BM-MNCs (2500, 5000, 10,000, or 20,000 cells added), and after treatment with BM-MNC-conditioned medium where A, B, C, and D represent 2500, 5000, 10,000, or 20,000 BM-MNCs, respectively, in (**C**) AIMV medium or (**D**) OptiMEM medium. Data points represent three technical replicates, and are presented as mean ± SEM. Non-significant via one-way ANOVA.

Quantification of the total tube length of HUVECs was performed after 8 h. The experiments were performed in triplicate with different BM-MNC isolates. The results did not show an increase in tube formation rate (Figure 4B and Supplementary Figure S6A). Figure 4B shows no differences in endothelial cell tube formation length of BM-MNCs in all dosages, which is confirmed in Figure S6A.

The indirect effects of BM-MNCs on endothelial cell tube formation length were studied by adding BM-MNC-conditioned medium to HUVECs. The conditioned medium was prepared in AIMV or OptiMEM medium, both suitable immune cell culture media to optimize the culturing conditions for the BM-MNCs. The results show no differences in the tube length (Figure 4C,D).

2.5. The Effect of BM-MNCs on Aortic Ring Sprouting

Aortic ring sprouting ex vivo is another very informative assay for the angiogenic potential of cells or factors. Explants of mouse aortas have the capacity to sprout and form branching microvessels ex vivo when embedded in gels of collagen. Angiogenesis in this system is driven by endogenous growth factors released by the aorta and its outgrowth in response to the injury of the dissection procedure [19]. The aortic ring assay offers many advantages over existing models of angiogenesis. Unlike isolated EC, the native endothelium of the aortic explants has not been modified by repeated passages in culture and retains its original properties. Angiogenic sprouting occurs in the presence of pericytes, macrophages, and fibroblasts, as seen during wound healing in vivo [20].

The incubation of murine aortic rings with increasing numbers of BM-MNC isolates (2500, 5000, 10,000, or 20,000 cells added) (Figure 5) did not result in any differences in the numbers of sprouts originating from the rings. The analysis was performed after 7 days. The experiment was repeated with three different BM-MNC isolates (Figure S7).

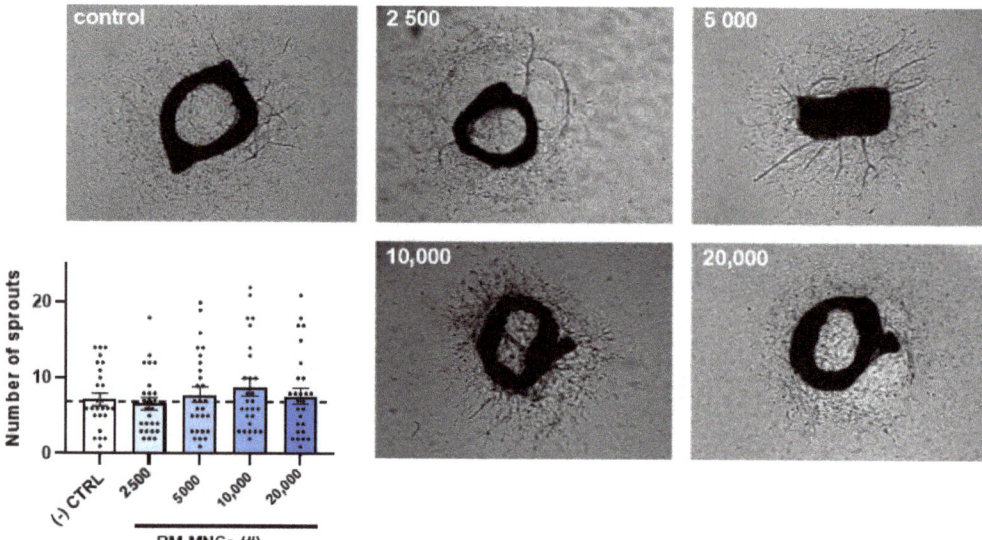

Figure 5. Quantification of neovessel sprouts of mice aortic rings after treatment with BM-MNCs (2500, 5000, 10,000, or 20,000 cells added). The graph is representative for 3 experiments performed with BM-MNCs isolated from 3 different bone marrow samples. Data are presented as mean ± SEM with data points in 30-fold. Non-significant via Kruskal-Wallis test.

2.6. BM-MNCs Release Angiogenic Cytokines

Thus far, no direct and paracrine effects of BM-MNCs on endothelial cells were observed. Therefore, we were interested in determining which cytokines and factors are released by BM-MNCs. To study factors excreted by BM-MNCs, the productions of IL-6, IL-8, MCP-1, and MMP-9 were determined. In OptiMEM, BM-MNCs produced 1.1 ng/mL of IL-6, 27.7 ng/mL of IL-8, 22.8 ng/mL of MCP-1, and 161.3 ng/mL of MMP-9. In AIMV,

BM-MNCs produced 1.5 ng/mL of IL-6, 33.3 ng/mL of IL-8, 31.2 ng/mL of MCP-1, and 203.9 ng/mL of MMP-9 (Figure 6).

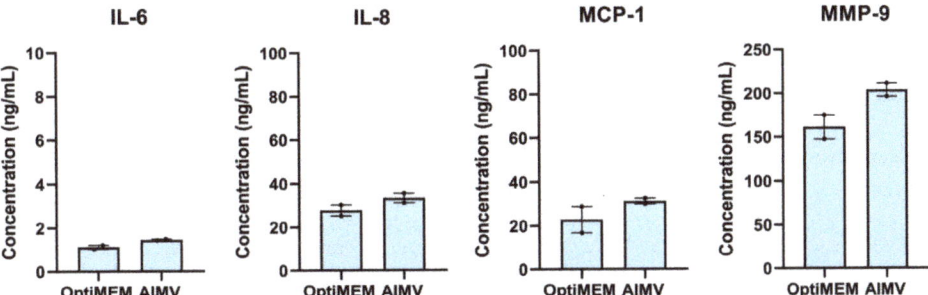

Figure 6. Quantification of the concentrations of IL-6, IL-8, MCP-1, and MMP-9 in OptiMEM or AIMV cell culture medium after 24 h incubation with 1.33×10^6 BM-MNCs/mL. Data points represent two measurements and are presented as mean ± SEM.

3. Discussion

The current study investigated whether bone marrow cell isolates, prepared according a strict protocol defined by Rojas-Torres et al. [18], have angiogenesis-stimulating potential. The effects of these bone marrow cell isolates on endothelial cell proliferation and endothelial cell migration were subsequently analyzed. Under none of the conditions tested could any stimulatory effects be observed with various concentrations, under normoxic or hypoxic conditions, or with direct contact or paracrine effect via conditioned medium exposure. Neither effects on Matrigel tube formation nor on aortic ring sprouting could be observed after incubation with different doses of cell isolates. Due to the lack of effects observed in these models, the effects were not evaluated in other ex vivo angiogenesis models such as spheroid cultures. In an attempt to unravel the mechanism of action of these specified BM-MNC isolates that showed promising results in clinical trials, it seems that the effect is most likely not due to an induction of angiogenesis per se [10].

The effect of bone marrow-derived mononuclear cells in patients with critical limb ischemia has been studied for a couple of decades, but its effectiveness remained unclear [21–23]. Clinical trials showed varying results, although there are plenty of studies that showed promising effects in patients with CLTI. However, the randomized controlled trials that reported beneficial effects of BM-MNCs are of relatively low quality. Thus far, the induction of neovascularization after BM-MNC therapy has not been convincingly demonstrated. Currently, a high-quality phase III randomized controlled trial is being conducted (NCT03174522) after reporting promising phase II trial (NCT00987363) results [10]. Despite these positive clinical trial results, the supposed mechanism of action by which these injected bone marrow cells induce neovascularization remains unclear.

Interestingly, most studies using bone marrow-derived cells as a therapy did not define the composition of cell types in the product nor analyze the product for quality assessment. In this study, quality requirements were set for the BM-MNC isolates, and each cell isolate was examined to confirm an adequate product quality. Hence, the cell composition of the BM-MNCs is known, and consists of multiple mononuclear cell types in certain proportions. Setting quality requirements, and thus assessing the proportions of different cell types in the product, is an important step in understanding the mechanism of action. Furthermore, it can help to acquire knowledge about why some patients do not respond to cell therapy, which may be related to the composition of the product. It is shown that 63.63–86.87% of the BM-MNC isolates consist of CD45+ leukocytes. However, the identity of the remaining 13–36% (CD45−) of the cells is still unknown. Previous research has shown that besides CD45+ hematopoietic stem cells, bone marrow also contains a population of heterogenous

CD45− nonhematopoietic tissue-committed stem cells [24]. In addition, CD45− cells in bone marrow cell fractions are of hematopoietic origin, and can be erythroid and lymphoid progenitors [25]. There is a substantial portion of CD34+ cells present in the our bone marrow cell isolates (Table 1), and CD34+ cell therapy has been shown to be one of the most promising approaches, most likely via the miR126 present in the condition medium of the CD34+ cells [26] that was reported to induce tube formation [27]. However, we were not able to demonstrate similar effects in our tube formation experiments.

Although no effects on angiogenesis were demonstrated, we showed IL-8, MCP-1, and MMP-9 to be present in the BM-MNC isolates, which are known proangiogenic factors. The role of IL-8 is widely researched in the oncologic field, where IL-8 promotes tumor angiogenesis by activating the VEGF pathway and enhancing MMP expression [13,28]. The presence of MMP-9 in the cell isolates suggests that extracellular matrix components can be degraded, which are key elements of the basement membrane surrounding blood vessels. Allowing the degradation of extracellular matrix allows endothelial cells to migrate into the surrounding tissue, starting new vessel formation [29]. MCP-1 is a chemokine that regulates the migration and infiltration of monocytes and macrophages to the site where it is released. Then, monocytes are able to differentiate into macrophages, which are important players in angiogenesis as they release factors including VEGF, MMPs, and enzymes promoting blood vessel growth by inducing endothelial cell proliferation and migration [30–32].

Understanding the absence of angiogenic effects of BM-MNCs on HUVECs, despite the proangiogenic chemokines that are excreted, is of great importance. A possible explanation for lacking angiogenic effects may be that the cell products were produced from bone marrow obtained from healthy donors. Although the BM-MNC isolates manufactured in this study fulfilled all quality criteria, BM-MNC isolates manufactured from CLTI patients (REX-001) suffering from type 2 diabetes mellitus may have a different composition and characteristics. Furthermore, the in vitro set-up only involved physiological HUVECs, while in the pathophysiologic situation of patients with CLTI dysfunctional endothelial cells are involved many more cells, chemokines, and inflammatory markers [33].

In this study, we studied the effects on HUVECs in normoxic and hypoxic environments, because endothelial cells in patients with CLTI suffer from hypoxia-induced endothelial cell dysfunction [34]. In addition, multiple culture media were used in the experiments to optimize the culturing conditions for both the BM-MNCs and the HUVECs together. Despite our efforts to unravel their effects on endothelial cells, one should bear in mind that other cell types, including smooth muscle cells, monocytes, and pericytes, are involved in angiogenesis and arteriogenesis. These cell types were not involved in our experiments, which is a limitation in our approach. However, REX-001 was studied in a murine model with the presence of all of the cell types involved, and improvement in revascularization and ischemic reperfusion was concluded [18]. Future fundamental biological studies should focus on identifying effects on these cell types. Moreover, there is a need for strategies to identify and augment the homing, survival, and effectiveness of the injected cells.

We believe that future clinical studies directed at cell therapeutic approaches to relieve CLI in patients should be based on a clear mechanism of action to avoid more disappointing clinical trial results.

4. Materials and Methods

4.1. BM-MNC Isolates Manufacturing

BM-MNCs were isolated from the heparinized bone marrow of healthy human donors (Hemacare, Charles-River) using a scaled-down density gradient. In short, the bone marrow was filtered using a 180 μm filter, and a sample of the filtered bone marrow was used for hematology analysis. The filtered bone marrow was then separated using Ficoll-Paque 1.077, and the upper layer, including the plasma and low-density cells, was isolated. These cells were washed twice with isotonic saline solution containing 2.5% human serum albumin.

Finally, the BM-MNCs were resuspended in Ringer's lactate solution containing 2.5% w/v glucose and 1% w/v HSA.

The filtered bone marrow and the final product were both measured in a Sysmex XP-300 hematology analyzer. The obtained amounts of leukocytes, erythrocytes, and thrombocytes were used to calculate the leukocyte recovery percentage, and the percentages of erythrocyte and thrombocyte depletion.

4.2. Flow Cytometric Analysis

Flow cytometry was performed on the isolated BM-MNC batches. The possible remaining erythrocytes were lysed in ACK lysis buffer (A1049201, Thermo Fisher, Waltham, MA, USA) and washed twice with PBS supplemented with 0.1% heat-inactivated fetal bovine serum. The conjugated antibodies to human CD45 (HI30, 1/200, 50 µg/mL, BioLegend, San Diego, CA, USA), CD3 (OKT3, 1/125, 30 µg/mL, BioLegend), CD4 (OKT4, 1/400, 150 µg/mL, BioLegend), CD8a (HIT8a, 1/300, 50 µg/mL, BioLegend), CD19 (HIB19, 1/200, 50 µg/mL, BioLegend), CD14 (M5E2, 1/400, 400 µg/mL, BioLegend), CD16 (3G8, 1/800, 0.5 mg/mL, BioLegend), CD56 (IgG κ, 1/200, 200 µg/mL, BioLegend), and CD34 (581, 1/100, 50 µg/mL, BioLegend) were incubated on ice for 30 min. 7-AAD viability staining solution (1/800, 50 µg/mL, BioLegend) was used as a viability marker. The flow cytometric acquisition was performed on an Aurora 3 Laser (Cytek, Fremont, CA, USA). The flow cytometry data were analyzed using FlowJo V10.1 software (BD).

4.3. Cell Culture of Human Umbilical Vein Endothelial Cells (HUVECs)

Human umbilical vein endothelial cells (C2519AS, Lonza, Basel, Switzerland) were cultured in plates coated with 0.1% gelatin in PBS in EBM-2 culture medium (CC-3156, Lonza, Basel, Switzerland) supplemented with EGM-2 SingleQuots (CC-4176, Lonza, Basel, Switzerland), and were between passages 3 and 4. The cells were incubated at 37 °C in a humidified 5% CO_2 environment.

4.4. BM-MNC Conditioned Medium

To prepare the conditioned media, 1.33×10^6 BM-MNCs/mL were incubated for 24 h in EBM-2, OptiMEM (Gibco, Billings, MT, USA), and AIMV (Gibco) culture media. The conditioned media were stored at −80 °C, and were thawed and diluted for use in the experiments.

4.5. MTT Assay

The cell proliferation ($n = 4$ experimental replicates) of the HUVECs was determined using MTT assays. A volume of 100 µL of HUVECs (4000 cells/well) were plated in 96-well plates and cultured until approximately 80% confluency was reached in complete endothelial cell culture medium. The medium was then replaced by endothelial cell low-serum medium containing 0.2% FBS for 24 h. Subsequently, the medium was replaced by BM-MNC treatments consisting of low-serum media containing 2500, 5000, 10,000, or 20,000 BM-MNCs. After 24 h of incubation, 10 µL of MTT (Thiazolyl, blue tetrazolium bromide, Sigma M5655) was added per well. The cells were incubated for 4 h, after which 75 µL of each well was discarded and replaced by 75 µL of isopropanol/0.1 M hydrogen chloride. The plates were incubated at room temperature on a platform shaker until dissolution of the formazan crystals was observed. Thereafter, the absorbance was read at 570 nm on a Cytation5 spectrophotometer (BioTek, Winooski, VT, USA), and the data were obtained using BioTek Gen5 software. The obtained mitochondrial metabolic activity data were quantified as a representative measure of cell proliferation.

4.6. Scratch Wound Healing Assay

For the scratch wound healing assays ($n = 6$ experimental replicates), HUVECs were plated on IncuCyte Imagelock 96-well plates (BA-04856, Sartorius AG, Goettingen, Germany) and cultured until approximately 90% confluence was reached in complete culture medium, as previously mentioned. The medium was then replaced by EBM-2 Basal medium supplemented with 2% FBS (SingleQuotsTM Supplements, CC-4176, Lonza) and 1% GA-1000 (SingleQuotsTM Supplements, CC-4176, Lonza). After 24 h, a scratch wound was introduced using the Incucyte Woundmaker Tool (4563, Sartorius AG, Goettingen, Germany), and different amounts of BM-MNCs were added in EBM-2 Basal medium supplemented with 2% FBS and 1% GA-1000. The plates were incubated in the IncuCyte S3, and pictures were taken after 12 h. The percentage of scratch wound closure after 12 h was calculated by measuring the difference in the wound surface at baseline and the wound surface after 12 h using the Wound Healing Tool of ImageJ.

4.7. Tube Formation Assay

HUVECs were seeded in a 6-well plate in EBM-2 culture medium supplemented with SingleQuots until they became confluent. The medium was replaced with low-serum medium for 24 h. Then, a 96-well plate was coated with 45 µL/well of Geltrex basement membrane matrix (A1413202, ThermoFisher, Waltham, MA, USA). Suspensions of HUVECs at a concentration of 250,000 cells/mL and different concentrations of BM-MNCs or BM-MNC-conditioned medium were prepared and seeded in the coated 96-well plate. The plate was incubated in IncuCyte S3, and pictures were taken every 2 h for 24 h. The analysis was performed using ImageJ at t = 8 h.

4.8. Aortic Ring Assay

The 8-week-old mice were sacrificed, the aortas were resected, and the surrounding fat and branching vessels were removed. The aortas were cut in <1 mm rings, and were overnight incubated at 37 °C in a humidified 5% CO_2 environment in OptiMEM supplemented with 1% penicillin/streptomycin. A 96-well plate was coated with 75 µL of collagen matrix (Collagen (Type 1, Merck Sigma-Aldrich, Millipore, Burlington, MA, USA) in DMEM (ThermoFisher, Waltham, MA, USA), pH adjusted with 5N NaOH), and then one aortic ring was added per well. After one hour, the collagen was solid and 150 µL of OptiMEM supplemented with 2.5% FBS, 1% penicillin–streptomycin solution (Cytiva, HyClone Laboratories, North Logan, UT, USA), 10 ng/mL of mouse VEGF (BioLegend, San Diego, CA, USA), and different amounts of BM-MNCs were added to each ring, with 20 or 30 rings per condition. After a total of 7 days of incubation at 37 °C in a humidified 5% CO_2 environment, with a medium replacement after 3 days, pictures of each aortic ring were taken using live phase-contrast microscopy (Axiovert 40C, Carl Zeiss, Oberkochen, Germany). The number of sprouts were counted manually.

4.9. ELISA

The bone marrow-derived mononuclear cells were plated at 1.33×10^6 cells/mL for 24 h to prepare the conditioned medium. After 24 h, the supernatant was stored at -20 °C. The IL-6, IL-8, MCP-1, and MMP-9 concentrations were determined via ELISA, according to the protocol (BD Biosciences, San Jose, CA, USA) in the supernatant of the BM-MNCs.

4.10. Statistical Analysis

Differences in the continuous variables between groups were statistically assessed using one-way ANOVA or Kruskal–Wallis tests in Graph Pad Prism 8 software. The data are represented as means ± SEM. The significance was set at $p < 0.05$.

5. Conclusions

In this study, no effect from human bone marrow cell isolates on the angiogenic behavior of experimental human endothelial cells (HUVEC) could be demonstrated. Our research holds significant relevance, as it addresses the shortage of supporting evidence regarding the effects of BM-MNCs on cultured endothelial cells.

Supplementary Materials: The following supporting information can be downloaded at: https://www.mdpi.com/article/10.3390/ijms241813822/s1.

Author Contributions: Conceptualization, J.A.H.M.P., A.S. and P.Q; methodology J.P, H.A.B.P., P.H.A.Q. and A.J.V.; analysis, H.P, J.A.H.M.P. and A.J.V.; resources, P.H.A.Q.; data curation, J.A.H.M.P. and H.A.B.P.; writing—original draft preparation, J.A.H.M.P., H.A.B.P. and P.H.A.Q.; writing—review and editing, J.A.H.M.P., A.S., J.F.H. and P.H.A.Q.; supervision, P.H.A.Q. and A.S.; funding acquisition, P.H.A.Q. All authors have read and agreed to the published version of the manuscript.

Funding: This research was funded by IXAKA. The APC was funded by LUMC.

Informed Consent Statement: Not applicable.

Data Availability Statement: Data can be made available upon request.

Acknowledgments: The help of Alwin de Jong with flow cytometry is highly appreciated.

Conflicts of Interest: The authors declare no conflict of interest.

References

1. Tateishi-Yuyama, E.; Matsubara, H.; Murohara, T.; Ikeda, U.; Shintani, S.; Masaki, H.; Amano, K.; Kishimoto, Y.; Yoshimoto, K.; Akashi, H.; et al. Therapeutic angiogenesis for patients with limb ischaemia by autologous transplantation of bone-marrow cells: A pilot study and a randomised controlled trial. *Lancet* **2002**, *360*, 427–435. [CrossRef]
2. Yusoff, F.M.; Kajikawa, M.; Matsui, S.; Hashimoto, H.; Kishimoto, S.; Maruhashi, T.; Chowdhury, M.; Noma, K.; Nakashima, A.; Kihara, Y.; et al. Review of the Long-term Effects of Autologous Bone-Marrow Mononuclear Cell Implantation on Clinical Outcomes in Patients with Critical Limb Ischemia. *Sci. Rep.* **2019**, *9*, 7711. [CrossRef]
3. Matoba, S.; Tatsumi, T.; Murohara, T.; Imaizumi, T.; Katsuda, Y.; Ito, M.; Saito, Y.; Uemura, S.; Suzuki, H.; Fukumoto, S.; et al. Long-term clinical outcome after intramuscular implantation of bone marrow mononuclear cells (Therapeutic Angiogenesis by Cell Transplantation [TACT] trial) in patients with chronic limb ischemia. *Am. Heart J.* **2008**, *156*, 1010–1018. [CrossRef] [PubMed]
4. Dash, N.R.; Dash, S.N.; Routray, P.; Mohapatra, S.; Mohapatra, P.C. Targeting nonhealing ulcers of lower extremity in human through autologous bone marrow-derived mesenchymal stem cells. *Rejuvenation Res.* **2009**, *12*, 359–366. [CrossRef] [PubMed]
5. Procházka, V.; Gumulec, J.; Jalůvka, F.; Salounová, D.; Jonszta, T.; Czerný, D.; Krajča, J.; Urbanec, R.; Klement, P.; Martinek, J.; et al. Cell therapy, a new standard in management of chronic critical limb ischemia and foot ulcer. *Cell Transplant.* **2010**, *19*, 1413–1424. [CrossRef]
6. Sharma, S.; Pandey, N.N.; Sinha, M.; Kumar, S.; Jagia, P.; Gulati, G.S.; Gond, K.; Mohanty, S.; Bhargava, B. Randomized, Double-Blind, Placebo-Controlled Trial to Evaluate Safety and Therapeutic Efficacy of Angiogenesis Induced by Intraarterial Autologous Bone Marrow-Derived Stem Cells in Patients with Severe Peripheral Arterial Disease. *J. Vasc. Interv. Radiol.* **2021**, *32*, 157–163. [CrossRef]
7. Rigato, M.; Monami, M.; Fadini, G.P. Autologous Cell Therapy for Peripheral Arterial Disease: Systematic Review and Meta-Analysis of Randomized, Nonrandomized, and Noncontrolled Studies. *Circ. Res.* **2017**, *120*, 1326–1340. [CrossRef]
8. Asahara, T.; Murohara, T.; Sullivan, A.; Silver, M.; van der Zee, R.; Li, T.; Witzenbichler, B.; Schatteman, G.; Isner, J.M. Isolation of putative progenitor endothelial cells for angiogenesis. *Science* **1997**, *275*, 964–967. [CrossRef] [PubMed]
9. Kobayashi, K.; Kondo, T.; Inoue, N.; Aoki, M.; Mizuno, M.; Komori, K.; Yoshida, J.; Murohara, T. Combination of in vivo angiopoietin-1 gene transfer and autologous bone marrow cell implantation for functional therapeutic angiogenesis. *Arterioscler. Thromb. Vasc. Biol.* **2006**, *26*, 1465–1472. [CrossRef]
10. Ruiz-Salmeron, R.; de la Cuesta-Diaz, A.; Constantino-Bermejo, M.; Pérez-Camacho, I.; Marcos-Sánchez, F.; Hmadcha, A.; Soria, B. Angiographic demonstration of neoangiogenesis after intra-arterial infusion of autologous bone marrow mononuclear cells in diabetic patients with critical limb ischemia. *Cell Transplant.* **2011**, *20*, 1629–1639. [CrossRef]
11. Folkman, J. Angiogenesis. *Annu. Rev. Med.* **2006**, *57*, 1–18. [CrossRef]
12. Potente, M.; Gerhardt, H.; Carmeliet, P. Basic and therapeutic aspects of angiogenesis. *Cell* **2011**, *146*, 873–887. [CrossRef]
13. Li, A.; Dubey, S.; Varney, M.L.; Dave, B.J.; Singh, R.K. IL-8 directly enhanced endothelial cell survival, proliferation, and matrix metalloproteinases production and regulated angiogenesis. *J. Immunol.* **2003**, *170*, 3369–3376. [CrossRef]
14. Ashtar Nakhaei, N.; Najarian, A.; Farzaei, M.H.; Norooznezhad, A.H. Endothelial dysfunction and angiogenesis: What is missing from COVID-19 and cannabidiol story? *J. Cannabis. Res.* **2022**, *4*, 21. [CrossRef]
15. Herbert, S.P.; Stainier, D.Y. Molecular control of endothelial cell behaviour during blood vessel morphogenesis. *Nat. Rev. Mol. Cell Biol.* **2011**, *12*, 551–564. [CrossRef] [PubMed]
16. Singh, S.; Anshita, D.; Ravichandiran, V. MCP-1: Function, regulation, and involvement in disease. *Int. Immunopharmacol.* **2021**, *101*, 107598. [CrossRef] [PubMed]
17. Avraham-Davidi, I.; Yona, S.; Grunewald, M.; Landsman, L.; Cochain, C.; Silvestre, J.S.; Mizrahi, H.; Faroja, M.; Strauss-Ayali, D.; Mack, M.; et al. On-site education of VEGF-recruited monocytes improves their performance as angiogenic and arteriogenic accessory cells. *J. Exp. Med.* **2013**, *210*, 2611–2625. [CrossRef]

18. Rojas-Torres, M.; Jiménez-Palomares, M.; Martín-Ramírez, J.; Beltrán-Camacho, L.; Sánchez-Gomar, I.; Eslava-Alcon, S.; Rosal-Vela, A.; Gavaldá, S.; Durán-Ruiz, M.C. REX-001, a BM-MNC Enriched Solution, Induces Revascularization of Ischemic Tissues in a Murine Model of Chronic Limb-Threatening Ischemia. *Front. Cell Dev. Biol.* **2020**, *8*, 602837. [CrossRef] [PubMed]
19. Nicosia, R.F.; Ottinetti, A. Growth of microvessels in serum-free matrix culture of rat aorta. A quantitative assay of angiogenesis in vitro. *Lab. Invest.* **1990**, *63*, 115–122.
20. Nowak-Sliwinska, P.; Alitalo, K.; Allen, E.; Anisimov, A.; Aplin, A.C.; Auerbach, R.; Augustin, H.G.; Bates, D.O.; van Beijnum, J.R.; Bender, R.H.F.; et al. Consensus guidelines for the use and interpretation of angiogenesis assays. *Angiogenesis* **2018**, *21*, 425–532. [CrossRef]
21. Samura, M.; Hosoyama, T.; Takeuchi, Y.; Ueno, K.; Morikage, N.; Hamano, K. Therapeutic strategies for cell-based neovascularization in critical limb ischemia. *J. Transl. Med.* **2017**, *15*, 49. [CrossRef]
22. Hart, C.A.; Tsui, J.; Khanna, A.; Abraham, D.J.; Baker, D.M. Stem cells of the lower limb: Their role and potential in management of critical limb ischemia. *Exp. Biol. Med.* **2013**, *238*, 1118–1126. [CrossRef] [PubMed]
23. Magenta, A.; Florio, M.C.; Ruggeri, M.; Furgiuele, S. Autologous cell therapy in diabetes-associated critical limb ischemia: From basic studies to clinical outcomes (Review). *Int. J. Mol. Med.* **2021**, *48*, 173. [CrossRef] [PubMed]
24. Kucia, M.; Reca, R.; Jala, V.R.; Dawn, B.; Ratajczak, J.; Ratajczak, M.Z. Bone marrow as a home of heterogenous populations of nonhematopoietic stem cells. *Leukemia* **2005**, *19*, 1118–1127. [CrossRef] [PubMed]
25. Boulais, P.E.; Mizoguchi, T.; Zimmerman, S.; Nakahara, F.; Vivié, J.; Mar, J.C.; van Oudenaarden, A.; Frenette, P.S. The Majority of $CD45^{(-)}$ $Ter119^{(-)}$ $CD31^{(-)}$ Bone Marrow Cell Fraction Is of Hematopoietic Origin and Contains Erythroid and Lymphoid Progenitors. *Immunity* **2018**, *49*, 627–639. [CrossRef]
26. Sahoo, S.; Klychko, E.; Thorne, T.; Misener, S.; Schultz, K.M.; Millay, M.; Ito, A.; Liu, T.; Kamide, C.; Agrawal, H.; et al. Exosomes from human $CD34^{(+)}$ stem cells mediate their proangiogenic paracrine activity. *Circ. Res.* **2011**, *109*, 724–728. [CrossRef]
27. van Solingen, C.; Seghers, L.; Bijkerk, R.; Duijs, J.M.; Roeten, M.K.; van Oeveren-Rietdijk, A.M.; Baelde, H.J.; Monge, M.; Vos, J.B.; de Boer, H.C.; et al. Antagomir-mediated silencing of endothelial cell specific microRNA-126 impairs ischemia-induced angiogenesis. *J. Cell Mol. Med.* **2009**, *13*, 1577–1585. [CrossRef]
28. Shi, J.; Wei, P.K. Interleukin-8: A potent promoter of angiogenesis in gastric cancer. *Oncol. Lett.* **2016**, *11*, 1043–1050. [CrossRef]
29. van Hinsbergh, V.W.; Engelse, M.A.; Quax, P.H. Pericellular proteases in angiogenesis and vasculogenesis. *Arterioscler. Thromb. Vasc. Biol.* **2006**, *26*, 716–728. [CrossRef]
30. Corliss, B.A.; Azimi, M.S.; Munson, J.M.; Peirce, S.M.; Murfee, W.L. Macrophages: An Inflammatory Link Between Angiogenesis and Lymphangiogenesis. *Microcirculation* **2016**, *23*, 95–121. [CrossRef]
31. Melincovici, C.S.; Boşca, A.B.; Şuşman, S.; Mărginean, M.; Mihu, C.; Istrate, M.; Moldovan, I.M.; Roman, A.L.; Mihu, C.M. Vascular endothelial growth factor (VEGF)—Key factor in normal and pathological angiogenesis. *Rom. J. Morphol. Embryol.* **2018**, *59*, 455–467. [PubMed]
32. Fields, G.B. Mechanisms of Action of Novel Drugs Targeting Angiogenesis-Promoting Matrix Metalloproteinases. *Front. Immunol.* **2019**, *10*, 1278. [CrossRef] [PubMed]
33. Kavurma, M.M.; Bursill, C.; Stanley, C.P.; Passam, F.; Cartland, S.P.; Patel, S.; Loa, J.; Figtree, G.A.; Golledge, J.; Aitken, S.; et al. Endothelial cell dysfunction: Implications for the pathogenesis of peripheral artery disease. *Front. Cardiovasc. Med.* **2022**, *9*, 1054576. [CrossRef] [PubMed]
34. Zhao, M.; Wang, S.; Zuo, A.; Zhang, J.; Wen, W.; Jiang, W.; Chen, H.; Liang, D.; Sun, J.; Wang, M. HIF-1α/JMJD1A signaling regulates inflammation and oxidative stress following hyperglycemia and hypoxia-induced vascular cell injury. *Cell. Mol. Biol. Lett.* **2021**, *26*, 40. [CrossRef]

Disclaimer/Publisher's Note: The statements, opinions and data contained in all publications are solely those of the individual author(s) and contributor(s) and not of MDPI and/or the editor(s). MDPI and/or the editor(s) disclaim responsibility for any injury to people or property resulting from any ideas, methods, instructions or products referred to in the content.

Article

Computational Fluid Dynamics (CFD) Model for Analysing the Role of Shear Stress in Angiogenesis in Rheumatoid Arthritis

Malaika K. Motlana [1] and Malebogo N. Ngoepe [1,2,*]

[1] Department of Mechanical Engineering, University of Cape Town, Rondebosch, Cape Town 7701, South Africa
[2] Centre for Research in Computational and Applied Mechanics (CERECAM), University of Cape Town, Rondebosch, Cape Town 7701, South Africa
* Correspondence: malebogo.ngoepe@uct.ac.za; Tel.: +27-2165-04444

Abstract: Rheumatoid arthritis (RA) is an autoimmune disease characterised by an attack on healthy cells in the joints. Blood flow and wall shear stress are crucial in angiogenesis, contributing to RA's pathogenesis. Vascular endothelial growth factor (VEGF) regulates angiogenesis, and shear stress is a surrogate for VEGF in this study. Our objective was to determine how shear stress correlates with the location of new blood vessels and RA progression. To this end, two models were developed using computational fluid dynamics (CFD). The first model added new blood vessels based on shear stress thresholds, while the second model examined the entire blood vessel network. All the geometries were based on a micrograph of RA blood vessels. New blood vessel branches formed in low shear regions (0.840–1.260 Pa). This wall-shear-stress overlap region at the junctions was evident in all the models. The results were verified quantitatively and qualitatively. Our findings point to a relationship between the development of new blood vessels in RA, the magnitude of wall shear stress and the expression of VEGF.

Keywords: rheumatoid arthritis (RA); vascular endothelial growth factor (VEGF); angiogenesis; wall shear stress; blood vessels; pathogenesis; computational fluid dynamics (CFD); CFD model

1. Introduction

Rheumatoid arthritis (RA) is a chronic, autoimmune disease in which the body's immune system attacks healthy tissue cells found in the lining of the joints [1–3]. RA impacts this lining, known as the synovium or synovial membrane. Owing to the involvement of many different variables, the pathophysiology of RA is still not fully understood [4]. However, angiogenesis is crucial to the progression of RA and results in the formation of new blood vessels [5–8]. Under physiological conditions, angiogenesis is governed by a complex, balanced network of chemical and mechanical cues, which maintain vital physiological functions [9]. In pathological cases such as RA, the altered environment and unbalanced angiogenetic processes contribute to disease progression [9,10].

Various sources have reported increased angiogenesis in RA patients, which can be detected at the point of clinical diagnosis [6]. RA produces a radically altered synovial composition with a reduced viscosity [11]. This enables pannus formation in the affected joint [3,10,12–16]. The environment in the pannus is hypoxic and inflammatory, necessitating increased angiogenesis to supply oxygen and nutrients [6,14,17]. Another view is that angiogenesis, a process natural to the body, drives RA through the development and maintenance of the pannus in an environment that upregulates vascular endothelial growth factor (VEGF) [7,10,12]. Although different, these perspectives elucidate the complexity of the disease and support the idea that RA angiogenesis arises from a misbalance between stimulating and inhibiting factors [10].

VEGF is essential to angiogenesis under physiological and pathological conditions [18–21]. Clinically, patients with RA present with increased VEGF levels in both serum and synovial fluid [22–24]. The role of VEGF-related genes has also been explored in these patients [7,23,25–28]. Elevated VEGF levels correlate with markers of inflammation and RA activity, such as increased C-reactive protein and an increase in swollen joints [29–31]. A wide range of factors, including hypoxia, hormones and signalling networks, all influence the expression of VEGF [32–36]. In addition, mechanical variables such as shear stress contribute to VEGF expression and angiogenesis [37].

Unlike genetics and chemical factors, the relationships between mechanical factors, angiogenesis, VEGF and RA have been less explored [37]. Shear stress has been shown to influence angiogenesis significantly, and the extracellular matrix's mechanical behaviour has been analysed [38–42]. From a fluid mechanics perspective, shear stress is pivotal when considering blood flow. In blood vessels, localised blood flow patterns influence a range of stress and stretch measures, including the wall shear stress experienced by blood vessel wall components [43–45]. These forces arise from many haemodynamic variables, including the pulsatile nature of blood flow and pressure [46]. Shear stress in the blood vessels can affect morphology, organisation of the endothelial cytoskeleton, the functioning of ion channels and gene expression within endothelial cells [47–49]. Given the wide-ranging impact, the effects of wall shear stress on VEGF and angiogenesis are varied [46].

A shear stress threshold of 1 Pa was found to trigger angiogenic sprouting in endothelial cells, which could then penetrate the underlying matrix [50]. Blood fluid shear stress has also been linked to the upregulation of VEGF gene expression [51]. One study hypothesised that shear stress generated by considerable differences between capillary, venous and arterial blood flow might influence the observed differences in VEGF expression [52]. In a murine model, VEGF was expressed exclusively by arterial endothelial cells. When laminar arterial shear stress was applied to human umbilical vein endothelial cells (HUVEC), the expression of VEGF increased, but the exact mechanism remained unclear. A more recent microfluidic study demonstrated that departure from the stabilised state, either in shear stress or VEGF concentration, led to neovascularisation [53]. A recent study estimated shear stress in endothelial cells in regions of neovascularisation [54]. The authors highlighted that values above 0.1 Pa were physiologically relevant. The expression of VEGF increased with both pulsatile flow and laminar shear stress [55,56]. While it is clear that shear stress and other mechanical variables play an essential role in VEGF regulation and angiogenesis, the application of these findings to RA remains limited [52].

Although many studies of angiogenesis and blood flow have been conducted in vitro, computational fluid dynamics (CFD) could augment these approaches. Significant advances in computing technology have improved the approximation of solutions for the Navier–Stokes equations, which describe fluid flow in a wide range of contexts [57–67]. As a result, CFD models have been developed for a plethora of applications, including biofluid flows [68–71]. The solvers compute pressure and velocity results, which can then be used to calculate quantities such as shear stress. In biofluid flows, CFD models provide a potential avenue for quantifying and exploring various mechanical variables, such as shear stress, in blood vessels. Countless cardiovascular CFD models have enabled the quantification of variables which would prove challenging to measure in other settings [59,72]. For example, a CFD angiogenesis model estimated shear stresses experienced by endothelial tip cells [54]. A limitation of CFD is computational cost, with very sophisticated models requiring significant computing resources and time [73]. Model developers must consider how different aspects of the model may be represented and which assumptions could enable reasonable simplification. Kretschmer et al. demonstrate that angiogenesis relies on mechanical and chemical factors [37]. They further argue that chemical cues can be translated into mechanical signals and vice versa. Together with Leblonde et al., they highlight that VEGF is one of the most critical drivers of RA angiogenesis and that VEGF plays a vital role in the mechanics of angiogenesis [3]. In a simplified CFD angiogenesis model, shear stress could be a proxy for VEGF [52,54,74].

This study explores the role of shear stress, as a surrogate for VEGF, in developing new blood vessels in RA. Using CFD, the study examines how shear stress emanating from the blood flow coincides with the location of new blood vessels and the progression of RA. This is achieved by analysing patterns of blood flow variability that may relate to where new blood vessels emerge within the blood vessel network. These findings are compared to a micrograph of blood vessel networks in RA. Understanding these patterns may be beneficial in gaining insight into the significance of the magnitude of shear stress and, by implication, the expression of VEGF.

2. Results

The results section begins with a brief note on mesh independence. The geometries from Section 2.2 were discretised to enable the numerical solution of the governing equations, and this process resulted in the development of a mesh. Greater detail regarding mesh independence is given in Section 2.1, followed by velocity and shear stress results for Models 1 and 2.

2.1. Mesh Independence

For a fluid flow solution, sufficient refinement must be achieved to ensure that the mesh does not have a negative impact on the numerical solution. A solution computed for an adequately refined mesh is considered mesh/grid independent. A less than 2% error was deemed an acceptable value for mesh independence in this study. The element size for the mesh independence was determined by modelling a single blood vessel represented by a solid cylinder of radius 0.001 m and length of 0.1 m. The mesh was generated with an initial element size of 0.0000334 m, and the blood velocity was set to 0.19 m/s at the inlets. The process was repeated twice, decreasing the element size in each subsequent run. Given that wall shear stress is calculated from velocity (and CFD computes velocity and pressure fields), velocities at the outlet of the flow field were compared, and the error (1.41%) was found to be sufficiently small for an element size of 0.0000334 m. Both the mesh and the mesh independence graph are illustrated in Figure 1.

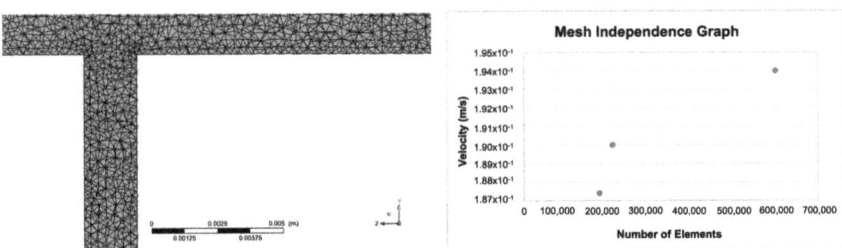

Figure 1. An image of the mesh generated from the original geometry and a graph illustrating grid independence for the model.

2.2. Model 1

Due to the wall shear stress being significantly higher than that prescribed for physiological arterial shear, the blood velocity was decreased from 0.19 m/s in the mesh independence study to 0.09 m/s in Models 1 and 2. The velocity was therefore reduced according to the Haagen–Poiseuille equation until the arterial shear range was satisfied.

The shear stress in wall 2 of Model 1A, shown in Figure 2, is outside the previously defined range. The velocity vectors indicate almost no flow within wall 2. The velocity was in the range of 0.00–0.055 m/s. This is a non-physiological finding because blood in the body constantly moves in the circulatory system. The position and number of inlets and outlets in Models 1B and 1C were adjusted to evaluate their effect on velocity. Compared to Model 1A, Model 1B showed a significant increase in the velocities (>0.22 m/s) of blood vessels 1 and 2. Parts of wall 1 had shear stresses greater than 4.2 Pa. Both wall shear stress and velocity lay outside the range of blood flow conditions in this model. Except for wall 2, which experienced the same shear stresses as Model 1A, Model 1C achieved plausible physiological velocity and wall shear stress values. Therefore, subsequent models were built from Model 1C. Junction 1 had a wall shear stress range of 0.420–1.260 Pa, and junction 2 had a range of 0.840–2.520 Pa. The two ranges intersected between 0.840–1.260 Pa. Based on this range, additional blood vessels were added in subsequent models. The junction overlap region for Model 1, shown in Table 1, is defined as the intersection between the wall shear stress range at all the junctions within a single model.

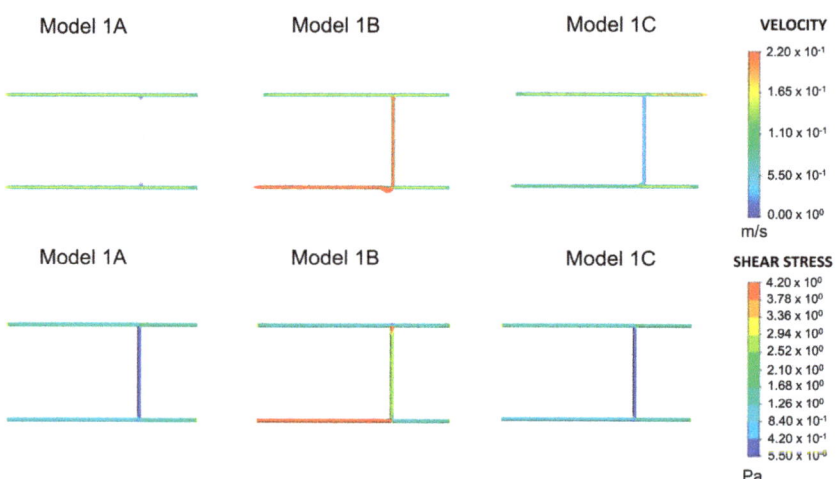

Figure 2. Velocity and shear stress results for Models 1A–1C. This was the same model with variations on inlets and outlets, as shown in Figure 2. Given the plausible velocity and shear stress ranges, Model 1C was chosen as the baseline model for all subsequent variations of Model 1.

Table 1. Wall shear stress and velocity ranges, and wall shear stress junction overlap regions at different locations in the blood vessel network for different variations of Model 1.

Model	Location of Wall Shear Stress	Wall Shear Stress Range (Pa)	Junction Overlap Region (Pa)	Velocity Range (m/s)
1C	Junction 1	0.420–1.260	0.840–1.260	0.055–0.11
	Junction 2	0.840–2.520		0.055–0.22
	Wall 1	0.840–1.680		0.11–0.165
	Wall 2	0.0000055–0.420		0.00–0.055
	Wall 3	0.840–2.10		0.055–0.22
1D	Junction 1	0.420–1.260	0.840–1.260	0.055–0.165
	Junction 2	0.840–1.680		0.055–0.11
	Junction 3	0.840–2.10		0.11–0.165
	Wall 1	0.420–1.680		0.11–0.165
	Wall 2	0.420–0.840		0.00–0.055
	Wall 3	0.0000055–1.680		0.055–0.165
	Wall 4	0.420–1.260		0.055–0.11
1E	Junction 1	0.420–1.260	0.840–1.260	0.055–0.165
	Junction 2	0.840–1.680		0.00–0.055
	Junction 3	0.840–2.10		0.055–0.11
	Junction 4	0.420–1.680		0.055–0.11
	Wall 1	0.840–1.260		0.11–0.165
	Wall 2	0.0000055–0.420		0.00–0.055
	Wall 3	0.840–2.10		0.11–0.22
	Wall 4	0.0000055–0.420		0.00–0.055
	Wall 5	0.840–1.680		0.11–0.165

Table 1. *Cont.*

Model	Location of Wall Shear Stress	Wall Shear Stress Range (Pa)	Junction Overlap Region (Pa)	Velocity Range (m/s)
1F	Junction 1	0.420–1.260	No overlap	0.00–0.055
	Junction 2	0.840–2.10		0.11–0.165
	Junction 3	1.260–4.20		0.11–0.165
	Wall 1	0.840–1.260		0.11–0.165
	Wall 2	0.0000055–0.420		0.00–0.055
	Wall 3	0.840–2.950		0.11–0.22
	Wall 4	0.840–1.680		0.11–0.165
1G	Junction 1	0.420–1.680	1.260–1.680	0.00–0.055
	Junction 2	0.840–4.20		0.055–0.11
	Junction 3	0.840–4.20		0.165–0.22
	Junction 4	1.260–4.20		0.11–0.22
	Wall 1	0.840–1.680		0.11–0.22
	Wall 2	0.0000055–0.420		0.00–0.055
	Wall 3	0.840–4.20		0.11–0.22
	Wall 4	1.680–2.940		0.165–0.22
	Wall 5	0.840–1.680		0.11–0.165

As previously described, models were built from Model 1C, as shown in Figure 3. Velocities in Model 1D fell predominantly within the desired range of 0.049–0.19 m/s. In regions where the velocity increased, the wall shear stress also increased, as predicted by the Hagen–Poiseuille equation. This directly proportional relationship also held where the velocity decreased. The overlap between junctions 1, 2, and 3 was 0.840–1.260 Pa. The next blood vessel was added to the network at wall 4, where the shear stress fell within this range. In Model 1E, the velocity and wall shear stress were lower in blood vessel walls 2 and 4 than in Model 1D. There were no considerable changes elsewhere in the blood vessel network. Consequently, the wall shear stress overlap at junctions 1–4 remained between 0.840–1.260 Pa.

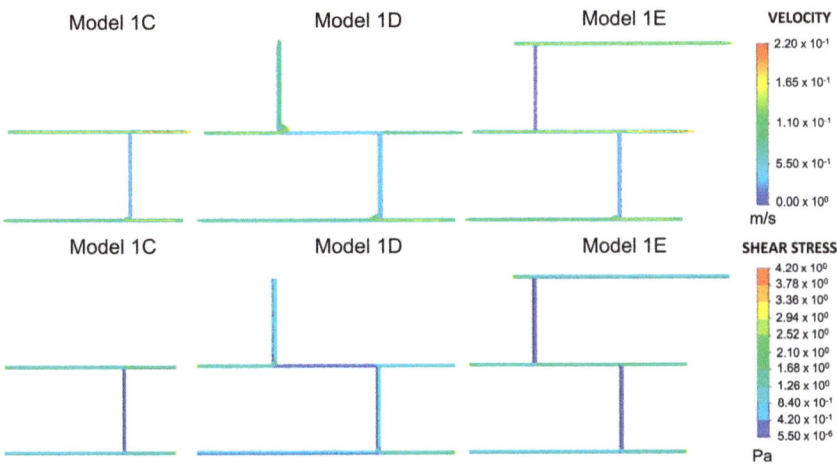

Figure 3. Velocity and shear stress results for Models 1C–1E. This was the same model with variations on inlets and outlets, as shown in Figure 2. Given the plausible velocity and shear stress ranges, Model 1C was chosen as the baseline model for all subsequent variations of Model 1.

Models 1F and 1G, shown in Figure 4, were used to analyse how the position and number of inlets and outlets would affect the junction overlap region. Models 1D and 1F had the same wall shear stress at junction 1. In Model 1F, junctions 2 and 3 had higher shear stress ranges than in Model 1D. Hence, no overlap occurred. The overlap region seen in Model 1G was higher than the overlap region seen in the other models, between 1.260–1.680 Pa. Therefore, it was regarded as an outlier. Compared to Model 1E, the shear stress at the junctions was higher and exceeded the limit for wall shear stress and velocity in wall 3.

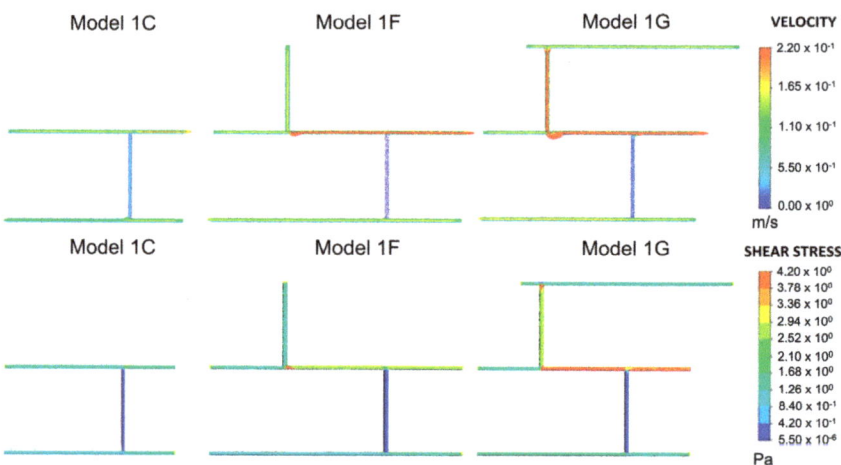

Figure 4. Model 1F and Model 1G are derived from Model 1C. These are similar to Model 1D and Model 1E, respectively, with the exception of inlet/outlet arrangements.

2.3. Model 2

For most of the vessels in Model 2A, shown in Figure 5, the wall shear stress lay within the range of the physiological arterial shear, and the velocity followed a similar pattern. Blood vessels 2, 3, and 5 were the exceptions to this observation. The results of Model 2A (Figure 5) were compared to Model 1E (Figure 3) as they were similar in shape. Although Model 2A had six blood vessels compared to five, the main difference between the two models lay in the direction of the highest-numbered blood vessel. In Model 2A, this blood vessel developed towards the left, whereas in Model 1E, it developed towards the right. Model 2A had the same wall shear as Model 1E at junction 1 and wall 1. The shear stress in wall 2 was also the same for both models. The wall shear for junctions 2, 5, and 6 in Model 2A, which were in a similar position to junctions 2, 3 and 4 in Model 1E, had similar values. Ultimately, the junction overlap region for Models 2A and 1E was identical. Model 2B investigated whether placing a blood vessel at an acute or obtuse angle would affect the wall shear and velocity. In blood vessel 2, the difference in the velocity was marginal compared to Model 2A; the wall shear and junction overlap remained the same. Thus, angling a blood vessel has little to no effect on the wall shear and velocity. Outlet 5, in Model 2A, was changed to inlet 5 in Model 2C (Figure 10). The velocity of blood vessels 1 and 4 increased as a result. Consequently, the wall shear was also higher in that region. The velocity and wall shear stress in blood vessels 2 and 3 were outside the already defined range of 0.049–0.19 m/s and 0.6–4 Pa, respectively. The junction overlap regions for Model 2 are shown in Table 2.

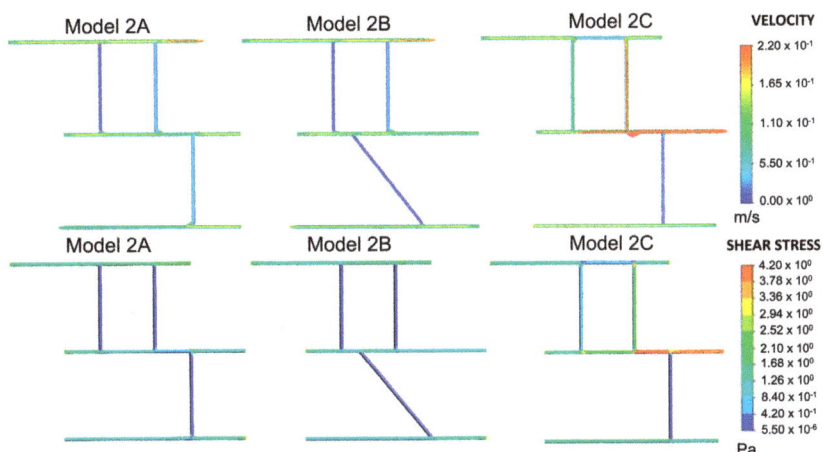

Figure 5. Velocity and shear stress plots for Models 2A, 2B and 2C. All three models represent the complete network.

Table 2. Wall shear stress and velocity ranges, and wall shear stress junction overlap regions at different locations in the blood vessel network for different variations of Model 2.

Model	Location of Wall Shear Stress	Wall Shear Stress Range (Pa)	Junction Overlap Region (Pa)	Velocity Range (m/s)
2A	Junction 1	0.420–1.260		0.055–0.11
	Junction 2	0.840–2.10		0.055–0.11
	Junction 3	0.420–1.260	0.840–1.260	0.055–0.11
	Junction 4	0.840–2.520		0.055–0.11
	Junction 5	0.420–1.260		0.00–0.11
	Junction 6	0.840–2.10		0.00–0.11
	Wall 1	0.840–1.260		0.11–0.165
	Wall 2	0.0000055–0.420		0.00–0.055
	Wall 3	0.420–1.680		0.11–0.165
	Wall 4	0.0000055–0.420		0.00–0.055
	Wall 5	0.0000055–0.420		0.00–0.055
	Wall 6	0.840–2.520		0.11–0.22
2B	Junction 1	0.420–1.260		0.055–0.11
	Junction 2	0.420–2.10		0.055–0.11
	Junction 3	0.420–1.680	0.840–1.260	0.055–0.11
	Junction 4	0.840–2.520		0.055–0.11
	Junction 5	0.840–1.260		0.00–0.11
	Junction 6	0.840–2.10		0.00–0.11
	Wall 1	0.840–1.680		0.11–0.165
	Wall 2	0.0000055–0.420		0.00–0.055
	Wall 3	0.840–1.680		0.11–0.165
	Wall 4	0.0000055–0.420		0.00–0.055
	Wall 5	0.0000055–0.420		0.00–0.055
	Wall 6	0.840–2.10		0.11–0.22
2C	Junction 1	0.420–1.680		0.055–0.11
	Junction 2	0.840–2.520		0.055–0.11
	Junction 3	0.840–4.20	No overlap	0.165–0.22
	Junction 4	1.680–3.780		0.165–0.22
	Junction 5	2.10–4.20		0.11–0.165
	Junction 6	0.840–2.940		0.11–0.165
	Wall 1	0.840–1.680		0.11–0.165
	Wall 2	0.0000055–0.420		0.00–0.055
	Wall 3	0.840–4.20		0.11–0.22
	Wall 4	1.680–2.10		0.165–0.22
	Wall 5	0.420–0.840		0.055–0.11
	Wall 6	0.0000055–1.680		0.055–0.165

2.4. Model Verification

The shear stress for the straight vessel was computed as 1.26 Pa, as shown in Figure 6A. On the basis of the analytical Hagen–Poiseuille formulation, we also calculated shear stress as 1.26 Pa. This verifies our result quantitatively. Qualitatively, Models 1E and 1G are very similar to the image shown in the micrograph (Figure 6B). Both models emerge from the initial geometry and emanate from the shear stress thresholds described above.

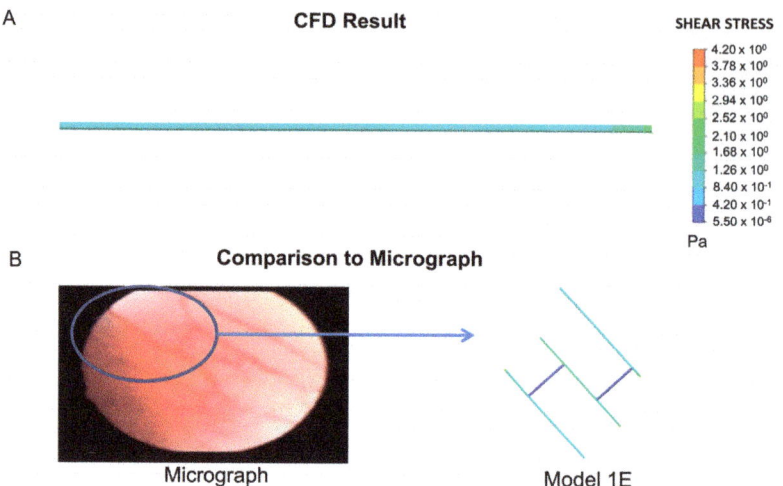

Figure 6. Quantitative and qualitative verification of the computational model. (**A**) Shear stress plot for a straight pipe based on the same parameters as those employed in Models 1 and 2. (**B**) Comparison between the geometry that emerged from Model 1 (Model 1E) and the original micrograph that was used to inform the initial model (permission obtained from RightsLink/Elsevier) [75].

3. Discussion

The objective of this study was to determine how blood flow influences angiogenesis in RA. Shear stress, a mechanical variable dependent on flow, was used as a surrogate for VEGF in determining the location of the new blood vessels. To achieve our objectives, two models of blood vessel networks in RA were built using CFD to analyse whether blood vessels would develop in low or high shear stress regions. Although several CFD studies have already been used to examine shear stress values in angiogenesis broadly, these have yet to be extended to RA [54,76]. Even in the in vitro angiogenesis studies, the exact role of wall shear stress remains controversial, with similar values having been shown to enhance and attenuate angiogenetic sprouting [46]. In this discussion, we consider our results in the context of other computational models and discuss the link between shear stress and VEGF.

The results of our two models, which are specific to RA, indicate that new blood vessels will form in areas of relatively low shear stress. Specifically, new branches formed between 0.840–1.260 Pa, which is in the lower half of the 0.6–4.20 Pa range. The overlap region for the wall shear stress at the junctions was the same for all models in our study, indicating a relationship between the emergence of new blood vessels and the magnitude of the wall shear stress. Our shear stress values are in the same order of magnitude as those in other computational studies that examined the effect of side branches [54,76–78]. In their CFD study on shear stress values along endothelial tip cells at the end of the capillary sprout in angiogenesis, Hu et al. considered a wall shear stress above 0.1 Pa to be physiologically relevant [54]. They found that tip cell shear stresses ranged from 0.019–0.465 Pa, and three out of eight cases achieved values above 0.1 Pa. Stapor et al. examined wall shear stress in angiogenetic capillary sprouts using CFD [76]. They found a local maximum wall shear stress value of 1.4 Pa at the base of a sprout with a non-permeable vessel wall. For larger blood vessels, such as the coronary arteries modelled by Wellnhofer et al. in a CFD model, a median wall shear stress value of 2.54 Pa was reported for steady state simulations with side branches [77]. In a fluid-structure interaction model using CFD to solve the fluid part, Ngoepe et al. found that for models with different side branch geometries, the peak wall shear stress varied from 0.7–2.3 Pa in arterial to venous anastomosis models [78]. The values from literature show that our shear stress values fall within the range of other CFD studies that include side branches. Wellnhofer et al. found that including side branches was necessary for wall shear stress estimation [77]. In particular, they found that the spread and distribution of wall shear stress, particularly for high and low values, was increased by including side branches. In our models, some blood vessels did not meet the arterial velocity or wall shear conditions, particularly the vertically orientated blood vessels. Wall shear stress in these vessels generally ranged between 5.5×10^{-6}–0.42 Pa, and the velocity was between 0.00–0.055 m/s. In their CFD model, Wellnhofer et al. found that very low wall shear stresses (i.e., less than 0.4 Pa) occurred in aneurysmatic coronary artery disease cases. In addition to these in silico findings, Galie et al. found that a shear stress threshold of 1 Pa triggered angiogenetic sprouting in an in vitro study [50].

Given that shear stress was used as a surrogate for VEGF, it is important to link our computed shear rate values to experimental observations of VEGF. In a microfluidic study examining the combined effect of shear stress and VEGF on neovascularisation, Zhao et al. found that shear stress plays a dominant role when VEGF is sufficient [53]. They found initiating neovascularisation under 1.5 Pa difficult, even with enough VEGF. Their threshold value is slightly higher than our maximum value for branch formation (1.260 Pa). Fey et al. examined the role of VEGF and shear stress on podosomes, which play a pivotal role in cell motility and are important for angiogenesis in endothelial cells [79]. In the absence of VEGF, changes in shear stress did not affect cell density, but higher shear stresses resulted in less podosome activity. When considering shear stress and VEGF together, it was found that high shear stress (1 Pa) increased podosome activity when there was sufficient VEGF in the system. This high shear stress value falls within our predicted angiogenetic range. Russo et al. considered how altering shear stress may change growth factor gene expression in endothelial cells [47]. A reduction from a physiological to pathological shear stress value (1.2 Pa to 0.4 Pa) increased VEGF gene expression. The pathological value fell outside our angiogenetic range, but direct comparison is somewhat challenging given that our model could not account for gene expression. Overall, our computed results fell in a range that supports VEGF expression and angiogenesis.

Our results demonstrate the importance of shear stress in RA angiogenesis and provide a tool for exploring the influence of haemodynamics. In cases where it is challenging to locate VEGF expression in newly developing blood vessels in RA, a haemodynamic simulation could map shear stress in the vascular network. Using wall shear stress as a surrogate for VEGF, researchers could identify parts of a blood vessel network where VEGF expression is likely to be highest. The application of this knowledge could contribute to the development of anti-VEGF biological therapies that inhibit expression. Some have even suggested that limiting angiogenesis by blocking the blood supply in the pannus may benefit patients [3]. Challenges to many therapeutic approaches would arise from patient variability. The main limitations of our work are the exclusion of chemical factors and the simplification of the blood flow. If developed further, the CFD model could account for patient-specificity factors such as VEGF and hypoxia, thereby enabling a coupled consideration of some of the most important variables for RA angiogenesis [3]. Other assumptions that should be revisited include modelling the blood flow as steady and laminar, with blood behaving as a Newtonian fluid. Furthermore, the role of the distensibility of the newly formed vessels should be explored.

4. Materials and Methods

The methods below describe the CFD simulations and blood vessel configurations used to analyse the relationship between the blood flow, shear stress, and the growth of new blood vessels. The experimental study employs steady-state conditions and makes several simplifying assumptions.

4.1. Fluid Flow Simulations

ANSYS Fluent Version 20.2.0 (ANSYS, Lebanon, NH, USA), a computational fluid dynamics simulation software, was used to model the blood flow in the respective geometries. ANSYS Fluent solves the Navier–Stokes equations by discretising the partial differential equations that govern the flow using the finite volume method (FVM),

$$\nabla \cdot U = 0, \tag{1}$$

$$\rho \frac{\partial U}{\partial t} + \rho U + \nabla U + \nabla P = \mu \nabla^2 U, \tag{2}$$

where U is velocity, ρ is the fluid density, μ is the dynamic fluid viscosity, t is time and P is pressure.

Boundary Conditions and Assumptions

Although a fair amount of pulsatility is experienced in blood vessels, steady-state conditions were applied. It has been shown that for more extended periods, such as when new vessel growth takes place, the baseline steady-state effect dominates the mechanical environment sensed by the cells [80]. In their CFD study, Wellnhofer et al. found that steady-state simulations were appropriate for a time-averaged wall shear stress [77]. Blood was assumed to be a Newtonian fluid with a density $\rho = 1060$ kg/m^3 and a constant viscosity of $\mu = 0.0035$ kg/m·s.

Even though angiogenesis is characterised by the formation of microvessels in the synovium [81], we were interested in exploring the role of arterial shear stresses as these have been implicated in marked VEGF expression [52]. As such, we had to balance two competing priorities: making the vessels sufficiently small while achieving arterial shear stress. The Haagen–Poiseuille equation was used to determine the maximum blood flow velocity and the vessel diameter that would result in shear stress inside the physiological arterial blood flow range (0.6–4 Pa) [52]. The arterial blood flow velocity was restricted to 0.049–0.19 m/s [82]. The blood vessel needed a diameter ranging from 0.1–10 mm [83]. All this information informed the selection of the blood vessel diameter and the maximum velocity. The blood vessels were assumed to be cylindrical and rigid, and a no-slip boundary condition was applied to the walls.

4.2. Geometries and Modelling Approach

The geometries are based on a micrograph of small blood vessels in the knee joint of an RA patient, which is presented in a study by Cañete et al. [75]. The vessels had a straight, branching pattern that is characteristic of RA. Although the exact dimensions of the vessels were not given, the micrographs were obtained using a 1.9 mm diameter or 2.7 mm diameter arthroscope.

Two different approaches, shown in Figure 7, were taken to develop models of blood vessel networks. The first approach (Model 1), shown in Figure 7, progressively added vessels based on shear stress thresholds. Given the strong link between shear stress and angiogenesis, we sought to find shear stress values that might support the development of new blood vessels. The process for determining the thresholds is described in the following paragraph and the threshold values are shown in Tables 1 and 2. The second approach (Model 2), also shown in Figure 7, modelled the complete vessel network as a starting point. Both models were used to analyse the relationship between wall shear stress and the development or positioning of new blood vessels. These examined how low shear and high shear regions influence blood vessel formation.

Model 1

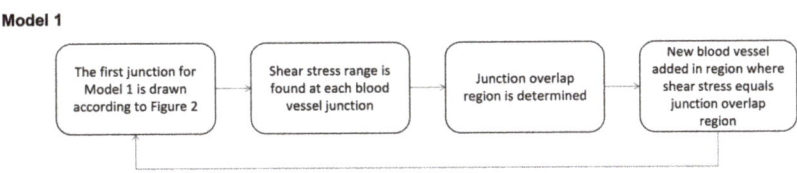

Model 2

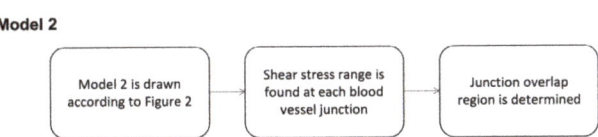

Figure 7. Two different approaches to developing the blood vessel network. Model 1 progressively adds vessels based on thresholds, while Model 2 begins with the entire vessel network.

The first approach, Model 1, began with a junction comprising two horizontal vessels connected by one vertical vessel. A three-vessel representation, comprising solid cylinders of radius 0.001 m, models a subsection of the blood vessels as depicted in Models 1A, 1B and 1C in Figure 8. Shear stress and velocity values were computed for these three different arrangements, where the positions of the inlets and outlets were varied. Model 1C was chosen as the final starter model as it achieved plausible physiological shear stresses and blood flow velocities.

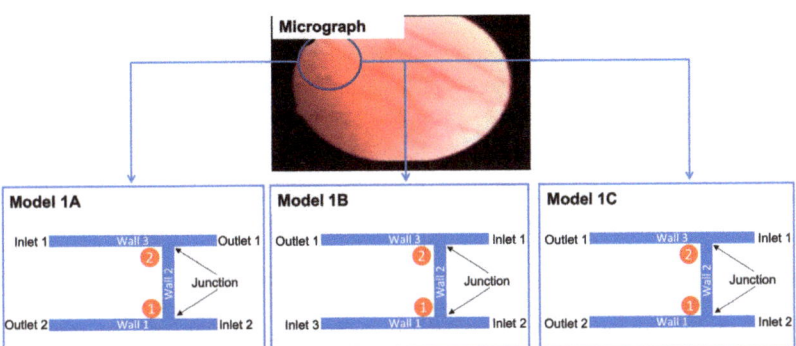

Figure 8. Development of models from realistic geometry for Model 1. The geometries for the initial geometry are derived from a micrograph of blood vessels in RA (permission obtained from RightsLink/Elsevier) [75]. The portion in the solid circle informs (B) the starting point for Model 1. The different variations (Model 1A, Model 1B and Model 1C) arise from rearranging inlets and outlets.

As described in Figure 8, subsequent geometries emerged from Model 1C, which determined where new blood vessels would form based on the range of shear stress values observed at junctions one and two. Shear stress was also calculated for wall three. A search for portions that achieved the range observed at junctions one and two was conducted. The part(s) of the wall that met this threshold were deemed capable of angiogenesis, and a vertical cylinder of length 0.05 m was constructed at these respective locations. This process was repeated twice, and the geometries which emerged from this process (Model 1D and Model 1E) are shown in Figure 9. Models 1F and 1G, also shown in Figure 9 and based on Models 1D and E, respectively, were included to analyse the effect of changing the numbers of inlets and outlets. The corresponding shear stress and velocity results for all the geometries are presented in the results section.

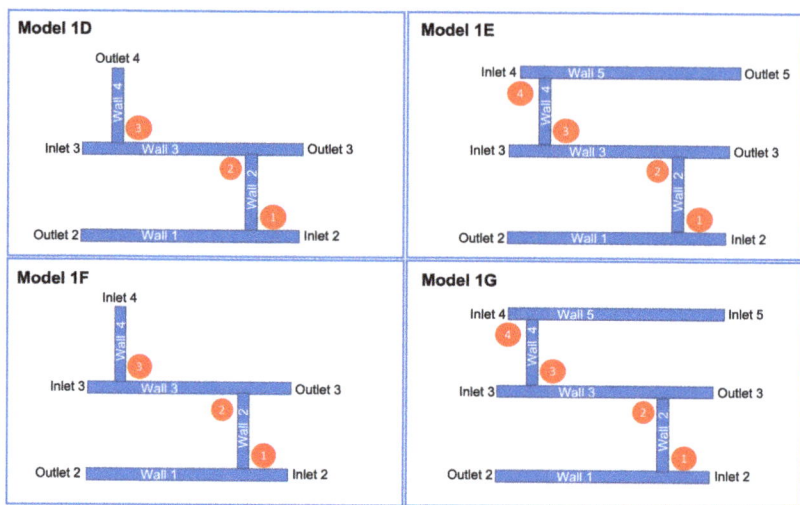

Figure 9. Geometries which emerge from vessels added to Model 1C. Models 1D and 1F are similar, except for inlet/outlet 4. Similarly, Models 1E and 1G differ only in so far as inlet/outlet 5 is concerned.

Figure 10 illustrates the designs for the second approach, Model 2. As described in Figure 7, this approach assumes the entire network as its starting point and examines shear stresses at the junctions of blood vessels. Variations on this basic model include changing the boundary conditions and placing one of the vertical vessels at an angle. These alterations are presented in Models 2A, 2B and 2C, shown in Figure 10 in the results section. The corresponding shear stress and velocity results for all the geometries are presented in the results section.

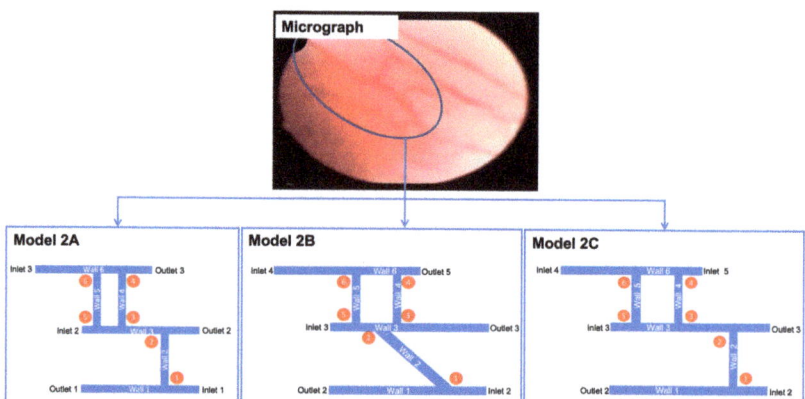

Figure 10. Development of models from realistic geometry for Model 2. The geometries for the initial geometry are derived from a micrograph of blood vessels in RA (permission obtained from RightsLink/Elsevier) [75]. The portion in the solid is the complete geometry for Model 2. The different variations (Model 2A, Model 2B and Model 2C) arise from rearranging inlets and outlets and placing one of the vessels at an angle.

4.3. Verification of Results

We verify our results quantitatively and qualitatively. For the former, we model blood flow through a straight pipe of radius 0.001 m, equivalent to the side branches in our other models. In this simple model, we use the same blood parameters and boundary conditions as those for Models 1 and 2. Once we have run the model, we compare the shear stress result to that calculated using the Hagen–Poiseuille formulation. This analytical solution enables us to calculate velocity and shear stress in straight pipes. The shear stress equation is given as

$$\tau = \frac{4\mu Q}{\pi r^3} \quad (3)$$

where τ is the shear stress, Q is the volumetric flow rate and r is the pipe radius. Qualitatively, we compare the final configuration for Model 1 with the original image that informed the initial model [75].

5. Conclusions

The findings of this study highlight the role of haemodynamics and wall shear stress in angiogenesis in rheumatoid arthritis. Specifically, we observed that new blood vessels are likely to develop in regions of low wall shear stress (0.840–1.260 Pa) in a computational fluid dynamics model. This highlights the role of mechanical factors in RA angiogenesis and provides a tool for further exploration of this phenomenon. Other angiogenesis-driven diseases, such as cancer, may also benefit from a similar modelling approach.

Author Contributions: Conceptualization, M.K.M. and M.N.N.; Data curation, M.K.M.; Investigation, M.K.M.; Methodology, M.K.M.; Project administration, M.N.N.; Supervision, M.N.N.; Visualization, M.N.N.; Writing—original draft, M.K.M. and M.N.N.; Writing—review and editing, M.K.M. and M.N.N. All authors have read and agreed to the published version of the manuscript.

Funding: The University of Cape Town paid for the software licence.

Institutional Review Board Statement: Not applicable.

Informed Consent Statement: Not applicable.

Data Availability Statement: Data are included in the tables presented in this paper.

Conflicts of Interest: The authors declare no conflict of interest.

References

1. Boissier, M.C.; Biton, J.; Semerano, L.; Decker, P.; Bessis, N. Origins of rheumatoid arthritis. *Jt. Bone Spine* **2020**, *87*, 301–306. [CrossRef]
2. Guo, Q.; Wang, Y.; Xu, D.; Nossent, J.; Pavlos, N.J.; Xu, J. Rheumatoid arthritis: Pathological mechanisms and modern pharmacologic therapies. *Bone Res.* **2018**, *6*, 15. [CrossRef]
3. Leblond, A.; Allanore, Y.; Avouac, J. Targeting synovial neoangiogenesis in rheumatoid arthritis. *Autoimmun. Rev.* **2017**, *16*, 594–601. [CrossRef]
4. Laurindo, L.F.; de Maio, M.C.; Barbalho, S.M.; Guiguer, E.L.; Araújo, A.C.; de Alvares Goulart, R.; Flato, U.A.; Júnior, E.B.; Detregiachi, C.R.; dos Santos Haber, J.F.; et al. Organokines in Rheumatoid Arthritis: A Critical Review. *Int. J. Mol. Sci.* **2022**, *23*, 6193. [CrossRef]
5. Hu, Y.; Yéléhé-Okouma, M.; Ea, H.K.; Jouzeau, J.Y.; Reboul, P. Galectin-3: A key player in arthritis. *Jt. Bone Spine* **2017**, *84*, 15–20. [CrossRef] [PubMed]
6. Khodadust, F.; Ezdoglian, A.; Steinz, M.M.; van Beijnum, J.R.; Zwezerijnen, G.J.C.; Jansen, G.; Tas, S.W.; van der Laken, C.J. Systematic Review: Targeted Molecular Imaging of Angiogenesis and Its Mediators in Rheumatoid Arthritis. *Int. J. Mol. Sci.* **2022**, *23*, 7071. [CrossRef]
7. MacDonald, I.; Liu, S.-C.; Su, C.-M.; Wang, Y.-H.; Tsai, C.-H.; Tang, C.-H. Implications of Angiogenesis Involvement in Arthritis. *Int. J. Mol. Sci.* **2018**, *19*, 2012. [CrossRef] [PubMed]
8. MacDonald, I.J.; Huang, C.-C.; Liu, S.-C.; Lin, Y.-Y.; Tang, C.-H. Targeting CCN Proteins in Rheumatoid Arthritis and Osteoarthritis. *Int. J. Mol. Sci.* **2021**, *22*, 4340. [CrossRef] [PubMed]
9. Akbarian, M.; Bertassoni, L.E.; Tayebi, L. Biological aspects in controlling angiogenesis: Current progress. *Cell. Mol. Life Sci.* **2022**, *79*, 349. [CrossRef]

10. Wang, Y.; Wu, H.; Deng, R. Angiogenesis as a potential treatment strategy for rheumatoid arthritis. *Eur. J. Pharmacol.* **2021**, *910*, 174500. [CrossRef]
11. Ben-Trad, L.; Matei, C.I.; Sava, M.M.; Filali, S.; Duclos, M.-E.; Berthier, Y.; Guichardant, M.; Bernoud-Hubac, N.; Maniti, O.; Landoulsi, A.; et al. Synovial Extracellular Vesicles: Structure and Role in Synovial Fluid Tribological Performances. *Int. J. Mol. Sci.* **2022**, *23*, 11998. [CrossRef]
12. Paleolog, E.M. The vasculature in rheumatoid arthritis: Cause or consequence? *Int. J. Exp. Pathol.* **2009**, *90*, 249–261. [CrossRef]
13. Alam, J.; Jantan, I.; Bukhari, S.N.A. Rheumatoid arthritis: Recent advances on its etiology, role of cytokines and pharmacotherapy. *Biomed. Pharmacother.* **2017**, *92*, 615–633. [CrossRef]
14. Marrelli, A.; Cipriani, P.; Liakouli, V.; Carubbi, F.; Perricone, C.; Perricone, R.; Giacomelli, R. Angiogenesis in rheumatoid arthritis: A disease specific process or a common response to chronic inflammation? *Autoimmun. Rev.* **2011**, *10*, 595–598. [CrossRef]
15. Paleolog, E.M. Angiogenesis in rheumatoid arthritis. *Arthritis Res.* **2002**, *4* (Suppl. S3), S81–S90. [CrossRef]
16. You, S.; Koh, J.H.; Leng, L.; Kim, W.; Bucala, R. The Tumor-like phenotype of rheumatoid synovium: Molecular profiling and prospects for precision medicine. *Arthritis Rheumatol.* **2018**, *70*, 637–652. [CrossRef]
17. Guo, X.; Chen, G. Hypoxia-Inducible Factor Is Critical for Pathogenesis and Regulation of Immune Cell Functions in Rheumatoid Arthritis. *Front. Immunol.* **2020**, *11*, 1668. [CrossRef]
18. Mabeta, P.; Steenkamp, V. The VEGF/VEGFR Axis Revisited: Implications for Cancer Therapy. *Int. J. Mol. Sci.* **2022**, *23*, 15585. [CrossRef]
19. Ahmad, A.; Nawaz, M.I. Molecular mechanism of VEGF and its role in pathological angiogenesis. *J. Cell. Biochem.* **2022**, *123*, 1938–1965. [CrossRef]
20. Melincovici, C.S.; Boşca, A.B.; Şuşman, S.; Mărginean, M.; Mihu, C.; Istrate, M.; Moldovan, I.M.; Roman, A.L.; Mihu, C.M. Vascular endothelial growth factor (VEGF)—Key factor in normal and pathological angiogenesis. *Rom. J. Morphol. Embryol.* **2018**, *59*, 455–467.
21. Abhinand, C.S.; Raju, R.; Soumya, S.J.; Arya, P.S.; Sudhakaran, P.R. VEGF-A/VEGFR2 signaling network in endothelial cells relevant to angiogenesis. *J. Cell Commun. Signal.* **2016**, *10*, 347–354. [CrossRef] [PubMed]
22. Lee, Y.H.; Bae, S.-C. Correlation between circulating VEGF levels and disease activity in rheumatoid arthritis: A meta-analysis. *Z. Rheumatol.* **2018**, *77*, 240–248. [CrossRef] [PubMed]
23. Paradowska-Gorycka, A.; Pawlik, A.; Romanowska-Prochnicka, K.; Haladyj, E.; Malinowski, D.; Stypinska, B.; Manczak, M.; Olesinska, M. Relationship between VEGF Gene Polymorphisms and Serum VEGF Protein Levels in Patients with Rheumatoid Arthritis. *PLoS ONE* **2016**, *11*, e0160769. [CrossRef] [PubMed]
24. Wu, Y.; Li, M.; Zeng, J.; Feng, Z.; Yang, J.; Shen, B.; Zeng, Y. Differential Expression of Renin-Angiotensin System-related Components in Patients with Rheumatoid Arthritis and Osteoarthritis. *Am. J. Med. Sci.* **2020**, *359*, 17–26. [CrossRef]
25. Li, Y.; Liu, Y.; Wang, C.; Xia, W.-R.; Zheng, J.-Y.; Yang, J.; Liu, B.; Liu, J.-Q.; Liu, L.-F. Succinate induces synovial angiogenesis in rheumatoid arthritis through metabolic remodeling and HIF-1α/VEGF axis. *Free Radic. Biol. Med.* **2018**, *126*, 1–14. [CrossRef]
26. Chen, Y.; Dawes, P.T.; Mattey, D.L. Polymorphism in the vascular endothelial growth factor A (VEGFA) gene is associated with serum VEGF-A level and disease activity in rheumatoid arthritis: Differential effect of cigarette smoking. *Cytokine* **2012**, *58*, 390–397. [CrossRef]
27. Afroz, S.; Giddaluru, J.; Vishwakarma, S.; Naz, S.; Khan, A.A.; Khan, N. A Comprehensive Gene Expression Meta-analysis Identifies Novel Immune Signatures in Rheumatoid Arthritis Patients. *Front. Immunol.* **2017**, *8*, 74. [CrossRef]
28. Ha, E.; Bae, S.-C.; Kim, K. Large-scale meta-analysis across East Asian and European populations updated genetic architecture and variant-driven biology of rheumatoid arthritis, identifying 11 novel susceptibility loci. *Ann. Rheum. Dis.* **2021**, *80*, 558–565. [CrossRef]
29. Badimon, L.; Peña, E.; Arderiu, G.; Padró, T.; Slevin, M.; Vilahur, G.; Chiva-Blanch, G. C-Reactive Protein in Atherothrombosis and Angiogenesis. *Front. Immunol.* **2018**, *9*, 430. [CrossRef]
30. Gümüş, A.; Coşkun, C.; Emre, H.Ö.; Temel, M.; İnal, B.B.; Seval, H.; Döventaş, Y.E.; Koldaş, M. Evaluation of vascular endothelial growth factor levels in rheumatoid arthritis patients, with and without joint swelling; a comparison with erythrocyte sedimentation rate, C-reactive protein, rheumatoid factor and anti-cyclic citruillnated protein. *Turkish J. Biochem.* **2018**, *43*, 76–82. [CrossRef]
31. Sakalyte, R.; Bagdonaite, L.; Stropuviene, S.; Naktinyte, S.; Venalis, A. VEGF Profile in Early Undifferentiated Arthritis Cohort. *Medicina* **2022**, *58*, 833. [CrossRef]
32. Konisti, S.; Kiriakidis, S.; Paleolog, E.M. Hypoxia—A key regulator of angiogenesis and inflammation in rheumatoid arthritis. *Nat. Rev. Rheumatol.* **2012**, *8*, 153–162. [CrossRef]
33. Chakraborty, D.; Sarkar, A.; Mann, S.; Agnihotri, P.; Saquib, M.; Malik, S.; Kumavat, R.; Mathur, A.; Biswas, S. Estrogen-mediated differential protein regulation and signal transduction in rheumatoid arthritis. *J. Mol. Endocrinol.* **2022**, *69*, R25–R43. [CrossRef]
34. Huang, C.-C.; Tseng, T.-T.; Liu, S.-C.; Lin, Y.-Y.; Law, Y.-Y.; Hu, S.-L.; Wang, S.-W.; Tsai, C.-H.; Tang, C.-H. S1P Increases VEGF Production in Osteoblasts and Facilitates Endothelial Progenitor Cell Angiogenesis by Inhibiting miR-16-5p Expression via the c-Src/FAK Signaling Pathway in Rheumatoid Arthritis. *Cells* **2021**, *10*, 2168. [CrossRef]
35. Amoroso, F.; Capece, M.; Rotondo, A.; Cangelosi, D.; Ferracin, M.; Franceschini, A.; Raffaghello, L.; Pistoia, V.; Varesio, L.; Adinolfi, E. The P2X7 receptor is a key modulator of the PI3K/GSK3β/VEGF signaling network: Evidence in experimental neuroblastoma. *Oncogene* **2015**, *34*, 5240–5251. [CrossRef]

36. Elshabrawy, H.A.; Chen, Z.; Volin, M.V.; Ravella, S.; Virupannavar, S.; Shahrara, S. The pathogenic role of angiogenesis in rheumatoid arthritis. *Angiogenesis* **2015**, *18*, 433–448. [CrossRef]
37. Kretschmer, M.; Rüdiger, D.; Zahler, S. Mechanical Aspects of Angiogenesis. *Cancers* **2021**, *13*, 4987. [CrossRef]
38. Augustin, H.G.; Koh, G.Y. Organotypic vasculature: From descriptive heterogeneity to functional pathophysiology. *Science* **2017**, *357*, eaal2379. [CrossRef]
39. Beech, D.J.; Kalli, A.C. Force sensing by piezo channels in cardiovascular health and disease. *Arterioscler. Thromb. Vasc. Biol.* **2019**, *39*, 2228–2239. [CrossRef]
40. Boldock, L.; Wittkowske, C.; Perrault, C.M. Microfluidic traction force microscopy to study mechanotransduction in angiogenesis. *Microcirculation* **2017**, *24*, e12361. [CrossRef]
41. Chatterjee, S.; Fujiwara, K.; Pérez, N.G.; Ushio-Fukai, M.; Fisher, A.B. Mechanosignaling in the vasculature: Emerging concepts in sensing, transduction and physiological responses. *Am. J. Physiol. Circ. Physiol.* **2015**, *308*, H1451–H1462. [CrossRef] [PubMed]
42. Marchand, M.; Monnot, C.; Muller, L.; Germain, S. Extracellular matrix scaffolding in angiogenesis and capillary homeostasis. *Semin. Cell Dev. Biol.* **2019**, *89*, 147–156. [CrossRef] [PubMed]
43. Urschel, K.; Tauchi, M.; Achenbach, S.; Dietel, B. Investigation of Wall Shear Stress in Cardiovascular Research and in Clinical Practice—From Bench to Bedside. *Int. J. Mol. Sci.* **2021**, *22*, 5635. [CrossRef] [PubMed]
44. Cecchi, E.; Giglioli, C.; Valente, S.; Lazzeri, C.; Gensini, G.F.; Abbate, R.; Mannini, L. Role of hemodynamic shear stress in cardiovascular disease. *Atherosclerosis* **2011**, *214*, 249–256. [CrossRef] [PubMed]
45. Davies, P.F. Hemodynamic shear stress and the endothelium in cardiovascular pathophysiology. *Nat. Clin. Pract. Cardiovasc. Med.* **2009**, *6*, 16–26. [CrossRef]
46. Barrasa-Ramos, S.; Dessalles, C.A.; Hautefeuille, M.; Barakat, A.I. Mechanical regulation of the early stages of angiogenesis. *J. R. Soc. Interface* **2022**, *19*, 20220360. [CrossRef]
47. Russo, T.A.; Banuth, A.M.M.; Nader, H.B.; Dreyfuss, J.L. Altered shear stress on endothelial cells leads to remodeling of extracellular matrix and induction of angiogenesis. *PLoS ONE* **2020**, *15*, e0241040. [CrossRef]
48. Inglebert, M.; Locatelli, L.; Tsvirkun, D.; Sinha, P.; Maier, J.A.; Misbah, C.; Bureau, L. The effect of shear stress reduction on endothelial cells: A microfluidic study of the actin cytoskeleton. *Biomicrofluidics* **2020**, *14*, 24115. [CrossRef]
49. Douguet, D.; Patel, A.; Xu, A.; Vanhoutte, P.M.; Honoré, E. Piezo Ion Channels in Cardiovascular Mechanobiology. *Trends Pharmacol. Sci.* **2019**, *40*, 956–970. [CrossRef]
50. Galie, P.A.; Nguyen, D.-H.T.; Choi, C.K.; Cohen, D.M.; Janmey, P.A.; Chen, C.S. Fluid shear stress threshold regulates angiogenic sprouting. *Proc. Natl. Acad. Sci. USA* **2014**, *111*, 7968–7973. [CrossRef]
51. Tsaryk, R.; Yucel, N.; Leonard, E.V.; Diaz, N.; Bondareva, O.; Odenthal-Schnittler, M.; Arany, Z.; Vaquerizas, J.M.; Schnittler, H.; Siekmann, A.F. Shear stress switches the association of endothelial enhancers from ETV/ETS to KLF transcription factor binding sites. *Sci. Rep.* **2022**, *12*, 4795. [CrossRef]
52. dela Paz, N.G.; Walshe, T.E.; Leach, L.L.; Saint-Geniez, M.; D'Amore, P.A. Role of shear-stress-induced VEGF expression in endothelial cell survival. *J. Cell Sci.* **2012**, *125*, 831–843. [CrossRef]
53. Zhao, P.; Liu, X.; Zhang, X.; Wang, L.; Su, H.; Wang, L.; He, N.; Zhang, D.; Li, Z.; Kang, H.; et al. Flow shear stress controls the initiation of neovascularization: Via heparan sulfate proteoglycans within a biomimetic microfluidic model. *Lab Chip* **2021**, *21*, 421–434. [CrossRef]
54. Hu, N.W.; Rodriguez, C.D.; Rey, J.A.; Rozenblum, M.J.; Courtney, C.P.; Balogh, P.; Sarntinoranont, M.; Murfee, W.L. Estimation of shear stress values along endothelial tip cells past the lumen of capillary sprouts. *Microvasc. Res.* **2022**, *142*, 104360. [CrossRef]
55. Goettsch, W.; Gryczka, C.; Korff, T.; Ernst, E.; Goettsch, C.; Seebach, J.; Schnittler, H.-J.; Augustin, H.G.; Morawietz, H. Flow-dependent regulation of angiopoietin-2. *J. Cell. Physiol.* **2008**, *214*, 491–503. [CrossRef]
56. Li, M.; Scott, D.E.; Shandas, R.; Stenmark, K.R.; Tan, W. High pulsatility flow induces adhesion molecule and cytokine mRNA expression in distal pulmonary artery endothelial cells. *Ann. Biomed. Eng.* **2009**, *37*, 1082–1092. [CrossRef]
57. Versteeg, H.K.; Malalasekera, W. *An Introduction to Computational Fluid Dynamics. The Finite Volume Method*; Pearson Education Limited: Harlow, UK, 2007; ISBN 978-0-13-127498-3.
58. Ferziger, J.H.; Peric, M. *Computational Methods for Fluid Dynamics*, 2nd ed.; Springer: Berlin/Heidelberg, Germany, 1999; ISBN 3-540-65373-2.
59. Morris, P.D.; Narracott, A.; von Tengg-Kobligk, H.; Soto, D.A.S.; Hsiao, S.; Lungu, A.; Evans, P.; Bressloff, N.W.; Lawford, P.V.; Hose, D.R.; et al. Computational fluid dynamics modelling in cardiovascular medicine. *Heart* **2016**, *102*, 18–28. [CrossRef]
60. Botha, E.; Malan, L.C.; Malan, A.G. Embedded One-Dimensional Orifice Elements for Slosh Load Calculations in Volume-Of-Fluid CFD. *Appl. Sci.* **2022**, *12*, 11909. [CrossRef]
61. Mani, M.; Dorgan, A.J. A Perspective on the State of Aerospace Computational Fluid Dynamics Technology. *Annu. Rev. Fluid Mech.* **2023**, *55*, 431–457. [CrossRef]
62. Liang, Y.Y.; Fletcher, D.F. Computational fluid dynamics simulation of forward osmosis (FO) membrane systems: Methodology, state of art, challenges and opportunities. *Desalination* **2023**, *549*, 116359. [CrossRef]
63. Aydin, E.S.; Yucel, O. Computational fluid dynamics study of hydrogen production using concentrated solar radiation as a heat source. *Energy Convers. Manag.* **2023**, *276*, 116552. [CrossRef]
64. Vinuesa, R.; Brunton, S.L. Enhancing computational fluid dynamics with machine learning. *Nat. Comput. Sci.* **2022**, *2*, 358–366. [CrossRef]

65. Micale, D.; Ferroni, C.; Uglietti, R.; Bracconi, M.; Maestri, M. Computational Fluid Dynamics of Reacting Flows at Surfaces: Methodologies and Applications. *Chemie Ing. Technol.* **2022**, *94*, 634–651. [CrossRef]
66. Rahimi-Gorji, M.; Debbaut, C.; Ghorbaniasl, G.; Cosyns, S.; Willaert, W.; Ceelen, W. Optimization of intraperitoneal aerosolized drug delivery using computational fluid dynamics (CFD) modeling. *Sci. Rep.* **2022**, *12*, 6305. [CrossRef]
67. Chen, T.B.; De Cachinho Cordeiro, I.M.; Yuen, A.C.; Yang, W.; Chan, Q.N.; Zhang, J.; Cheung, S.C.P.; Yeoh, G.H. An Investigation towards Coupling Molecular Dynamics with Computational Fluid Dynamics for Modelling Polymer Pyrolysis. *Molecules* **2022**, *27*, 292. [CrossRef]
68. Peach, T.; Cornhill, J.F.; Nguyen, A.; Riina, H.; Ventikos, Y. The "Sphere": A Dedicated Bifurcation Aneurysm Flow-Diverter Device. *Cardiovasc. Eng. Technol.* **2014**, *5*, 334–347. [CrossRef]
69. Peach, T.W.; Ricci, D.; Ventikos, Y. A Virtual Comparison of the eCLIPs Device and Conventional Flow-Diverters as Treatment for Cerebral Bifurcation Aneurysms. *Cardiovasc. Eng. Technol.* **2019**, *10*, 508–519. [CrossRef]
70. Jimoh-Taiwo, Q.; Haffejee, R.; Ngoepe, M. A Mechano-Chemical Computational Model of Deep Vein Thrombosis. *Front. Phys.* **2022**, *10*, 481. [CrossRef]
71. Guo, L.; Vardakis, J.C.; Chou, D.; Ventikos, Y. A multiple-network poroelastic model for biological systems and application to subject-specific modelling of cerebral fluid transport. *Int. J. Eng. Sci.* **2020**, *147*, 103204. [CrossRef]
72. Vardhan, M.; Randles, A. Application of physics-based flow models in cardiovascular medicine: Current practices and challenges. *Biophys. Rev.* **2021**, *2*, 11302. [CrossRef]
73. Zhong, L.; Zhang, J.-M.; Su, B.; Tan, R.S.; Allen, J.C.; Kassab, G.S. Application of Patient-Specific Computational Fluid Dynamics in Coronary and Intra-Cardiac Flow Simulations: Challenges and Opportunities. *Front. Physiol.* **2018**, *9*, 742. [CrossRef]
74. Miller, B.; Sewell-Loftin, M.K. Mechanoregulation of Vascular Endothelial Growth Factor Receptor 2 in Angiogenesis. *Front. Cardiovasc. Med.* **2022**, *8*, 804934. [CrossRef]
75. Cañete, J.D.; Rodríguez, J.R.; Salvador, G.; Gómez-Centeno, A.; Muñoz-Gómez, J.; Sanmartí, R. Diagnostic usefulness of synovial vascular morphology in chronic arthritis. A systematic survey of 100 cases. *Semin. Arthritis Rheum.* **2003**, *32*, 378–387. [CrossRef] [PubMed]
76. Stapor, P.C.; Wang, W.; Murfee, W.L.; Khismatullin, D.B. The Distribution of Fluid Shear Stresses in Capillary Sprouts. *Cardiovasc. Eng. Technol.* **2011**, *2*, 124–136. [CrossRef]
77. Wellnhofer, E.; Osman, J.; Kertzscher, U.; Affeld, K.; Fleck, E.; Goubergrits, L. Flow simulation studies in coronary arteries-Impact of side-branches. *Atherosclerosis* **2010**, *213*, 475–481. [CrossRef]
78. Ngoepe, M.N.; Reddy, B.D.; Kahn, D.; Meyer, C.; Zilla, P.; Franz, T. A Numerical Tool for the Coupled Mechanical Assessment of Anastomoses of PTFE Arterio-venous Access Grafts. *Cardiovasc. Eng. Technol.* **2011**, *2*, 160–172. [CrossRef]
79. Fey, T.; Schubert, K.M.; Schneider, H.; Fein, E.; Kleinert, E.; Pohl, U.; Dendorfer, A. Impaired endothelial shear stress induces podosome assembly via VEGF up-regulation. *FASEB J.* **2016**, *30*, 2755–2766. [CrossRef]
80. Papaioannou, T.G.; Stefanadis, C. Vascular wall shear stress: Basic principles and methods. *Hell. J. Cardiol.* **2005**, *46*, 9–15.
81. Semerano, L.; Clavel, G.; Assier, E.; Denys, A.; Boissier, M.C. Blood vessels, a potential therapeutic target in rheumatoid arthritis? *Jt. Bone Spine* **2011**, *78*, 118–123. [CrossRef]
82. Klarhöfer, M.; Csapo, B.; Balassy, C.; Szeles, J.C.; Moser, E. High-resolution blood flow velocity measurements in the human finger. *Magn. Reson. Med.* **2001**, *45*, 716–719. [CrossRef]
83. Renner, J.; Gårdhagen, R.; Ebbers, T.; Heiberg, E.; Länne, T.; Karlsson, M. A method for subject specific estimation of aortic wall shear stress. *WSEAS Trans. Biol. Biomed.* **2009**, *6*, 49–57.

Disclaimer/Publisher's Note: The statements, opinions and data contained in all publications are solely those of the individual author(s) and contributor(s) and not of MDPI and/or the editor(s). MDPI and/or the editor(s) disclaim responsibility for any injury to people or property resulting from any ideas, methods, instructions or products referred to in the content.

Review

Building a Scaffold for Arteriovenous Fistula Maturation: Unravelling the Role of the Extracellular Matrix

Suzanne L. Laboyrie [1], Margreet R. de Vries [2,3], Roel Bijkerk [1] and Joris I. Rotmans [1,*]

1. Department of Internal Medicine, Leiden University Medical Centre, 2333 ZA Leiden, The Netherlands; s.l.laboyrie@lumc.nl (S.L.L.); r.bijkerk@lumc.nl (R.B.)
2. Department of Surgery and the Heart and Vascular Center, Brigham & Women's Hospital and Harvard Medical School, Boston, MA 02115, USA; m.r.de_vries@lumc.nl
3. Department of Vascular Surgery, Leiden University Medical Centre, 2333 ZA Leiden, The Netherlands
* Correspondence: j.i.rotmans@lumc.nl

Abstract: Vascular access is the lifeline for patients receiving haemodialysis as kidney replacement therapy. As a surgically created arteriovenous fistula (AVF) provides a high-flow conduit suitable for cannulation, it remains the vascular access of choice. In order to use an AVF successfully, the luminal diameter and the vessel wall of the venous outflow tract have to increase. This process is referred to as AVF maturation. AVF non-maturation is an important limitation of AVFs that contributes to their poor primary patency rates. To date, there is no clear overview of the overall role of the extracellular matrix (ECM) in AVF maturation. The ECM is essential for vascular functioning, as it provides structural and mechanical strength and communicates with vascular cells to regulate their differentiation and proliferation. Thus, the ECM is involved in multiple processes that regulate AVF maturation, and it is essential to study its anatomy and vascular response to AVF surgery to define therapeutic targets to improve AVF maturation. In this review, we discuss the composition of both the arterial and venous ECM and its incorporation in the three vessel layers: the tunica intima, media, and adventitia. Furthermore, we examine the effect of chronic kidney failure on the vasculature, the timing of ECM remodelling post-AVF surgery, and current ECM interventions to improve AVF maturation. Lastly, the suitability of ECM interventions as a therapeutic target for AVF maturation will be discussed.

Keywords: extracellular matrix; arteriovenous fistula; vascular remodelling; AVF maturation

Citation: Laboyrie, S.L.; de Vries, M.R.; Bijkerk, R.; Rotmans, J.I. Building a Scaffold for Arteriovenous Fistula Maturation: Unravelling the Role of the Extracellular Matrix. *Int. J. Mol. Sci.* **2023**, *24*, 10825. https://doi.org/10.3390/ijms241310825

Academic Editors: Elisabeth Deindl, Paul Quax and Ewa K. Szczepanska-Sadowska

Received: 28 April 2023
Revised: 20 June 2023
Accepted: 27 June 2023
Published: 28 June 2023

Copyright: © 2023 by the authors. Licensee MDPI, Basel, Switzerland. This article is an open access article distributed under the terms and conditions of the Creative Commons Attribution (CC BY) license (https://creativecommons.org/licenses/by/4.0/).

1. Introduction

End-stage kidney disease (ESKD) patients receive renal replacement therapy through kidney transplantation or dialysis treatment. During haemodialysis (HD), vascular access (VA) connects the patient's blood supply to the dialysis machine. VA can be achieved through a central venous catheter (CVC), arteriovenous graft (AVG), or an arteriovenous fistula (AVF). The radiocephalic AVF was the first arteriovenous configuration, introduced in 1966 [1]. This provided a stepping stone for a high-flow vascular access design for HD, served as inspiration for other AVF configurations, and resulted in the AVF being the gold standard in VA [2].

The National Kidney Foundation's Kidney Disease Outcomes Quality Initiative (KDOQI) guidelines recommend AV access (AVF or AVG) in patients requiring HD [3]. AVFs have better longevity and reduced complications when compared to AVG and CVC [4–7]. However, non-maturation rates are around 9–31%, depending on AVF type, compared to 9% functional failure in AVGs, which remains a hurdle in the lifeline to HD therapy [8,9]. The role of different cell types, such as endothelial cells (ECs), inflammatory cells, and vascular smooth muscle cells (VSMCs), is being actively investigated, but to date, there is no adequate intervention to promote AVF maturation. Although thoroughly reviewed in models of vascular disease, such as atherosclerosis plaque rupture, aneurysm formation, hypertension, and vascular calcification [10–12], the extracellular matrix (ECM) is an active but frequently overlooked participant in the process of AVF maturation, and reviews regarding the role of the ECM in AVF maturation are lacking. As the ECM gives structural and mechanical strength to the vessel and provides communication to the vascular cells to regulate their differentiation and proliferation, it is involved in multiple processes that regulate AVF maturation. The ECM is also essential during the only healthy vascular remodelling comparable to AVF maturation, namely remodelling of the uterine vasculature during pregnancy [13]. Uterine arterial blood flow increases eight-fold over the course of 36 weeks of pregnancy, partially due to ECM-remodelling-induced diameter expansion [13,14].

The importance of the ECM in AVF remodelling is underscored in a recent study by Martinez et al. [15], who performed bulk RNA sequence analysis of pre-access veins and pair-matched AVFs of ESKD patients. They found that ECM proteins were amongst the most differentially expressed genes, with 87% of them upregulated in the AVF. Pathway enrichment analysis showed that collagen remodelling, both degradation and production, was strongly increased. Moreover, clusters of differentially expressed ECM components enabled the crude separation of failed and matured AVFs. These findings demonstrate the importance of ECM remodelling to facilitate increased blood flow in the AVF.

In the present review on the role of the ECM in AVF maturation, AVF failure and the differences in vascular anatomy of arteries and veins will be discussed, followed by the temporal regulation of ECM remodelling post-AVF creation. Factors regulating ECM remodelling will be explored, followed by future perspectives.

2. AVF Maturation Failure

As the venous vascular wall differs from the arterial vessel wall, a non-physiological pressure is created in the venous outflow tract. The venous outflow tract thus has to undergo arterialisation: the vessel wall has to thicken to endure the increase in pressure and tensile stress and expand its luminal diameter to facilitate both enhanced blood flow and adapt to the increase in shear stress [16,17]. As this is an intricate process, the venous outflow tract or juxta-anastomotic region of the AVF is often the culprit in AVF maturation and luminal narrowing due to thrombosis and intimal hyperplasia (IH) occurring at the AVF venous outflow tract. IH, together with the degree of enlargement of the vessel diameter, defined as outward remodelling (OR), defines the luminal diameter of the vessel and, thereby, the ability of the AVF to facilitate the increase in blood flow [18]. A disruption in this balance leads to AVF failure, where stenosis decreases the blood flow throughout the AVF and hinders efficient dialysis. To understand the difference between the arterial and venous vessel walls and what happens during AVF remodelling, we will give an overview of the anatomical differences between major arteries and veins and how their mechanical tasks are supported by their differences in the ECM scaffold.

3. The Arterial and Venous Vessel Wall and Its ECM Components

3.1. Vascular Identity: Phenotypic Differences between Arteries and Veins

Arteries and veins are functionally and anatomically distinct, whereby their arterial-venous cell fate is not only determined by hemodynamic differences but also by genetic mechanisms [19–21]. The most well-known arterial and venous markers belong to the Ephrin family [22], which consists of Ephrin ligands and Eph protein-tyrosine kinases receptors. Ligand ephrinB2 marks arterial cells [23], while its receptor Eph-B4 is expressed in venous ECs and VSMCs [24]. Protack et al. [25] showed that murine AVFs adopt a dual arteriovenous identity, with increased expression of both Eph-B4 and Ephrin-B2 in the venous outflow tract shortly after AVF creation. This supports the hypothesis that the venous part of the AVF has to undergo 'arterialisation'. To understand the process of arterialisation of the AVF, the distinction between the venous and arterial anatomy has to be elucidated.

AVFs are created from native arteries and veins, which, despite their expressional differences, comprise three layers: the tunica intima, tunica media, and tunica adventitia, sometimes also referred to as tunica externa—three layers with different functions in haemostasis, yet closely working together. The vascular ECM is composed of numerous different macromolecules, proteoglycans, glycoproteins, and cell adhesion proteins that are dispersed throughout the three tunicae and together resist compressive forces, provide adhesive surfaces, and give tensile strength to the vessel. These ECM components are produced by different cell types dispersed throughout the matrix. Figure 1 shows an overview of the anatomy of arteries versus veins and the differences in their ECM structure.

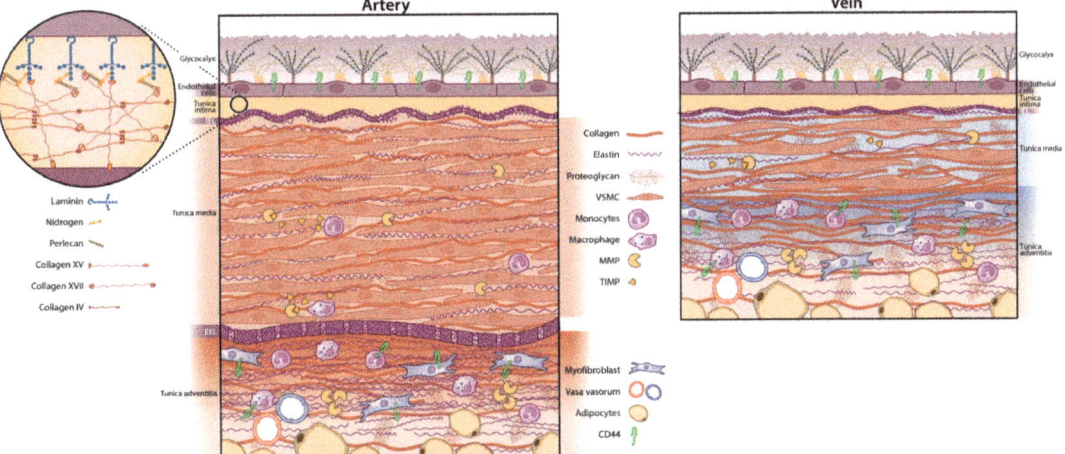

Figure 1. Overview of the arterial versus venous vessel layers and their respective ECM. The luminal side of both vessel types contains an intimal layer with endothelial cells (ECs) covered by the glycocalyx. The basal lamina is a mesh-like structure situated underneath the ECs. The internal elastic lamina (IEL) separates the tunica intima and media. The arterial (**left panel**) tunica media is thicker and more elastic than its venous counterpart (**right panel**) with more vascular smooth muscle cells (VSMCs), organised with elastin into a contractile-elastic unit. Arteries have a prominent external elastic lamina (EEL). Veins are less muscular with a lower elastin-to-collagen ratio. The tunica adventitia contains collagen-producing (myo) fibroblasts, surrounded by perivascular adipose tissue and the vasa vasorum: a capillary network of minor blood vessels. Matrix metalloproteinases (MMPs) are proteinases that regulate ECM degradation and extend into the media and adventitia. They are inhibited by tissue inhibitors of metalloproteinases (TIMPs). CD44 is a cell-surface glycoprotein receptor found in the venous and arterial adventitial layer and venous tunica intima. CD44 is expressed by ECs and can bind ECM components, such as collagen, fibronectin, MMPs, and hyaluronic acid. ECM = extracellular matrix.

3.2. The Tunica Intima

The inner layer of the vasculature is lined by endothelial cells (ECs), which require the ECM for their adhesion, migration, and proliferation. The luminal side of ECs is covered by the glycocalyx: a pericellular matrix composed of glycosaminoglycans (GAGs), proteoglycans, glycolipids, and glycoproteins [26,27]. GAGs, including heparan sulphate and hyaluronan, line the luminal side of the vessel and interact with plasma proteins, chemokines, and growth factors [28]. A healthy glycocalyx is essential for vascular haemostasis as, in addition to interacting with both ECs and plasma-produced molecules, it aids in mechanotransduction, vascular permeability, and, thereby, inflammation [27,29]. The glycocalyx senses and translates mechanical forces into intracellular signals, thereby mediating endothelial nitric oxide (NO) production [30]. Increased shear stress leads to reduced NO, which stimulates heparan sulphate synthesis [31]. In the healthy individual, glycocalyx synthesis and degradation are regulated to maintain endothelial function and adapt to environmental changes. Unfortunately, ESKD patients, and those receiving HD in particular, have elevated serum markers of glycocalyx damage and endothelial dysfunction [32]. This suggests that the vasculature of ESKD patients has reduced capability to adequately sense and respond to a change in shear stress.

In addition to the glycocalyx, ECs also produce and deposit components of the basal lamina and internal elastic lamina (IEL). The basal lamina, also referred to as the basement membrane, is composed of collagens type IV, XV, and XVIII as well as laminin, fibronectin, and perlecan dispersed throughout a mesh-like structure, providing structural support and anchor sites to the ECs [33–35]. When the basement membrane is exposed, it forms a bed for activated platelets to adhere [36], which in turn can induce VSMC proliferation and migration to aid in intimal hyperplasia [37,38]. The aforementioned IEL separates the tunica intima from the tunica media.

3.3. The Tunica Media

The tunica media is the most prominent vessel layer and highlights the different arterial and venous functions and how their mechanic requirements are translated into their anatomical composition. As arteries are exposed to higher blood pressure, the tunica media is thicker and more elastic than its venous counterpart. Their VSMCs are diagonally organised, whereas venous VSMCs are irregularly organised to make the vessels more dilatable [39].

As indicated by its nomenclature, the IEL is formed by elastin, one of the main components of the vascular ECM. VSMCs produce elastin, proteoglycans, and collagen and disperse these components in between elastic fibres and into the IEL. Elastic fibres are composed of elastin on a microfibril scaffold of fibrillins, microfibril-associated glycoproteins (MAGPs), microfibrillar-associated proteins (MFAPs), and fibulins [40,41]. As VSMCs are dispersed throughout the elastin fibres, they create a contractile-elastic unit and connect adjacent elastic laminae [42]. Veins have lower elastin expression, resulting in reduced transmission of tension throughout the vessel wall and diminished contractability compared to arteries. Arterial biomechanical tasks include dampening pulsation due to cardiac output and blood pressure differences throughout the vasculature [43]. Arterial VSMCs and elastic laminae are organised circumferentially [39], resulting in enhanced elasticity, constriction, and expansion capacity.

Loss of arterial elastin expression increases distensibility both longitudinally and circumferentially [44]. Veins, on the other hand, require less constriction capacity and have a bigger diameter and reduced blood pressure. Therefore, the venous tunica media contains less elastin and more collagen and smooth muscle fibres. Arteries have more abundant elastin expression compared to veins and, thus, a lower collagen-to-elastin ratio [45]. At low pressures, the arterial pressure curve strongly relates to elastin, while at high pressures, the elastic modulus of collagen is more important, indicating the importance of venous collagen deposition after AVF creation [46]. Long-term circumferential strain enhances the expression of both collagen type III and elastin [47]. Collagen limits wall distension by forming a triple helix of elongated fibrils. Collagen type I, III, IV, V, and VI are most abundant in the vasculature and are involved in forming its structure and cell signalling. Fibril-forming collagens—i.e., collagen I, III, and IV—are produced by VSMCs and provide vessel wall strength, as illustrated by the pathology in vascular Ehlers-Danlos syndrome where collagen IIIα1 is mutated, leading to fragile blood vessels and increased incidence of vessel rupture [47,48]. Furthermore, collagens provide binding sites for other ECM components, such as fibronectin and thrombospondin, which regulate vascular remodelling and activity of MMP-2 and MMP-9 [33,49,50].

In vascular homeostasis, the degradation of ECM components is of vital importance to facilitate elastin and collagen turnover. Matrix metalloproteinases (MMPs) produced by VSMCs, monocytes, and macrophages that extend into the media or adventitia are proteinases that partially regulate this ECM degradation [51]. MMPs are specifically inhibited by tissue inhibitors of metalloproteinases (TIMPs). Excessive TIMP production can result in vascular fibrosis, while abundant MMP expression and ECM degradation can lead to aneurysm formation, a form of AVF failure where the vessel wall expands abnormally, resulting in an increased risk of vessel rupture [52]. The formation of both elastin and collagen is dependent on lysyl oxidase (LOX), as LOX induces extracellular catalysis of lysine and hydroxylysine residues for collagen and elastin cross-linking [53,54]. Interventions in elastin and collagen production and turnover to promote AVF functioning will be discussed later.

3.4. The Tunica Adventitia

The tunica adventitia is separated from the tunica media by the external elastic lamina (EEL). Overall, the adventitia of large vessels is characterised by collagen-producing (myo)fibroblasts forming fibrous connective tissue, which is surrounded by perivascular adipose tissue (PVAT) and the vasa vasorum: a capillary network of minor blood vessels supplying major blood vessels with oxygen and nutrients [33,55]. Increased vasa vasorum is often observed in pathological processes [33]. In major arteries, the vasa vasorum spans from the adventitia to the outer layer of the media, while in large muscular veins, the vasa vasorum extends deeply into the tunica media [56]. However, vascularisation of the pre-access vein or venous AVF outflow tract is not associated with maturation outcomes [57].

Myofibroblasts in the tunica adventitia are major collagen producers, which they secrete into the tunica media to aggregate into fibrils to provide mechanical strength and prevent extreme vasodilation. The majority of the adventitial ECM is formed by elastin, fibronectin, proteoglycans, and collagen type I to prevent vessel rupture. Moreover, it serves as storage for molecules such as growth factors and MMPs. CD44 is a cell-surface glycoprotein receptor found in the venous and arterial adventitial layer and venous tunica intima [58]. CD44 is expressed by inflammatory and vascular cells such as ECs and fibroblasts, which can bind ECM components, such as collagen, fibronectin, MMPs, and hyaluronic acid [59]. As the adventitia contains ECM adhesion proteins such as CD44 and fibroblasts, which produce a majority of vascular ECM components, injury of the adventitia can radically alter the vascular ECM [33,60].

4. The Effect of Chronic Kidney Disease on the Vasculature

Kidney failure and cardiovascular pathology often go hand in hand, and reduced glomerular filtration rate (GFR) is an independent risk factor for cardiovascular events [61]. When creating an AVF, patients are at the ESKD stage; thus, the vessels are affected by an uremic environment and exposed to high blood pressure, as the majority of ESKD patients have hypertension. These factors affect vascular health and contribute to arterial stiffening, vascular calcification, increased risk of atherosclerosis, and impaired vascular repair [62]. Wali et al. [63] described that cephalic veins of ESKD patients show morphologically altered VSMCs and increased irregular deposition of collagen and elastin in between VSMCs when compared to healthy controls. Native veins of ESKD patients pre-AVF surgery contain significant collagen deposition in the tunica media and an enlarged intima mostly consisting of proteoglycans and collagen [64]. Furthermore, there is increased phagocytic activity of elastin and collagen in medial VSMCs [63]. Kidney failure is often paired with endothelial dysfunction, increased oxidative stress, VSMC proliferation, and peripheral vascular dysfunction [65–67]. Previous research in rats with kidney failure revealed that the increased oxidative stress and enhanced NO resistance hinder AVF maturation [68]. It has been shown that antioxidant selenium can modulate oxidative stress in CKD patients and reduce inflammation and oxidative stress markers [69].

However, when studying the vascular remodelling of AVFs, it is important to keep in mind that kidney failure impacts vascular health.

5. ECM Remodelling in the AVF: A Timely Matter

AVF maturation research is mostly focused on the venous outflow tract and its adaptation to the newly created arterial environment. Recently, there has been an increasing interest in modulating the ECM to promote AVF maturation and functionality. Hall et al. [70] elegantly studied temporal regulation of venous ECM remodelling after AVF creation in an aortocaval AVF model in C57BL/6J mice.

Murine AVF maturation encompasses three distinct phases of ECM remodelling: early ECM degradation, followed by a transition phase through reorganisation of the collagen and elastin scaffold, and lastly, rebuilding of the matrix with non-collagenous proteins and glycoproteins. Hence, timely expression of factors regulating the ECM is especially important after AVF creation.

Early MMP expression was observed in murine AVFs that would mature: *MMP-9* mRNA was increased maximally on day 1, while *MMP-2* RNA and protein expression increased on day 7 [70]. *TIMP1* RNA expression was also elevated on day 1 until 21 days post-AVF creation, whereas expression of *TIMP2*, *TIMP3* and *TIMP4* was upregulated later in the remodelling process at day 21. *Collagen III* mRNA showed upregulation on day 3 post-AVF creation, while *collagen I, IV, VIII,* and *XVIII* showed increased expression on day 7. Collagen III deposition was mainly observed in the adventitia and collagen I within the layers of the remodelling venous outflow tract. Total collagen protein expression was upregulated at day 21. *Elastin* mRNA expression was increased at 3 and 7 days post-AVF creation, and elastin protein deposition was elevated at day 28.

As MMP expression is regulated post-transcriptionally [71], data about mRNA expression should be interpreted with caution. However, the work of Hall et al. [70] indicates that the balance between vessel wall degradation and remodelling post-AVF creation is a fine line. Figure 2 gives a timeline overview of balanced degradation and reconstruction of the ECM after AVF creation. Next, studies on the ECM in AVF maturation will be discussed in the sequence of events occurring in AVF vascular remodelling: (i) ECM degradation regulated by MMPs and TIMPs, (ii) regulation of collagen and elastin deposition and LOX and (iii) the effects of TGF-β and inflammation on ECM remodelling.

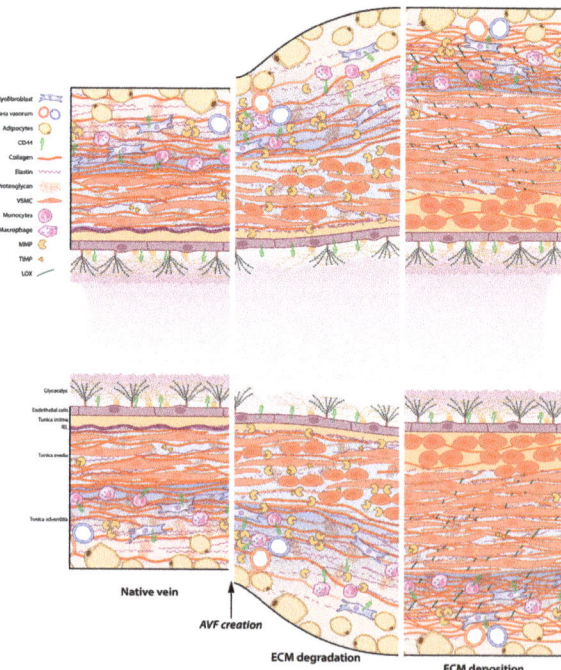

Figure 2. ECM remodelling after AVF creation. The native vein (**left panel**) requires rebuilding its vascular scaffold to facilitate AVF maturation. The venous outflow tract undergoes ECM degradation

(**middle panel**) and rebuilding of the ECM framework through ECM deposition (**right panel**). After AVF creation, the intimal layer and glycocalyx are damaged, and increased wall shear stress induces outward remodelling. Degradation of the ECM is facilitated by increased MMP (matrix metalloprotease) production, degrading collagen and elastin. Proliferating VSMCs ensure outward remodelling and wall thickening. Eventually, VSMCs migrate into the intima, where they proliferate and form intimal hyperplasia. During ECM deposition, collagen is produced by VSMCs and myofibroblasts. LOX (lysyl oxidase) cross-links elastin and collagen into their respective fibre formations.

6. Rebuilding the Vascular Framework: ECM Remodelling during AVF Maturation

6.1. ECM Degradation: The Role of MMPs and TIMPs in the AVF

ECM degradation, due to MMP activation or TIMP inhibition, is an essential process early on in AVF maturation. Rat AVFs showed an increase in both blood flow and intimal and medial area post-AVF creation compared to sham-operated controls, accompanied by an increase in MMP-2 and MMP-9 and downregulation of TIMP-4, which resulted in collagen degradation and an increased collagen I/III ratio [72]. At the time of AVF creation, elevated serum levels of MMP-2/TIMP-2 are measured in patients that would have a matured AVF compared to those who would experience AVF failure, with border-significant elevated levels of MMP-9/TIMP-4 as well ($p = 0.06$) [73]. Another study verified that patients with veins that would become matured AVFs had increased pre-operative expression of TIMP-2, MMP-2 activator MT1-MMP (Membrane type-1-MMP), and MMP-2 itself when compared to veins that would become failed AVFs [74]. Misra et al. [75] show that in rats, increased expression of MMP-2 and MMP-9 at a later time point is associated with venous stenosis, similar to ESKD patient data on increased MMP-9 expression in stenotic AVF lesions [76]. This is in line with the increased deposition of pro MMP 9 in AVFs that had to undergo surgical revision due to thrombosis or stenosis [77]. Many clinical descriptive studies have focused on the role of MMPs in AVF vascular remodelling but encompass findings based on systemic plasma levels or patient tissue of AVFs that need to undergo revision and are, thus, AVFs that failed to mature. This raises the question: what is the role of MMPs and TIMPs locally in AVF maturation? Therefore, interventions in MMP and TIMP expression are essential to see if they could be a therapeutic target, and what the therapeutic timing should be.

Doxycycline has been used as a specific MMP inhibitor as it directly inactivates MMPs through their zinc sites and indirectly through binding to inactive calcium sites [78]. Nath et al. [79] administered doxycycline chronically in their murine carotid-jugular AVF model. They verified suppression of MMP-9 expression in the venous outflow tract of the AVF, which did not affect AVF patency. Nonetheless, doxycycline is proven effective in preventing IH in a mouse model of arterial intimal hyperplasia and vein graft thickening [80]. Conditional forward logistical analysis was used with patients undergoing chronic maintenance haemodialysis to assess the likelihood of vascular access aneurysm formation after doxycycline treatment. Here, doxycycline treatment was compared to other antibiotic treatments. Patients that had received doxycycline and thus MMP inhibition appeared to have a decrease in aneurysm formation: a long-outcome effect rather than short-term maturation [78].

Shih et al. [81] studied the direct effect of MMP-9 deficiency on AVF remodelling using a murine knockout model with chronic kidney injury. MMP-9 deficient mice showed reduced VSMCs and collagen, accompanied by an increase in the AVF-venous luminal area and a decrease in IH when compared to wild-type (WT) mice. This was due to a reduction in vascular inflammation, demonstrated by reduced CD44 protein expression. After a femoral artery wire injury, MMP-9 and MMP-2 deficient mice showed reduced IH formation at two weeks post-injury and at four weeks post-injury [82]. As for patients studies, Lin et al. [83] state that the variant and accompanying transcriptional activity of MMPs is associated with AVF outcome in patients: genetic variants that lead to reduced transcription of MMP-1, MMP-3 and MMP-9 are associated with increased risk of AVF failure and stenosis, probably due to reduced proteolytic degradation of the ECM, resulting in ECM accumulation. Taken together, there is a solid foundation to state that MMPs play a significant role in AVF remodelling in animal models, which can possibly be translated to patients.

6.2. Strengthening the ECM Framework: Macro-Proteins Elastin and Collagen

Collagen and elastin are major ECM components that give structure to the vessel. Vascular pathologies emphasise the importance of the maintenance of these ECM proteins. Loss of both elastin-network integrity, as seen in Marfan syndrome [40,84], and loss-of-function mutations in collagen [48,85], both result in a weakened vessel wall and increased aneurysm formation. Excessive collagen deposition, however, is also a common cause of vascular clinical manifestations, seen in vessel fibrosis and vascular stiffness [86]. Figure 3 shows patient AVFs with different gradients of ECM deposition.

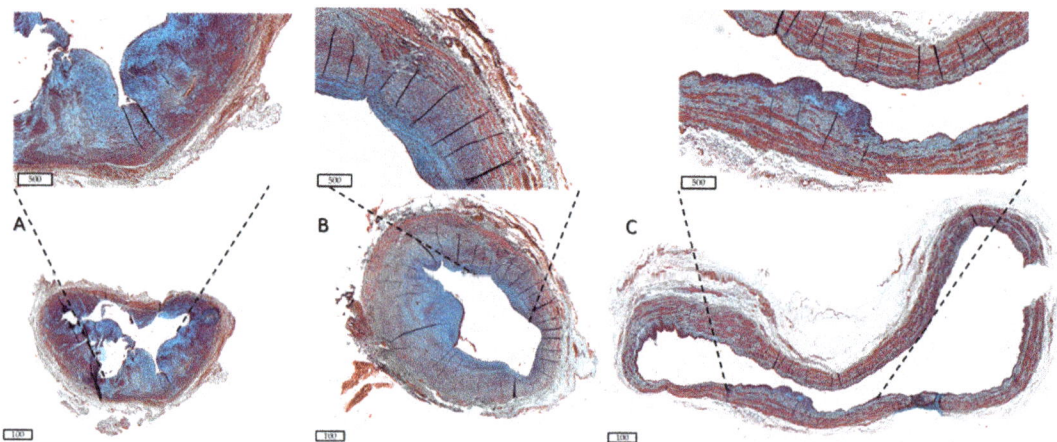

Figure 3. Differential ECM deposition in human AVFs. Samples were stained with Masson Trichrome. ECM is shown in blue, and smooth muscle cells are stained red. Scale bars are 100 μm in the bottom images and 500 μm in the inlay. (**A**) shows an AVF with little outward remodelling and a lot of ECM deposition in the IH, as shown in the inlay, indicative of a fibrotic AVF. (**B**) AVF with wall thickening and some ECM deposition amongst well-organised muscle fibres. (**C**) AVF with excessive OR and little ECM deposition, indicative of aneurysm formation. AVF = arteriovenous fistula, ECM = extracellular matrix, IH = intimal hyperplasia, OR = outward remodelling.

It has been established that mechanical factors, such as the increase in flow and wall shear stress stretching the elastic fibres, can increase the binding of elastase and the density of its binding sites along the fibre [87]. After AVF surgery, degeneration of the internal elastic lamina indeed occurs both arterially and at sites of the venous outflow tract that experience the biggest hemodynamic changes proximal to the anastomosis [88–90]. Rabbit AVF models show flow and shear stress-induced elastin fragmentation post AVF creation in the afferent artery [90–94], preceded by increased mRNA expression of both MMP-2, MMP-9, and MT1-MMP in ECs and VSMCs [92]. These fragmentations of the IEL contain VMSC deposition in between the laminae, facilitating OR and, thus, luminal enlargement [90,94]. Elastin haplodeficiency in a murine carotid-jugular AVF model resulted in accelerated OR of the venous outflow tract, no differential development of IH, and, eventually, an increase in the AVF venous luminal area [95]. Meanwhile, increased elastin deposition and a reduction in elastase activity hinder elastin degradation and impair OR and murine AVF maturation [96].

Two recent randomised, double-blind, placebo-controlled interventions using recombinant human elastase PRT-201 were performed in patients receiving a brachio- or radiocephalic AVF [97,98]. In these studies, PRT-201 was applied to the inflow artery, anastomosis, and outflow vein. Usage was proven safe, and there was no increase in adverse events; however, there was no increase in AVF venous diameter, stenosis, blood flow, or successful maturation compared to placebo-treated patients. PRT-201 was also not proven effective in AVGs [99]. Low-dosage PRT-201 was associated with improved unassisted maturation [100]. A larger prospective trial again showed the safety of using PRT-201 but no clinical value to surgical outcomes of radiocephalic fistulas nor secondary patency [97]. This indicates that increasing elastin degradation through temporary supplementation of an elastase might not be of therapeutic value to improve the maturation and patency of AVFs. It is hypothesised that the degradation of elastin has to be guided by the repair of elastic fibres [70,101].

Case reports of Alport's syndrome patients, a renal disease caused by a systemic disorder of collagen type IV (COL4A5), show a high prevalence of aneurysmal AVFs [102,103]. An increase in collagen content in the venous AVF is essential [104]; however, excessive collagen deposition leads to fibrosis. Interestingly, pre-existing arterial medial fibrosis is positively associated with AVF diameter, blood flow, and AVF maturation, while pre-existing venous medial fibrosis did not correlate with AVF functionality [105]. Excessive fibrotic remodelling in the form of circumferential alignment of collagenous fibres along the venous AVF outflow tract is associated with the non-maturation of brachiobasilic AVFs [106]. Bulk RNA sequence analysis of pre-access veins and pair-matched two-stage brachiobasilic AVFs of ESKD patients showed increased expression levels of fibrillar collagens I and II in the AVF, whereby increased COL8A1 and decreased MMP-9 and MMP-19 expression marked failed AVFs [15]. Upregulated COL8A1 results in increased collagen VIII, which forms a hexagonal network instead of fibres surrounding VSMCs in the tunica media. Martinez et al. hypothesise that the increase of collagen VIII in the medial layer might induce enhanced TGF-β and collagen I and collagen III production, resulting in vascular stiffness and AVF failure [15].

Little is known about the effect of collagen modulation post-AVF creation, as only a few interventional studies have been performed. Inhibiting LOX is the most common method used to study the role of collagen expression in vascular remodelling post-AVF, as it aids in intra- and intermolecular covalent cross-linking of collagen and facilitates the formation of collagen's triple helix conformation [107]. This is often performed by administering β-aminopropionitrile (BAPN), an irreversible LOX inhibitor. Hernandez et al. [108] inhibited LOX both locally and systemically through intraperitoneal BAPN injection or a BAPN-loaded scaffold around the venous outflow tract of an end-to-side rat AVF of the epigastric vein and femoral artery. The scaffold was degraded in 60 days, and systemic delivery was given from 2 days prior to 21 days post-AVF creation. Both treatments decreased fibrosis and shear stress and improved flow volume and distensibility in the AVF. A follow-up rat study showed that a BAPN-loaded PGLA nanofibre scaffold around the AVF promotes OR, prevents adventitial fibrosis, and improves vascular compliance [109]. BAPN and LOX, however, are both involved in cross-linking of collagen as well as elastin fibres and are, therefore, not exclusively related to collagen [110]. In ESKD patients, pre-access veins that would fail had higher LOX expression compared to veins that would mature. In failed AVFs, increased collagen cross-linking was observed [108]. Thus, modulation of LOX activity, and, thereby, elastin and collagen, seems to be a promising therapeutic target. However, as we can conclude from clinical trials administering PRT-201, interfering with elastin fragmentation is a delicate matter in AVF remodelling. Similar to the modulation of MMP and TIMP activity, inducing collagen and elastin degradation to promote AVF functionality is a timely matter and should be balanced by a certain degree of ECM production, orchestrated into a balanced sequence of events.

6.3. The Effect of TGF-β on ECM Remodelling in the AVF

In addition to regulating elastin and collagen cross-linking, LOX also influences the activation of the TGF-β pathway. TGF-β can be activated through the canonical pathway (Smad-activation) or non-canonical pathway (non-Smad-regulated) and generally results in myofibroblast activation, ECM production, and prevention of ECM degradation [111]. LOX suppresses Smad3 phosphorylation, a signalling intermediate normally activated by TGF-β stimulation to induce VSMC proliferation [112]. TGF-β, in turn, enhances LOX expression in rat aortic VSMCs [113,114]. TGF-β has been proven a key player in both kidney and vascular fibrosis and an important determinant for AVF functioning, as demonstrated by interventional in vivo studies summarised here.

Inhibiting TGF-β receptor-I through adventitial administration of SB431542-loaded nanoparticles reduces AVF wall thickness, collagen deposition, and VSMC proliferation [115]. Cell-specific TGF-β inhibition in ECs or VSMCs results in a reduction of collagen density, and EC-specific modulation of TGF-β signalling affects wall thickness, OR, and AVF patency [115]. Murine and matured patient AVFs show non-canonical activation of TGF-β signalling through TGF-β-activated kinase 1 (TAK1): reducing TAK1 function from 7 days prior to 7 days post-surgery using 5Z-7-oxozeaenol (OZ) results in decreased fibronectin, collagen I, wall thickness, and vessel diameter, which are increased when overexpressing TAK1 through periadventitial lentiviral transduction [116]. Differential regulation of TGF-β signalling might also underly the sex difference that is observed in AVF maturation, where female ESKD patients have decreased AVF maturation and patency [117,118]. Cai et al. [119] observed decreased gene expression of *BMP7* and *IL17Rb* and increased *Tgf-β1* and *Tgfβ-r1* in female murine AVFs, which resulted in venous fibrosis and negative vascular remodelling. Percutaneous transluminal angioplasty (PTA), a common intervention to salvage AVF functionality, increased TGF-β signalling in female but not male murine AVFs. This resulted in a reduction in the diameter of the venous outflow tract, luminal area, peak systolic velocity and an increased intima-to-media ratio [120]. In addition to murine studies, TGF-β is shown to play a role in human AVF maturation as well. Heine et al. [121] observed that polymorphisms that cause increased TGF-β1 expression were associated with a decreased 12-month patency rate of 62.4%, while intermediate production of TGF-β1 was associated with a patency of 81.2%. More research is needed to evaluate if TGF-β modulation is a valuable therapeutic remedy to reduce fibrosis and collagen deposition in the AVF.

6.4. Inflammation Influencing ECM Remodelling

Inflammation is an inherent process of AVF creation. Several inflammatory markers have been related to poor AVF patency outcome or AVF complications, including elevated systemic C-reactive protein (CRP) levels, neutrophil-to-lymphocyte ratio (NLR), platelet-to-lymphocyte ratio (PLR) and systemic inflammatory index (SII) as well as the local deposition of IL-6, TNF-α, and MCP-1 in the AVF [122–127]. Administering liposomal prednisolone inhibited inflammatory markers IL-6, TNF-α, and MCP-1 in a murine AVF model [128]. Liposomal prednisolone was proven safe in humans but has not been studied yet in a large randomised control trial [129]. Similar to ECM remodelling, inflammation requires a balanced approach between pro-inflammatory and anti-inflammatory signalling to facilitate OR and wall thickening. The influence of inflammation on AVF vascular remodelling is thoroughly reviewed [130–135].

Here, we uncover the relationship between inflammation and ECM remodelling in an AVF model, whereby inflammation can increase both ECM degradation and deposition. In the context of this review, we broadly uncover two categories in the relationship between inflammation and ECM remodelling in the AVF. Namely, inflammation-influencing MMP production by VSMCs and macrophages and inflammation-enhancing CD44 activity.

Inflammatory cells, such as macrophages and T-cells, produce numerous factors and cytokines that affect the vascular ECM, such as TGF-β and MMPs, by regulatory T cells and macrophages [136,137]. Pro-inflammatory M1 macrophages secrete pro-inflammatory cytokines such as TNF-α, IFN-γ, and IL-1 [134]. These local inflammatory stimuli can enhance regional upregulation of MMPs, induce elastin breaks, and thereby facilitate VSMC migration and degradation of the vascular ECM structure. IL-1 or TNF-α stimulation of VSMCs can lead to MMP-1, 3, and 9 production, but not TIMPs, and IL-1 is essential to facilitate OR [138,139]. Thrombosed AVFs are characterised by increased inflammation at the luminal site through MMP-9 expressing macrophages [124]. Inflammatory stimuli through CD44 activation, however, lead to the upregulation of M2 macrophages, which produce IL-10 and TGF-β, inducing ECM deposition, and leading to venous wall thickening, thereby promoting AVF maturation [134,140]. Administering cyclosporine, a T-cell inhibitor, resulted in reduced pro and anti-inflammatory macrophage presence in AVFs, hindering wall thickening, and promoting OR [141]. The relationship between inflammation-induced ECM remodelling, OR, and wall thickening versus IH formation due to VSMC proliferation and thrombi is complex and temporal and needs to be unravelled further to design therapeutic agents for AVF maturation.

7. Interventions Creating an ECM Framework Supporting Arteriovenous Fistulas

A few interventional studies have been performed in animal models using an ECM-mimicking framework to wrap around the AVF and possibly enhance AVF maturation.

CorMatrix, a decellularised matrix containing mostly collagen I fibres, GAGs, and glycoproteins, was wrapped around the venous outflow tract of carotid-jugular AVFs in immunodeficient mice. CorMatrix had beneficial outcomes regarding luminal outflow area and OR [142]. It is hypothesised that CorMatrix traps adventitial fibroblasts and prevents their differentiation and inwards migration, thereby reducing IH. CorMatrix has been applied in patient AVFs that needed reconstruction, and the data suggests that CorMatrix is safe to use, yet a high incidence of IH occurred, and there was no control group [143]. One case report shows the use of CorMatrix to repair an aneurysmal AVF, which resulted in complication-free clinical patency for the duration of a follow-up of four months [144]. A clinical study, however, has to be performed to investigate the effectiveness of the CorMatrix application peri-operatively. Natural Vascular Scaffolding (NVS) Therapy is another ECM intervention studied in rats. A small molecule (4-amino-1,8-naphtalamide) was administered perivascular at the AVF anastomosis and activated with a laser, which induces covalent binding of collagen and elastin, resulting in a framework for the vascular wall [145]. This resulted in an increase in AVF luminal area compared to controls, with decreased MMP-2 and MMP-9 expression at four weeks. Less collagen deposition was observed, although not significant, and the collagen fibres were organised more perpendicular [145], which seems favourable in AVFs [106]. The same treatment was also tested in sheep after cephalic veins were dilated through a balloon catheter [146]. This led to an increase in wall thickness and luminal diameter. These ECM interventions applying or creating an ECM-wrap peri-operatively during AVF surgery in animal models suggest that these remedies are safe and may have therapeutic potential by facilitating OR and luminal expansion to enhance AVF maturation.

8. Future Directions

In this review, we have summarised the influence of the ECM in AVF remodelling. Restructuring the ECM after AVF creation is a complex and timely process, with a dynamic balance between ECM degradation and ECM synthesis, mostly occurring in the venous tract. ECM degradation is essential for OR and venous wall thickening. However, excessive ECM degradation or insufficient repair results in a weak aneurysmatic venous outflow tract. Most novel therapeutic interventions focused on the ECM have solely been studied in animal models of AVF failure, and the success of those interventions is primarily based on (end-point) histological and morphometric parameters. To unravel how timely ECM turnover influences AVF outcome, new research methods should be employed to study ECM turnover longitudinally.

Currently, there are several live-imaging modalities available to track AVF flow, ECM remodelling and wall thickening in vivo, including magnetic resonance imaging (MRI), ultrasound analysis (US) and photoacoustics. Increased MRI specificity can be achieved through the use of contrast agents, for example, targeting elastin or tropoelastin [84,147,148]. US analysis can both determine AVF flow [149] and track wall thickening, total vessel wall area and OR [150,151]. By determining the optical absorption contrast of the target area, photoacoustics can characterise different tissues and distinguish collagen from its surroundings [152]. Live tissue imaging can also be accompanied by labelling agents targeting collagen [153–155]. MMPSense is a protease-activatable fluorescent agent activated by MMP-2, -3, -9, -12, and 13, which allows real-time imaging of local MMP activation and, thereby, ECM turnover [156]. These imaging techniques are suitable for studying ECM turnover post-AVF creation in vivo and should be incorporated into future research.

Furthermore, it is important to approach vascular AVF remodelling as a system tightly regulated by both vascular cells and the ECM that they produce. AVF maturation is an interplay between regulated inflammation and non-excessive cell proliferation to facilitate re-endothelisation, wall thickening, and OR and ECM turnover. Therapeutic interventions should, therefore, have time-dependent delivery. Post-surgery, peri-vascular scaffolds can easily be applied when constructing an AVF to target early vascular remodelling. During this timeframe, OR is the most prominent process, while IH develops at a later time point. The scaffolds could contain slow-release gels, anti-inflammatory cytokines, small molecules, or gene therapy. Such an example has already been described above, with Natural Vascular Scaffolding Therapy as an ECM intervention that interlinks collagen and elastin via photoactivation of a locally delivered small molecule (4-amino-1,8-naphtalamide) [145]. Another small molecule that showed promise in preventing AVF failure is MCC950. This NLRP3 inflammasome inhibitor was shown to repress Smad2/3 phosphorylation and suppress CKD–promoted AVF failure [157]. Another study showed that AVF failure due to IH and subsequent venous stenosis in a porcine AVF model could be improved by using $1\alpha,25(OH)2D3$-encapsulated nanoparticles. Perivascular AVF release of this molecule from poly(lactic-co-glycolic acid) nanoparticles that were embedded in a pluronic F127 hydrogel resulted in better AVF flow and hemodynamics, while it also reduced inflammation and fibrosis [158]. Vascular scaffolds could also be implemented to study the effect of redox regulation on ECM components. Both cross-linking and microfibril-assembly of collagen, elastin, and fibrillin are known to be influenced by intrinsic redox regulation, affecting the protein's cysteine residues and MMP-2 and MMP-9 [159]. Moreover, adenoviral-delivered gene therapy is a well-studied delivery method in vein grafts and AVFs [160–162] and could be employed to target LOX, MMPs, or TIMPs.

An alternative approach can be found in the capacity of non-coding RNA, such as microRNAs (miRNAs), to post-transcriptionally regulate ECM composition and AVF outcome. For instance, miR-29a/b regulates ECM mRNAs, such as collagen (type I and II), fibronectin, and elastin, and inhibits their translation [163]. As such, miR-29 inhibitors can stimulate elastin formation and LOX expression [164,165], although this could also induce a fibrotic response and vascular calcification [163,166]. Possibly, combination therapy with LOX-inhibition might achieve synergistic favourable effects on AVF maturation, especially when locally applied and in a timely manner. Interestingly, miR-21 also directly affects the synthesis of several collagen species, and miR-21 inhibition improved AVF patency by reducing IH and VSMCs and myofibroblast presence in the vessel wall [167].

In conclusion, the ECM is a vital scaffold for vessels, and its remodelling post-AVF creation is a balance between degradation and deposition, enabling outward remodelling of the AVF, followed by supporting and strengthening it. This requires timely therapeutic delivery, taking into account the interplay between the vascular cells and the ECM they both produce and reside in.

Author Contributions: S.L.L. drafted the manuscript and prepared the schematic figures; M.R.d.V. provided the patient material S.L.L., M.R.d.V. and J.I.R. designed the concept. R.B. provided critical input. All authors have read and agreed to the published version of the manuscript.

Funding: This research received no external funding.

Institutional Review Board Statement: Anonymous arteriovenous fistula specimens were obtained in accordance with guidelines set out by the 'Code for Proper Secondary Use of Human Tissue' of the Dutch Federation of Biomedical Scientific Societies (Federa) and conform with the principles outlined in the Declaration of Helsinki.

Informed Consent Statement: Not applicable.

Data Availability Statement: Not applicable.

Acknowledgments: The authors would like to express their gratitude to Manon Zuurmond, who helped design the schematic figures.

Conflicts of Interest: The authors declare no conflict of interest.

References

1. Brescia, M.J.; Cimino, J.E.; Appel, K.; Hurwich, B.J. Chronic hemodialysis using venipuncture and a surgically created arteriovenous fistula. *N. Engl. J. Med.* **1966**, *275*, 1089–1092. [CrossRef] [PubMed]
2. Schmidli, J.; Widmer, M.K.; Basile, C.; de Donato, G.; Gallieni, M.; Gibbons, C.P.; Haage, P.; Hamilton, G.; Hedin, U.; Kamper, L.; et al. Editor's Choice—Vascular Access: 2018 Clinical Practice Guidelines of the European Society for Vascular Surgery (ESVS). *Eur. J. Vasc. Endovasc. Surg.* **2018**, *55*, 757–818. [CrossRef] [PubMed]
3. Lok, C.E.; Huber, T.S.; Lee, T.; Shenoy, S.; Yevzlin, A.S.; Abreo, K.; Allon, M.; Asif, A.; Astor, B.C.; Glickman, M.H.; et al. KDOQI Clinical Practice Guideline for Vascular Access: 2019 Update. *Am. J. Kidney Dis.* **2020**, *75*, S1–S164. [CrossRef] [PubMed]
4. Astor, B.C.; Eustace, J.A.; Powe, N.R.; Klag, M.J.; Fink, N.E.; Coresh, J.; Josef Coresh for the CHOICE Study. Type of Vascular Access and Survival among Incident Hemodialysis Patients: The Choices for Healthy Outcomes in Caring for ESRD (CHOICE) Study. *J. Am. Soc. Nephrol.* **2005**, *16*, 1449–1455. [CrossRef]
5. Banerjee, T.; Kim, S.J.; Astor, B.; Shafi, T.; Coresh, J.; Powe, N.R. Vascular Access Type, Inflammatory Markers, and Mortality in Incident Hemodialysis Patients: The Choices for Healthy Outcomes in Caring for End-Stage Renal Disease (CHOICE) Study. *Am. J. Kidney Dis.* **2014**, *64*, 954–961. [CrossRef]
6. Dhingra, R.K.; Young, E.W.; Hulbert-Shearon, T.E.; Leavey, S.F.; Port, F.K. Type of vascular access and mortality in U.S. hemodialysis patients. *Kidney Int.* **2001**, *60*, 1443–1451. [CrossRef]
7. Choi, J.; Ban, T.H.; Choi, B.S.; Baik, J.H.; Kim, B.S.; Kim, Y.O.; Park, C.W.; Yang, C.W.; Jin, D.C.; Park, H.S. Comparison of vascular access patency and patient survival between native arteriovenous fistula and synthetic arteriovenous graft according to age group. *Hemodial Int.* **2020**, *24*, 309–316. [CrossRef]
8. Voorzaat, B.M.; van der Bogt, K.E.A.; Janmaat, C.J.; van Schaik, J.; Dekker, F.W.; Rotmans, J.I.; Voorzaat, B.M.; van der Bogt, K.E.A.; Janmaat, C.J.; van Schaik, J.; et al. Arteriovenous Fistula Maturation Failure in a Large Cohort of Hemodialysis Patients in the Netherlands. *World J. Surg.* **2018**, *42*, 1895–1903. [CrossRef]
9. Arhuidese, I.J.; Orandi, B.J.; Nejim, B.; Malas, M. Utilization, patency, and complications associated with vascular access for hemodialysis in the United States. *J. Vasc. Surg.* **2018**, *68*, 1166–1174. [CrossRef]
10. Ponticos, M.; Smith, B.D. Extracellular matrix synthesis in vascular disease: Hypertension, and atherosclerosis. *J. Biomed. Res.* **2014**, *28*, 25–39.
11. Chistiakov, D.A.; Sobenin, I.A.; Orekhov, A.N. Vascular Extracellular Matrix in Atherosclerosis. *Cardiol. Rev.* **2013**, *21*, 270–288. [CrossRef]
12. Stepien, K.L.; Bajdak-Rusinek, K.; Fus-Kujawa, A.; Kuczmik, W.; Gawron, K. Role of Extracellular Matrix and Inflammation in Abdominal Aortic Aneurysm. *Int. J. Mol. Sci.* **2022**, *23*, 11078. [CrossRef]
13. Osol, G.; Moore, L.G. Maternal Uterine Vascular Remodeling During Pregnancy. *Microcirculation* **2014**, *21*, 38–47. [CrossRef]
14. Palmer, S.K.; Zamudio, S.; Coffin, C.; Parker, S.; Stamm, E.; Moore, L.G. Quantitative estimation of human uterine artery blood flow and pelvic blood flow redistribution in pregnancy. *Obstet. Gynecol.* **1992**, *80*, 1000–1006.
15. Martinez, L.; Rojas, M.G.; Tabbara, M.; Pereira-Simon, S.; Falcon, N.S.; Rauf, M.A.; Challa, A.S.; Zigmond, Z.M.; Griswold, A.J.; Duque, J.C.; et al. The Transcriptomics of the Human Vein Transformation After Arteriovenous Fistula Anastomosis Uncovers Layer-Specific Remodeling and Hallmarks of Maturation Failure. *Kidney Int. Rep.* **2023**, *8*, 837–850. [CrossRef]
16. Hu, H.; Patel, S.; Hanisch, J.J.; Santana, J.M.; Hashimoto, T.; Bai, H.; Kudze, T.; Foster, T.R.; Guo, J.; Yatsula, B.; et al. Future research directions to improve fistula maturation and reduce access failure. *Semin. Vasc. Surg.* **2016**, *29*, 153–171. [CrossRef]
17. Lu, D.Y.; Chen, E.Y.; Wong, D.J.; Yamamoto, K.; Protack, C.D.; Williams, W.T.; Assi, R.; Hall, M.R.; Sadaghianloo, N.; Dardik, A. Vein graft adaptation and fistula maturation in the arterial environment. *J. Surg. Res.* **2014**, *188*, 162–173. [CrossRef]
18. Rothuizen, T.C.; Wong, C.; Quax, P.H.A.; van Zonneveld, A.J.; Rabelink, T.J.; Rotmans, J.I. Arteriovenous access failure: More than just intimal hyperplasia? *Nephrol. Dial. Transplant.* **2013**, *28*, 1085–1092. [CrossRef]
19. Adams, R.H. Molecular control of arterial–venous blood vessel identity. *J. Anat.* **2003**, *202*, 105–112. [CrossRef]
20. Lawson, N.D.; Scheer, N.; Pham, V.N.; Kim, C.-H.; Chitnis, A.B.; Campos-Ortega, J.A.; Weinstein, B.M. Notch signaling is required for arterial-venous differentiation during embryonic vascular development. *Development* **2001**, *128*, 3675–3683. [CrossRef]

21. le Noble, F.; Moyon, D.; Pardanaud, L.; Yuan, L.; Djonov, V.; Matthijsen, R.; Bréant, C.; Fleury, V.; Eichmann, A. Flow regulates arterial-venous differentiation in the chick embryo yolk sac. *Development* **2004**, *131*, 361–375. [CrossRef] [PubMed]
22. Wang, H.U.; Chen, Z.F.; Anderson, D.J. Molecular distinction and angiogenic interaction between embryonic arteries and veins revealed by ephrin-B2 and its receptor Eph-B4. *Cell* **1998**, *93*, 741–753. [CrossRef] [PubMed]
23. Gale, N.W.; Baluk, P.; Pan, L.; Kwan, M.; Holash, J.; DeChiara, T.M.; McDonald, D.M.; Yancopoulos, G.D. Ephrin-B2 selectively marks arterial vessels and neovascularization sites in the adult, with expression in both endothelial and smooth-muscle cells. *Dev. Biol.* **2001**, *230*, 151–160. [CrossRef] [PubMed]
24. Steinle, J.J.; Meininger, C.J.; Forough, R.; Wu, G.; Wu, M.H.; Granger, H.J. Eph B4 receptor signaling mediates endothelial cell migration and proliferation via the phosphatidylinositol 3-kinase pathway. *J. Biol. Chem.* **2002**, *277*, 43830–43835. [CrossRef]
25. Protack, C.D.; Foster, T.R.; Hashimoto, T.; Yamamoto, K.; Lee, M.Y.; Kraehling, J.R.; Bai, H.; Hu, H.; Isaji, T.; Santana, J.M.; et al. Eph-B4 regulates adaptive venous remodeling to improve arteriovenous fistula patency. *Sci. Rep.* **2017**, *7*, 15386. [CrossRef]
26. Rix, D.A.; Douglas, M.S.; Talbot, D.; Dark, J.H.; Kirby, J.A. Role of glycosaminoglycans (GAGs) in regulation of the immunogenicity of human vascular endothelial cells. *Clin. Exp. Immunol.* **1996**, *104*, 60–65. [CrossRef]
27. Wang, G.; Tiemeier, G.L.; van den Berg, B.M.; Rabelink, T.J. Endothelial Glycocalyx Hyaluronan: Regulation and Role in Prevention of Diabetic Complications. *Am. J. Pathol.* **2020**, *190*, 781–790. [CrossRef]
28. Reitsma, S.; Slaaf, D.W.; Vink, H.; van Zandvoort, M.A.M.J.; oude Egbrink, M.G.A. The endothelial glycocalyx: Composition, functions, and visualization. *Pflügers Arch.—Eur. J. Physiol.* **2007**, *454*, 345–359. [CrossRef]
29. Tarbell, J.M.; Simon, S.I.; Curry, F.R. Mechanosensing at the vascular interface. *Annu. Rev. Biomed. Eng.* **2014**, *16*, 505–532. [CrossRef]
30. Bartosch, A.M.W.; Mathews, R.; Tarbell, J.M. Endothelial Glycocalyx-Mediated Nitric Oxide Production in Response to Selective AFM Pulling. *Biophys. J.* **2017**, *113*, 101–108. [CrossRef]
31. Giantsos-Adams, K.M.; Koo, A.J.-A.; Song, S.; Sakai, J.; Sankaran, J.; Shin, J.H.; Garcia-Cardena, G.; Dewey, C.F. Heparan Sulfate Regrowth Profiles Under Laminar Shear Flow Following Enzymatic Degradation. *Cell. Mol. Bioeng.* **2013**, *6*, 160–174. [CrossRef]
32. Liew, H.; Roberts, M.A.; Pope, A.; McMahon, L.P. Endothelial glycocalyx damage in kidney disease correlates with uraemic toxins and endothelial dysfunction. *BMC Nephrol.* **2021**, *22*, 21. [CrossRef]
33. Halper, J. Chapter Four—Basic Components of Vascular Connective Tissue and Extracellular Matrix. In *Advances in Pharmacology*; Khalil, R.A., Ed.; Academic Press: Cambridge, MA, USA, 2018; pp. 95–127.
34. Kalluri, R. Basement membranes: Structure, assembly and role in tumour angiogenesis. *Nat. Rev. Cancer* **2003**, *3*, 422–433. [CrossRef]
35. Tomono, Y.; Naito, I.; Ando, K.; Yonezawa, T.; Sado, Y.; Hirakawa, S.; Arata, J.; Okigaki, T.; Ninomiya, Y. Epitope-defined Monoclonal Antibodies against Multiplexin Collagens Demonstrate that Type XV and XVIII Collagens are Expressed in Specialized Basement Membranes. *Cell Struct. Funct.* **2002**, *27*, 9–20. [CrossRef]
36. Massberg, S.; Grüner, S.; Konrad, I.; Garcia Arguinzonis, M.I.; Eigenthaler, M.; Hemler, K.; Kersting, J.; Schulz, C.; Müller, I.; Besta, F.; et al. Enhanced in vivo platelet adhesion in vasodilator-stimulated phosphoprotein (VASP)–deficient mice. *Blood* **2004**, *103*, 136–142. [CrossRef]
37. Shi, Y.; Field, D.J.; Long, X.; Mickelsen, D.; Ko, K.-a.; Ture, S.; Korshunov, V.A.; Miano, J.M.; Morrell, C.N. Platelet factor 4 mediates vascular smooth muscle cell injury responses. *Blood* **2013**, *121*, 4417–4427. [CrossRef]
38. Fingerle, J.; Johnson, R.; Clowes, A.W.; Majesky, M.W.; Reidy, M.A. Role of platelets in smooth muscle cell proliferation and migration after vascular injury in rat carotid artery. *Proc. Natl. Acad. Sci. USA* **1989**, *86*, 8412–8416. [CrossRef]
39. O'Connell, M.K.; Murthy, S.; Phan, S.; Xu, C.; Buchanan, J.; Spilker, R.; Dalman, R.L.; Zarins, C.K.; Denk, W.; Taylor, C.A. The three-dimensional micro- and nanostructure of the aortic medial lamellar unit measured using 3D confocal and electron microscopy imaging. *Matrix Biol.* **2008**, *27*, 171–181. [CrossRef]
40. Bunton, T.E.; Biery, N.J.; Myers, L.; Gayraud, B.; Ramirez, F.; Dietz, H.C. Phenotypic Alteration of Vascular Smooth Muscle Cells Precedes Elastolysis in a Mouse Model of Marfan Syndrome. *Circ. Res.* **2001**, *88*, 37–43. [CrossRef]
41. Pereira, L.; Andrikopoulos, K.; Tian, J.; Lee, S.Y.; Keene, D.R.; Ono, R.; Reinhardt, D.P.; Sakai, L.Y.; Biery, N.J.; Bunton, T.; et al. Targetting of the gene encoding fibrillin-1 recapitulates the vascular aspect of Marfan syndrome. *Nat. Genet.* **1997**, *17*, 218–222. [CrossRef]
42. Davis, E.C. Smooth muscle cell to elastic lamina connections in developing mouse aorta. Role in aortic medial organization. Laboratory investigation. *J. Tech. Methods Pathol.* **1993**, *68*, 89–99.
43. Niklason, L.; Dai, G. Arterial Venous Differentiation for Vascular Bioengineering. *Annu. Rev. Biomed. Eng.* **2018**, *20*, 431–447. [CrossRef] [PubMed]
44. Chow, M.J.; Choi, M.; Yun, S.H.; Zhang, Y. The effect of static stretch on elastin degradation in arteries. *PLoS ONE* **2013**, *8*, e81951. [CrossRef] [PubMed]
45. Basu, P.; Sen, U.; Tyagi, N.; Tyagi, S.C. Blood flow interplays with elastin: Collagen and MMP: TIMP ratios to maintain healthy vascular structure and function. *Vasc. Health Risk Manag.* **2010**, *6*, 215–228.
46. Berry, C.L.; Greenwald, S.E.; Rivett, J.F. Static mechanical properties of the developing and mature rat aorta. *Cardiovasc. Res.* **1975**, *9*, 669–678. [CrossRef]
47. Wanjare, M.; Agarwal, N.; Gerecht, S. Biomechanical strain induces elastin and collagen production in human pluripotent stem cell-derived vascular smooth muscle cells. *Am. J. Physiol. Cell Physiol.* **2015**, *309*, C271–C281. [CrossRef]

48. Schwarze, U.; Schievink, W.I.; Petty, E.; Jaff, M.R.; Babovic-Vuksanovic, D.; Cherry, K.J.; Pepin, M.; Byers, P.H. Haploinsufficiency for One COL3A1 Allele of Type III Procollagen Results in a Phenotype Similar to the Vascular Form of Ehlers-Danlos Syndrome, Ehlers-Danlos Syndrome Type IV. *Am. J. Hum. Genet.* **2001**, *69*, 989–1001. [CrossRef]
49. Krady, M.M.; Zeng, J.; Yu, J.; MacLauchlan, S.; Skokos, E.A.; Tian, W.; Bornstein, P.; Sessa, W.C.; Kyriakides, T.R. Thrombospondin-2 modulates extracellular matrix remodeling during physiological angiogenesis. *Am. J. Pathol.* **2008**, *173*, 879–891. [CrossRef]
50. Chiang, H.Y.; Korshunov, V.A.; Serour, A.; Shi, F.; Sottile, J. Fibronectin is an important regulator of flow-induced vascular remodeling. *Arterioscler. Thromb. Vasc. Biol.* **2009**, *29*, 1074–1079. [CrossRef]
51. Nagase, H.; Woessner, J.F. Matrix Metalloproteinases. *J. Biol. Chem.* **1999**, *274*, 21491–21494. [CrossRef]
52. Páramo, J.A. New mechanisms of vascular fibrosis: Role of lysyl oxidase. *Clínica Investig. Arterioscler. Engl. Ed.* **2017**, *29*, 166–167. [CrossRef]
53. Pinnell, S.R.; Martin, G.R. The cross-linking of collagen and elastin: Enzymatic conversion of lysine in peptide linkage to alpha-aminoadipic-delta-semialdehyde (allysine) by an extract from bone. *Proc. Natl. Acad. Sci. USA* **1968**, *61*, 708–716. [CrossRef]
54. Csiszar, K. Lysyl oxidases: A novel multifunctional amine oxidase family. In *Progress in Nucleic Acid Research and Molecular Biology*; Academic Press: Cambridge, MA, USA, 2001; pp. 1–32.
55. Takaoka, M.; Nagata, D.; Kihara, S.; Shimomura, I.; Kimura, Y.; Tabata, Y.; Saito, Y.; Nagai, R.; Sata, M. Periadventitial Adipose Tissue Plays a Critical Role in Vascular Remodeling. *Circ. Res.* **2009**, *105*, 906–911. [CrossRef]
56. Dashwood, M.R.; Anand, R.; Loesch, A.; Souza, D.S.R. Hypothesis: A Potential Role for the Vasa Vasorum in the Maintenance of Vein Graft Patency. *Angiology* **2004**, *55*, 385–395. [CrossRef]
57. Duque, J.C.; Martinez, L.; Tabbara, M.; Parikh, P.; Paez, A.; Selman, G.; Salman, L.H.; Velazquez, O.C.; Vazquez-Padron, R.I. Vascularization of the arteriovenous fistula wall and association with maturation outcomes. *J. Vasc. Access* **2020**, *21*, 161–168. [CrossRef]
58. Hellström, M.; Engström-Laurent, A.; Hellström, S. Expression of the CD44 Receptor in the Blood Vessel System: An Experimental Study in Rat. *Cells Tissues Organs* **2005**, *179*, 102–108. [CrossRef]
59. Goodison, S.; Urquidi, V.; Tarin, D. CD44 cell adhesion molecules. *Mol. Pathol.* **1999**, *52*, 189–196. [CrossRef]
60. Stenmark, K.R.; Yeager, M.E.; El Kasmi, K.C.; Nozik-Grayck, E.; Gerasimovskaya, E.V.; Li, M.; Riddle, S.R.; Frid, M.G. The adventitia: Essential regulator of vascular wall structure and function. *Annu. Rev. Physiol.* **2013**, *75*, 23–47. [CrossRef]
61. Go, A.S.; Chertow, G.M.; Fan, D.; McCulloch, C.E.; Hsu, C.-y. Chronic Kidney Disease and the Risks of Death, Cardiovascular Events, and Hospitalization. *N. Engl. J. Med.* **2004**, *351*, 1296–1305. [CrossRef]
62. Brunet, P.; Gondouin, B.; Duval-Sabatier, A.; Dou, L.; Cerini, C.; Dignat-George, F.; Jourde-Chiche, N.; Argiles, A.; Burtey, S. Does Uremia Cause Vascular Dysfunction. *Kidney Blood Press. Res.* **2011**, *34*, 284–290. [CrossRef]
63. Wali, M.A.; Eid, R.A.; Al-Homrany, M.A. Smooth Muscle Changes in the Cephalic Vein of Renal Failure Patients before Use as an Arteriovenous Fistula (AVF). *J. Smooth Muscle Res.* **2002**, *38*, 75–85. [CrossRef] [PubMed]
64. Alpers, C.E.; Imrey, P.B.; Hudkins, K.L.; Wietecha, T.A.; Radeva, M.; Allon, M.; Cheung, A.K.; Dember, L.M.; Roy-Chaudhury, P.; Shiu, Y.-T.; et al. Histopathology of Veins Obtained at Hemodialysis Arteriovenous Fistula Creation Surgery. *J. Am. Soc. Nephrol.* **2017**, *28*, 3076–3088. [CrossRef] [PubMed]
65. Martens, C.R.; Edwards, D.G. Peripheral vascular dysfunction in chronic kidney disease. *Cardiol. Res. Pract.* **2011**, *2011*, 267257. [CrossRef] [PubMed]
66. Günthner, T.; Jankowski, V.; Kretschmer, A.; Nierhaus, M.; Van Der Giet, M.; Zidek, W.; Jankowski, J. Progress in Uremic Toxin Research: Endothelium and Vascular Smooth Muscle Cells in the Context of Uremia. *Semin. Dial.* **2009**, *22*, 428–432. [CrossRef]
67. Moody, W.E.; Edwards, N.C.; Madhani, M.; Chue, C.D.; Steeds, R.P.; Ferro, C.J.; Townend, J.N. Endothelial dysfunction and cardiovascular disease in early-stage chronic kidney disease: Cause or association? *Atherosclerosis* **2012**, *223*, 86–94. [CrossRef]
68. Geenen, I.L.; Kolk, F.F.; Molin, D.G.; Wagenaar, A.; Compeer, M.G.; Tordoir, J.H.; Schurink, G.W.; De Mey, J.G.; Post, M.J. Nitric Oxide Resistance Reduces Arteriovenous Fistula Maturation in Chronic Kidney Disease in Rats. *PLoS ONE* **2016**, *11*, e0146212. [CrossRef]
69. Stockler-Pinto, M.B.; Mafra, D.; Moraes, C.; Lobo, J.; Boaventura, G.T.; Farage, N.E.; Silva, W.S.; Cozzolino, S.F.; Malm, O. Brazil Nut (*Bertholletia excelsa*, H.B.K.) Improves Oxidative Stress and Inflammation Biomarkers in Hemodialysis Patients. *Biol. Trace Elem. Res.* **2014**, *158*, 105–112. [CrossRef]
70. Hall, M.R.; Yamamoto, K.; Protack, C.D.; Tsuneki, M.; Kuwahara, G.; Assi, R.; Brownson, K.E.; Bai, H.; Madri, J.A.; Dardik, A. Temporal regulation of venous extracellular matrix components during arteriovenous fistula maturation. *J. Vasc. Access* **2015**, *16*, 93–106. [CrossRef]
71. Fanjul-Fernández, M.; Folgueras, A.R.; Cabrera, S.; López-Otín, C. Matrix metalloproteinases: Evolution, gene regulation and functional analysis in mouse models. *Biochim. Biophys. Acta BBA—Mol. Cell Res.* **2010**, *1803*, 3–19. [CrossRef]
72. Chan, C.-Y.; Chen, Y.-S.; Ma, M.-C.; Chen, C.-F. Remodeling of experimental arteriovenous fistula with increased matrix metalloproteinase expression in rats. *J. Vasc. Surg.* **2007**, *45*, 804–811. [CrossRef]
73. Lee, E.S.; Shen, Q.; Pitts, R.L.; Guo, M.; Wu, M.H.; Sun, S.C.; Yuan, S.Y. Serum metalloproteinases MMP-2, MMP-9, and metalloproteinase tissue inhibitors in patients are associated with arteriovenous fistula maturation. *J. Vasc. Surg.* **2011**, *54*, 454–459. [CrossRef]
74. Lee, E.S.; Shen, Q.; Pitts, R.L.; Guo, M.; Wu, M.H.; Yuan, S.Y. Vein tissue expression of matrix metalloproteinase as biomarker for hemodialysis arteriovenous fistula maturation. *Vasc. Endovasc. Surg.* **2010**, *44*, 674–679. [CrossRef]

75. Misra, S.; Fu, A.A.; Anderson, J.L.; Sethi, S.; Glockner, J.F.; McKusick, M.A.; Bjarnason, H.; Woodrum, D.A.; Mukhopadhyay, D. The rat femoral arteriovenous fistula model: Increased expression of matrix metalloproteinase-2 and -9 at the venous stenosis. *J. Vasc. Interv. Radiol.* **2008**, *19*, 587–594. [CrossRef]
76. Ruan, L.; Yao, X.; Li, W.; Zhang, L.; Yang, H.; Sun, J.; Li, A. Effect of galectin-3 in the pathogenesis of arteriovenous fistula stenosis formation. *Ren. Fail.* **2021**, *43*, 566–576. [CrossRef]
77. Misra, S.; Fu, A.A.; Rajan, D.K.; Juncos, L.A.; McKusick, M.A.; Bjarnason, H.; Mukhopadhyay, D. Expression of Hypoxia Inducible Factor–1α, Macrophage Migration Inhibition Factor, Matrix Metalloproteinase–2 and −9, and Their Inhibitors in Hemodialysis Grafts and Arteriovenous Fistulas. *J. Vasc. Interv. Radiol.* **2008**, *19*, 252–259. [CrossRef]
78. Diskin, C.; Stokes, T.J.; Dansby, L.M.; Radcliff, L.; Carter, T.B. Doxycycline may reduce the incidence of aneurysms in haemodialysis vascular accesses. *Nephrol. Dial. Transplant.* **2005**, *20*, 959–961. [CrossRef]
79. Nath, K.A.; Grande, J.P.; Kang, L.; Juncos, J.P.; Ackerman, A.W.; Croatt, A.J.; Katusic, Z.S. ß-Catenin is markedly induced in a murine model of an arteriovenous fistula: The effect of metalloproteinase inhibition: The effect of metalloproteinase inhibition. *Am. J. Physiol. Renal Physiol.* **2010**, *299*, F1270–F1277. [CrossRef]
80. Lardenoye, J.H.; de Vries, M.R.; Deckers, M.; van Lent, N.; Hanemaaijer, R.; van Bockel, J.H.; Quax, P.H. Inhibition of intimal hyperplasia by the tetracycline derived mmp inhibitor doxycycline in vein graft disease in vitro and in vivo. *EuroIntervention* **2005**, *1*, 236–243.
81. Shih, Y.-C.; Chen, P.-Y.; Ko, T.-M.; Huang, P.-H.; Ma, H.; Tarng, D.-C. MMP-9 Deletion Attenuates Arteriovenous Fistula Neointima through Reduced Perioperative Vascular Inflammation. *Int. J. Mol. Sci.* **2021**, *22*, 5448. [CrossRef]
82. Guo, L.; Ning, W.; Tan, Z.; Gong, Z.; Li, X. Mechanism of matrix metalloproteinase axis-induced neointimal growth. *J. Mol. Cell. Cardiol.* **2014**, *66*, 116–125. [CrossRef]
83. Lin, C.-C.; Yang, W.-C.; Chung, M.-Y.; Lee, P.-C. Functional Polymorphisms in Matrix Metalloproteinases-1, -3, -9 are Associated with Arteriovenous Fistula Patency in Hemodialysis Patients. *Clin. J. Am. Soc. Nephrol.* **2010**, *5*, 1805. [CrossRef] [PubMed]
84. Okamura, H.; Pisani, L.J.; Dalal, A.R.; Emrich, F.; Dake, B.A.; Arakawa, M.; Onthank, D.C.; Cesati, R.R.; Robinson, S.P.; Milanesi, M.; et al. Assessment of Elastin Deficit in a Marfan Mouse Aneurysm Model Using an Elastin-Specific Magnetic Resonance Imaging Contrast Agent. *Circ. Cardiovasc. Imaging* **2014**, *7*, 690–696. [CrossRef]
85. Vouyouka, A.G.; Pfeiffer, B.J.; Liem, T.K.; Taylor, T.A.; Mudaliar, J.; Phillips, C.L. The role of type I collagen in aortic wall strength with a homotrimeric [α1(I)]3 collagen mouse model. *J. Vasc. Surg.* **2001**, *33*, 1263–1270. [CrossRef] [PubMed]
86. Wynn, T.A. Cellular and molecular mechanisms of fibrosis. *J. Pathol.* **2008**, *214*, 199–210. [CrossRef] [PubMed]
87. Suki, B.; Jesudason, R.; Sato, S.; Parameswaran, H.; Araujo, A.D.; Majumdar, A.; Allen, P.G.; Bartolák-Suki, E. Mechanical failure, stress redistribution, elastase activity and binding site availability on elastin during the progression of emphysema. *Pulm. Pharmacol. Ther.* **2012**, *25*, 268–275. [CrossRef] [PubMed]
88. Davis, P.F.; Ryan, P.A.; Osipowicz, J.; Anderson, M.J.; Sweeney, A.; Stehbens, W.E. The biochemical composition of hemodynamically stressed vascular tissue: The insoluble elastin of experimental arteriovenous fistulae. *Exp. Mol. Pathol.* **1989**, *51*, 103–110. [CrossRef]
89. Chang, C.-J.; Chen, C.-C.; Hsu, L.-A.; Chang, G.-J.; Ko, Y.-H.; Chen, C.-F.; Chen, M.-Y.; Yang, S.-H.; Pang, J.-H.S. Degradation of the Internal Elastic Laminae in Vein Grafts of Rats with Aortocaval Fistulae: Potential Impact on Graft Vasculopathy. *Am. J. Pathol.* **2009**, *174*, 1837–1846. [CrossRef]
90. Masuda, H.; Zhuang, Y.-J.; Singh, T.M.; Kawamura, K.; Murakami, M.; Zarins, C.K.; Glagov, S. Adaptive Remodeling of Internal Elastic Lamina and Endothelial Lining During Flow-Induced Arterial Enlargement. *Arterioscler. Thromb. Vasc. Biol.* **1999**, *19*, 2298–2307. [CrossRef]
91. Greenhill, N.S.; Stehbens, W.E. Scanning electron microscopic investigation of the afferent arteries of experimental femoral arteriovenous fistulae in rabbits. *Pathology* **1987**, *19*, 22–27. [CrossRef]
92. Sho, E.; Sho, M.; Singh, T.M.; Nanjo, H.; Komatsu, M.; Xu, C.; Masuda, H.; Zarins, C.K. Arterial Enlargement in Response to High Flow Requires Early Expression of Matrix Metalloproteinases to Degrade Extracellular Matrix. *Exp. Mol. Pathol.* **2002**, *73*, 142–153. [CrossRef]
93. Tronc, F.; Wassef, M.; Esposito, B.; Henrion, D.; Glagov, S.; Tedgui, A. Role of NO in Flow-Induced Remodeling of the Rabbit Common Carotid Artery. *Arterioscler. Thromb. Vasc. Biol.* **1996**, *16*, 1256–1262. [CrossRef]
94. Jones, G.T.; Stehbens, W.E.; Martin, B.J. Ultrastructural changes in arteries proximal to short-term experimental carotid-jugular arteriovenous fistulae in rabbits. *Int. J. Exp. Pathol.* **1994**, *75*, 225–232. [CrossRef]
95. Wong, C.Y.; Rothuizen, T.C.; de Vries, M.R.; Rabelink, T.J.; Hamming, J.F.; van Zonneveld, A.J.; Quax, P.H.; Rotmans, J.I. Elastin is a key regulator of outward remodeling in arteriovenous fistulas. *Eur. J. Vasc. Endovasc. Surg.* **2015**, *49*, 480–486. [CrossRef]
96. Bezhaeva, T.; de Vries, M.R.; Geelhoed, W.J.; van der Veer, E.P.; Versteeg, S.; van Alem, C.M.A.; Voorzaat, B.M.; Eijkelkamp, N.; van der Bogt, K.E.; Agoulnik, A.I.; et al. Relaxin receptor deficiency promotes vascular inflammation and impairs outward remodeling in arteriovenous fistulas. *FASEB J.* **2018**, *32*, 6293–6304. [CrossRef]
97. Peden, E.K.; Lucas, J.F.; Browne, B.J.; Settle, S.M.; Scavo, V.A.; Bleyer, A.J.; Ozaki, C.K.; Teruya, T.H.; Wilson, S.E.; Mishler, R.E.; et al. PATENCY-2 trial of vonapanitase to promote radiocephalic fistula use for hemodialysis and secondary patency. *J. Vasc. Access* **2022**, *23*, 265–274. [CrossRef]

98. Bleyer, A.J.; Scavo, V.A.; Wilson, S.E.; Browne, B.J.; Ferris, B.L.; Ozaki, C.K.; Lee, T.; Peden, E.K.; Dixon, B.S.; Mishler, R.; et al. A randomized trial of vonapanitase (PATENCY-1) to promote radiocephalic fistula patency and use for hemodialysis. *J. Vasc. Surg.* **2019**, *69*, 507–515. [CrossRef]
99. Dwivedi, A.J.; Roy-Chaudhury, P.; Peden, E.K.; Browne, B.J.; Ladenheim, E.D.; Scavo, V.A.; Gustafson, P.N.; Wong, M.D.; Magill, M.; Lindow, F.; et al. Application of Human Type I Pancreatic Elastase (PRT-201) to the venous Anastomosis of Arteriovenous Grafts in Patients with Chronic Kidney Disease. *J. Vasc. Access* **2014**, *15*, 376–384. [CrossRef]
100. Hye, R.J.; Peden, E.K.; O'Connor, T.P.; Browne, B.J.; Dixon, B.S.; Schanzer, A.S.; Jensik, S.C.; Dember, L.M.; Jaff, M.R.; Burke, S.K. Human type I pancreatic elastase treatment of arteriovenous fistulas in patients with chronic kidney disease. *J. Vasc. Surg.* **2014**, *60*, 454–461.e1. [CrossRef]
101. Andraska, E.; Skirtich, N.; McCreary, D.; Kulkarni, R.; Tzeng, E.; McEnaney, R. Simultaneous Upregulation of Elastolytic and Elastogenic Factors Are Necessary for Regulated Collateral Diameter Expansion. *Front. Cardiovasc. Med.* **2022**, *8*, 762094. [CrossRef]
102. Field, M.A.; McGrogan, D.G.; Tullet, K.; Inston, N.G. Arteriovenous fistula aneurysms in patients with Alport's. *J. Vasc. Access* **2013**, *14*, 397–399. [CrossRef]
103. Klüsch, V.; Aper, T.; Sonnenschein, K.; Becker, L.S.; Umminger, J.; Haverich, A.; Rustum, S. A Hyperdynamic Arteriovenous Fistula Aneurysm After Long Time Renal Transplantation. *Vasc. Endovasc. Surg.* **2022**, *57*, 182–185. [CrossRef] [PubMed]
104. Smith, R.A.; Stehbens, W.E.; Weber, P. Hemodynamically-induced increase in soluble collagen in the anastomosed veins of experimental arteriovenous fistulae. *Atherosclerosis* **1976**, *23*, 429–436. [CrossRef]
105. Shiu, Y.T.; Litovsky, S.H.; Cheung, A.K.; Pike, D.B.; Tey, J.C.; Zhang, Y.; Young, C.J.; Robbin, M.; Hoyt, K.; Allon, M. Preoperative Vascular Medial Fibrosis and Arteriovenous Fistula Development. *Clin. J. Am. Soc. Nephrol.* **2016**, *11*, 1615–1623. [CrossRef] [PubMed]
106. Martinez, L.; Duque, J.C.; Tabbara, M.; Paez, A.; Selman, G.; Hernandez, D.R.; Sundberg, C.A.; Tey, J.C.S.; Shiu, Y.T.; Cheung, A.K.; et al. Fibrotic Venous Remodeling and Nonmaturation of Arteriovenous Fistulas. *J. Am. Soc. Nephrol.* **2018**, *29*, 1030–1040. [CrossRef] [PubMed]
107. Siegel, R.C. Lysyl Oxidase. In *International Review of Connective Tissue Research*; Hall, D.A., Jackson, D.S., Eds.; Elsevier: Amsterdam, The Netherlands, 1979; pp. 73–118.
108. Hernandez, D.R.; Applewhite, B.; Martinez, L.; Laurito, T.; Tabbara, M.; Rojas, M.G.; Wei, Y.; Selman, G.; Knysheva, M.; Velazquez, O.C.; et al. Inhibition of Lysyl Oxidase with β aminopropionitrile Improves Venous Adaptation after Arteriovenous Fistula Creation. *Kidney360* **2021**, *2*, 270–278. [CrossRef]
109. Applewhite, B.; Gupta, A.; Wei, Y.; Yang, X.; Martinez, L.; Rojas, M.G.; Andreopoulos, F.; Vazquez-Padron, R.I. Periadventitial β-aminopropionitrile-loaded nanofibers reduce fibrosis and improve arteriovenous fistula remodeling in rats. *Front. Cardiovasc. Med.* **2023**, *10*, 1124106. [CrossRef]
110. Wilmarth, K.R.; Froines, J.R. In vitro and in vivo inhibition of lysyl oxidase byaminopropionitriles. *J. Toxicol. Environ. Health* **1992**, *37*, 411–423. [CrossRef]
111. Meng, X.-M.; Nikolic-Paterson, D.J.; Lan, H.Y. TGF-β: The master regulator of fibrosis. *Nat. Rev. Nephrol.* **2016**, *12*, 325–338. [CrossRef]
112. Suwanabol, P.A.; Seedial, S.M.; Shi, X.; Zhang, F.; Yamanouchi, D.; Roenneburg, D.; Liu, B.; Kent, K.C. Transforming growth factor-β increases vascular smooth muscle cell proliferation through the Smad3 and extracellular signal-regulated kinase mitogen-activated protein kinases pathways. *J. Vasc. Surg.* **2012**, *56*, 446–454.e1. [CrossRef]
113. Atsawasuwan, P.; Mochida, Y.; Katafuchi, M.; Kaku, M.; Fong, K.S.; Csiszar, K.; Yamauchi, M. Lysyl oxidase binds transforming growth factor-beta and regulates its signaling via amine oxidase activity. *J. Biol. Chem.* **2008**, *283*, 34229–34240. [CrossRef]
114. Gacheru, S.N.; Thomas, K.M.; Murray, S.A.; Csiszar, K.; Smith-Mungo, L.I.; Kagan, H.M. Transcriptional and post-transcriptional control of lysyl oxidase expression in vascular smooth muscle cells: Effects of TGF-beta 1 and serum deprivation. *J. Cell. Biochem.* **1997**, *65*, 395–407. [CrossRef]
115. Taniguchi, R.; Ohashi, Y.; Lee, J.S.; Hu, H.; Gonzalez, L.; Zhang, W.; Langford, J.; Matsubara, Y.; Yatsula, B.; Tellides, G.; et al. Endothelial Cell TGF-β (Transforming Growth Factor-Beta) Signaling Regulates Venous Adaptive Remodeling to Improve Arteriovenous Fistula Patency. *Arterioscler. Thromb. Vasc. Biol.* **2022**, *42*, 868–883. [CrossRef]
116. Hu, H.; Lee, S.R.; Bai, H.; Guo, J.; Hashimoto, T.; Isaji, T.; Guo, X.; Wang, T.; Wolf, K.; Liu, S.; et al. TGFβ (Transforming Growth Factor-Beta)-Activated Kinase 1 Regulates Arteriovenous Fistula Maturation. *Arterioscler. Thromb. Vasc. Biol.* **2020**, *40*, e203–e213. [CrossRef]
117. Bashar, K.; Zafar, A.; Elsheikh, S.; Healy, D.A.; Clarke-Moloney, M.; Casserly, L.; Burke, P.E.; Kavanagh, E.G.; Walsh, S.R. Predictive Parameters of Arteriovenous Fistula Functional Maturation in a Population of Patients with End-Stage Renal Disease. *PLoS ONE* **2015**, *10*, e0119958. [CrossRef]
118. Peterson, W.J.; Barker, J.; Allon, M. Disparities in Fistula Maturation Persist Despite Preoperative Vascular Mapping. *Clin. J. Am. Soc. Nephrol.* **2008**, *3*, 437–441. [CrossRef]
119. Cai, C.; Kilari, S.; Singh, A.K.; Zhao, C.; Simeon, M.L.; Misra, A.; Li, Y.; Misra, S. Differences in Transforming Growth Factor-β1/BMP7 Signaling and Venous Fibrosis Contribute to Female Sex Differences in Arteriovenous Fistulas. *J. Am. Heart Assoc.* **2020**, *9*, e017420. [CrossRef]

120. Cai, C.; Zhao, C.; Kilari, S.; Sharma, A.; Singh, A.K.; Simeon, M.L.; Misra, A.; Li, Y.; Misra, S. Effect of sex differences in treatment response to angioplasty in a murine arteriovenous fistula model. *Am. J. Physiol.-Ren. Physiol.* **2020**, *318*, F565–F575. [CrossRef]
121. Heine, G.H.; Ulrich, C.; Sester, U.; Sester, M.; Köhler, H.; Girndt, M. Transforming growth factor β1 genotype polymorphisms determine AV fistula patency in hemodialysis patients. *Kidney Int.* **2003**, *64*, 1101–1107. [CrossRef]
122. Stirbu, O.; Gadalean, F.; Pitea, I.V.; Ciobanu, G.; Schiller, A.; Grosu, I.; Nes, A.; Bratescu, R.; Olariu, N.; Timar, B.; et al. C-reactive protein as a prognostic risk factor for loss of arteriovenous fistula patency in hemodialyzed patients. *J. Vasc. Surg.* **2019**, *70*, 208–215. [CrossRef]
123. Kaller, R.; Arbănași, E.M.; Mureșan, A.V.; Voidăzan, S.; Arbănași, E.M.; Horváth, E.; Suciu, B.A.; Hosu, I.; Halmaciu, I.; Brinzaniuc, K.; et al. The Predictive Value of Systemic Inflammatory Markers, the Prognostic Nutritional Index, and Measured Vessels' Diameters in Arteriovenous Fistula Maturation Failure. *Life* **2022**, *12*, 1447. [CrossRef]
124. Chang, C.-J.; Ko, Y.-S.; Ko, P.-J.; Hsu, L.-A.; Chen, C.-F.; Yang, C.-W.; Hsu, T.-S.; Pang, J.-H.S. Thrombosed arteriovenous fistula for hemodialysis access is characterized by a marked inflammatory activity. *Kidney Int.* **2005**, *68*, 1312–1319. [CrossRef] [PubMed]
125. Sener, E.F.; Taheri, S.; Korkmaz, K.; Zararsiz, G.; Serhatlioglu, F.; Unal, A.; Emirogullari, O.N.; Ozkul, Y. Association of TNF-α −308 G > A and ACE I/D gene polymorphisms in hemodialysis patients with arteriovenous fistula thrombosis. *Int. Urol. Nephrol.* **2014**, *46*, 1419–1425. [CrossRef] [PubMed]
126. Juncos, J.P.; Grande, J.P.; Kang, L.; Ackerman, A.W.; Croatt, A.J.; Katusic, Z.S.; Nath, K.A. MCP-1 Contributes to Arteriovenous Fistula Failure. *J. Am. Soc. Nephrol.* **2011**, *22*, 43–48. [CrossRef] [PubMed]
127. Marrone, D.; Pertosa, G.; Simone, S.; Loverre, A.; Capobianco, C.; Cifarelli, M.; Memoli, B.; Schena, F.P.; Grandaliano, G. Local Activation of Interleukin 6 Signaling Is Associated with Arteriovenous Fistula Stenosis in Hemodialysis Patients. *Am. J. Kidney Dis.* **2007**, *49*, 664–673. [CrossRef]
128. Wong, C.; Bezhaeva, T.; Rothuizen, T.C.; Metselaar, J.M.; de Vries, M.R.; Verbeek, F.P.R.; Vahrmeijer, A.L.; Wezel, A.; van Zonneveld, A.-J.; Rabelink, T.J.; et al. Liposomal prednisolone inhibits vascular inflammation and enhances venous outward remodeling in a murine arteriovenous fistula model. *Sci. Rep.* **2016**, *6*, 30439. [CrossRef]
129. Voorzaat, B.M.; van der Bogt, K.E.A.; Bezhaeva, T.; van Schaik, J.; Eefting, D.; van der Putten, K.; van Nieuwenhuizen, R.C.; Groeneveld, J.O.; Hoogeveen, E.K.; van der Meer, I.M.; et al. A Randomized Trial of Liposomal Prednisolone (LIPMAT) to Enhance Radiocephalic Fistula Maturation: A Pilot Study. *Kidney Int. Rep.* **2020**, *5*, 1327–1332. [CrossRef]
130. Penn, D.L.; Witte, S.R.; Komotar, R.J.; Sander Connolly, E. The role of vascular remodeling and inflammation in the pathogenesis of intracranial aneurysms. *J. Clin. Neurosci.* **2014**, *21*, 28–32. [CrossRef]
131. Nguyen, M.; Thankam, F.G.; Agrawal, D.K. Sterile inflammation in the pathogenesis of maturation failure of arteriovenous fistula. *J. Mol. Med.* **2021**, *99*, 729–741. [CrossRef]
132. Samra, G.; Rai, V.; Agrawal, D.K. Innate and adaptive immune cells associate with arteriovenous fistula maturation and failure. *Can. J. Physiol. Pharmacol.* **2022**, *100*, 716–727. [CrossRef]
133. Satish, M.; Gunasekar, P.; Agrawal, D.K. Pro-inflammatory and pro-resolving mechanisms in the immunopathology of arteriovenous fistula maturation. *Expert Rev. Cardiovasc. Ther.* **2019**, *17*, 369–376. [CrossRef]
134. Matsubara, Y.; Kiwan, G.; Fereydooni, A.; Langford, J.; Dardik, A. Distinct subsets of T cells and macrophages impact venous remodeling during arteriovenous fistula maturation. *JVS Vasc. Sci.* **2020**, *1*, 207–218. [CrossRef]
135. Chan, S.M.; Weininger, G.; Langford, J.; Jane-Wit, D.; Dardik, A. Sex Differences in Inflammation During Venous Remodeling of Arteriovenous Fistulae. *Front. Cardiovasc. Med.* **2021**, *8*, 715114. [CrossRef]
136. Zhao, X.; Chen, J.; Sun, H.; Zhang, Y.; Zou, D. New insights into fibrosis from the ECM degradation perspective: The macrophage-MMP-ECM interaction. *Cell Biosci.* **2022**, *12*, 117. [CrossRef]
137. Sukhova, G.K.; Schönbeck, U.; Rabkin, E.; Schoen, F.J.; Poole, A.R.; Billinghurst, R.C.; Libby, P. Evidence for Increased Collagenolysis by Interstitial Collagenases-1 and -3 in Vulnerable Human Atheromatous Plaques. *Circulation* **1999**, *99*, 2503–2509. [CrossRef]
138. Galis, Z.S.; Muszynski, M.; Sukhova, G.K.; Simon-Morrissey, E.; Unemori, E.N.; Lark, M.W.; Amento, E.; Libby, P. Cytokine-stimulated human vascular smooth muscle cells synthesize a complement of enzymes required for extracellular matrix digestion. *Circ. Res.* **1994**, *75*, 181–189. [CrossRef]
139. Alexander, M.R.; Moehle, C.W.; Johnson, J.L.; Yang, Z.; Lee, J.K.; Jackson, C.L.; Owens, G.K. Genetic inactivation of IL-1 signaling enhances atherosclerotic plaque instability and reduces outward vessel remodeling in advanced atherosclerosis in mice. *J. Clin. Investig.* **2012**, *122*, 70–79. [CrossRef]
140. Kuwahara, G.; Hashimoto, T.; Tsuneki, M.; Yamamoto, K.; Assi, R.; Foster, T.R.; Hanisch, J.J.; Bai, H.; Hu, H.; Protack, C.D.; et al. CD44 Promotes Inflammation and Extracellular Matrix Production During Arteriovenous Fistula Maturation. *Arterioscler. Thromb. Vasc. Biol.* **2017**, *37*, 1147–1156. [CrossRef]
141. Matsubara, Y.; Kiwan, G.; Liu, J.; Gonzalez, L.; Langford, J.; Gao, M.; Gao, X.; Taniguchi, R.; Yatsula, B.; Furuyama, T.; et al. Inhibition of T-Cells by Cyclosporine A Reduces Macrophage Accumulation to Regulate Venous Adaptive Remodeling and Increase Arteriovenous Fistula Maturation. *Arterioscler. Thromb. Vasc. Biol.* **2021**, *41*, e160–e174. [CrossRef]
142. Yang, B.; Kilari, S.; Brahmbhatt, A.; McCall, D.L.; Torres, E.N.; Leof, E.B.; Mukhopadhyay, D.; Misra, S. CorMatrix Wrapped Around the Adventitia of the Arteriovenous Fistula Outflow Vein Attenuates Venous Neointimal Hyperplasia. *Sci. Rep.* **2017**, *7*, 14298. [CrossRef]

143. Leskovar, B.; Furlan, T.; Poznic, S.; Hrastelj, M.; Adamlje, A. Using CorMatrix for partial and complete (re)construction of arteriovenous fistulas in haemodialysis patients: (Re)construction of arteriovenous fistulas with CorMatrix. *J. Vasc. Access* **2019**, *20*, 597–603. [CrossRef]
144. DuBose, J.J.; Azizzadeh, A. Utilization of a Tubularized CorMatrix Extracellular Matrix for Repair of an Arteriovenous Fistula Aneurysm. *Ann. Vasc. Surg.* **2015**, *29*, e1–e366. [CrossRef] [PubMed]
145. Shiu, Y.T.; He, Y.; Tey, J.C.S.; Knysheva, M.; Anderson, B.; Kauser, K. Natural Vascular Scaffolding Treatment Promotes Outward Remodeling During Arteriovenous Fistula Development in Rats. *Front. Bioeng. Biotechnol.* **2021**, *9*, 622617. [CrossRef] [PubMed]
146. He, Y.; Anderson, B.; Hu, Q.; Hayes, R.; Huff, K.; Isaacson, J.; Warner, K.S.; Hauser, H.; Greenberg, M.; Chandra, V.; et al. Photochemically Aided Arteriovenous Fistula Creation to Accelerate Fistula Maturation. *Int. J. Mol. Sci.* **2023**, *24*, 7571. [CrossRef] [PubMed]
147. Protti, A.; Lavin, B.; Dong, X.; Lorrio, S.; Robinson, S.; Onthank, D.; Shah, A.M.; Botnar, R.M. Assessment of Myocardial Remodeling Using an Elastin/Tropoelastin Specific Agent with High Field Magnetic Resonance Imaging (MRI). *J. Am. Heart Assoc.* **2015**, *4*, e001851. [CrossRef] [PubMed]
148. Brangsch, J.; Reimann, C.; Kaufmann, J.O.; Adams, L.C.; Onthank, D.C.; Thöne-Reineke, C.; Robinson, S.P.; Buchholz, R.; Karst, U.; Botnar, R.M.; et al. Concurrent Molecular Magnetic Resonance Imaging of Inflammatory Activity and Extracellular Matrix Degradation for the Prediction of Aneurysm Rupture. *Circ. Cardiovasc. Imaging* **2019**, *12*, e008707. [CrossRef]
149. Laboyrie, S.L.; Vries, M.R.d.; Jong, A.d.; Boer, H.C.d.; Lalai, R.A.; Martinez, L.; Vazquez-Padron, R.I.; Rotmans, J.I. von Willebrand Factor: A Central Regulator of Arteriovenous Fistula Maturation Through Smooth Muscle Cell Proliferation and Outward Remodeling. *J. Am. Heart Assoc.* **2022**, *11*, e024581. [CrossRef]
150. Murphy, G.J.; Angelini, G.D. Insights into the pathogenesis of vein graft disease: Lessons from intravascular ultrasound. *Cardiovasc. Ultrasound* **2004**, *2*, 8. [CrossRef]
151. de Jong, A.; Sier, V.Q.; Peters, H.A.B.; Schilder, N.K.M.; Jukema, J.W.; Goumans, M.J.T.H.; Quax, P.H.A.; de Vries, M.R. Interfering in the ALK1 Pathway Results in Macrophage-Driven Outward Remodeling of Murine Vein Grafts. *Front. Cardiovasc. Med.* **2022**, *8*, 784980. [CrossRef]
152. Wu, M.; Awasthi, N.; Rad, N.M.; Pluim, J.P.W.; Lopata, R.G.P. Advanced Ultrasound and Photoacoustic Imaging in Cardiology. *Sensors* **2021**, *21*, 7947. [CrossRef]
153. Muzard, J.; Sarda-Mantel, L.; Loyau, S.; Meulemans, A.; Louedec, L.; Bantsimba-Malanda, C.; Hervatin, F.; Marchal-Somme, J.; Michel, J.B.; Le Guludec, D.; et al. Non-invasive molecular imaging of fibrosis using a collagen-targeted peptidomimetic of the platelet collagen receptor glycoprotein VI. *PLoS ONE* **2009**, *4*, e5585. [CrossRef]
154. De Jong, S.; van Middendorp, L.B.; Hermans, R.H.A.; de Bakker, J.M.T.; Bierhuizen, M.F.A.; Prinzen, F.W.; van Rijen, H.V.M.; Losen, M.; Vos, M.A.; van Zandvoort, M.A.M.J. Ex Vivo and In Vivo Administration of Fluorescent CNA35 Specifically Marks Cardiac Fibrosis. *Mol. Imaging* **2014**, *13*, 1–9. [CrossRef]
155. Adams, L.C.; Brangsch, J.; Reimann, C.; Kaufmann, J.O.; Buchholz, R.; Karst, U.; Botnar, R.M.; Hamm, B.; Makowski, M.R. Simultaneous molecular MRI of extracellular matrix collagen and inflammatory activity to predict abdominal aortic aneurysm rupture. *Sci. Rep.* **2020**, *10*, 15206. [CrossRef]
156. Bezhaeva, T.; Wong, C.; de Vries, M.R.; van der Veer, E.P.; van Alem, C.M.A.; Que, I.; Lalai, R.A.; van Zonneveld, A.-J.; Rotmans, J.I.; Quax, P.H.A. Deficiency of TLR4 homologue RP105 aggravates outward remodeling in a murine model of arteriovenous fistula failure. *Sci. Rep.* **2017**, *7*, 10269. [CrossRef]
157. Ding, X.; Chen, J.; Wu, C.; Wang, G.; Zhou, C.; Chen, S.; Wang, K.; Zhang, A.; Ye, P.; Wu, J.; et al. Nucleotide-Binding Oligomerization Domain-Like Receptor Protein 3 Deficiency in Vascular Smooth Muscle Cells Prevents Arteriovenous Fistula Failure Despite Chronic Kidney Disease. *J. Am. Heart Assoc.* **2019**, *8*, e011211. [CrossRef]
158. Singh, A.K.; Cai, C.; Kilari, S.; Zhao, C.; Simeon, M.L.; Takahashi, E.; Edelman, E.R.; Kong, H.; Macedo, T.; Singh, R.J.; et al. 1α,25-Dihydroxyvitamin D3 Encapsulated in Nanoparticles Prevents Venous Neointimal Hyperplasia and Stenosis in Porcine Arteriovenous Fistulas. *J. Am. Soc. Nephrol.* **2021**, *32*, 866–885. [CrossRef]
159. Tanaka, L.Y.; Laurindo, F.R.M. Vascular remodeling: A redox-modulated mechanism of vessel caliber regulation. *Free. Radic. Biol. Med.* **2017**, *109*, 11–21. [CrossRef]
160. George, S.J.; Wan, S.; Hu, J.; MacDonald, R.; Johnson, J.L.; Baker, A.H. Sustained Reduction of Vein Graft Neointima Formation by Ex Vivo TIMP-3 Gene Therapy. *Circulation* **2011**, *124* (Suppl. 1), S135–S142. [CrossRef]
161. Ballmann, M.Z.; Raus, S.; Engelhart, R.; Kaján, G.L.; Beqqali, A.; Hadoke, P.W.F.; van der Zalm, C.; Papp, T.; John, L.; Khan, S.; et al. Human AdV-20-42-42, a Promising Novel Adenoviral Vector for Gene Therapy and Vaccine Product Development. *J. Virol.* **2021**, *95*, e0038721. [CrossRef]
162. Rotmans, J.I.; Verhagen, H.J.; Velema, E.; de Kleijn, D.P.; van den Heuvel, M.; Kastelein, J.J.; Pasterkamp, G.; Stroes, E.S. Local overexpression of C-type natriuretic peptide ameliorates vascular adaptation of porcine hemodialysis grafts. *Kidney Int.* **2004**, *65*, 1897–1905. [CrossRef]
163. van Rooij, E.; Sutherland, L.B.; Thatcher, J.E.; DiMaio, J.M.; Naseem, R.H.; Marshall, W.S.; Hill, J.A.; Olson, E.N. Dysregulation of microRNAs after myocardial infarction reveals a role of miR-29 in cardiac fibrosis. *Proc. Natl. Acad. Sci. USA* **2008**, *105*, 13027–13032. [CrossRef]

164. Zhang, P.; Huang, A.; Ferruzzi, J.; Mecham, R.P.; Starcher, B.C.; Tellides, G.; Humphrey, J.D.; Giordano, F.J.; Niklason, L.E.; Sessa, W.C. Inhibition of MicroRNA-29 Enhances Elastin Levels in Cells Haploinsufficient for Elastin and in Bioengineered Vessels—Brief Report. *Arterioscler. Thromb. Vasc. Biol.* **2012**, *32*, 756–759. [CrossRef] [PubMed]
165. Rothuizen, T.C.; Kemp, R.; Duijs, J.M.; de Boer, H.C.; Bijkerk, R.; van der Veer, E.P.; Moroni, L.; van Zonneveld, A.J.; Weiss, A.S.; Rabelink, T.J.; et al. Promoting Tropoelastin Expression in Arterial and Venous Vascular Smooth Muscle Cells and Fibroblasts for Vascular Tissue Engineering. *Tissue Eng. Part C Methods* **2016**, *22*, 923–931. [CrossRef] [PubMed]
166. Sudo, R.; Sato, F.; Azechi, T.; Wachi, H. MiR-29-mediated elastin down-regulation contributes to inorganic phosphorus-induced osteoblastic differentiation in vascular smooth muscle cells. *Genes Cells* **2015**, *20*, 1077–1087. [CrossRef] [PubMed]
167. Kilari, S.; Cai, C.; Zhao, C.; Sharma, A.; Chernogubova, E.; Simeon, M.; Wu, C.C.; Song, H.L.; Maegdefessel, L.; Misra, S. The Role of MicroRNA-21 in Venous Neointimal Hyperplasia: Implications for Targeting miR-21 for VNH Treatment. *Mol. Ther.* **2019**, *27*, 1681–1693. [CrossRef]

Disclaimer/Publisher's Note: The statements, opinions and data contained in all publications are solely those of the individual author(s) and contributor(s) and not of MDPI and/or the editor(s). MDPI and/or the editor(s) disclaim responsibility for any injury to people or property resulting from any ideas, methods, instructions or products referred to in the content.

Article

Induction of Angiogenesis by Genetically Modified Human Umbilical Cord Blood Mononuclear Cells

Dilara Z. Gatina [1], Ilnaz M. Gazizov [2], Margarita N. Zhuravleva [1], Svetlana S. Arkhipova [1], Maria A. Golubenko [1], Marina O. Gomzikova [1], Ekaterina E. Garanina [1], Rustem R. Islamov [2], Albert A. Rizvanov [1] and Ilnur I. Salafutdinov [1,2,*]

[1] Institute of Fundamental Medicine and Biology, Kazan (Volga Region) Federal University, 420008 Kazan, Russia
[2] Department of Medical Biology and Genetics, Kazan State Medical University, 420012 Kazan, Russia
* Correspondence: sal.ilnur@gmail.com

Abstract: Stimulating the process of angiogenesis in treating ischemia-related diseases is an urgent task for modern medicine, which can be achieved through the use of different cell types. Umbilical cord blood (UCB) continues to be one of the attractive cell sources for transplantation. The goal of this study was to investigate the role and therapeutic potential of gene-engineered umbilical cord blood mononuclear cells (UCB-MC) as a forward-looking strategy for the activation of angiogenesis. Adenovirus constructs Ad-VEGF, Ad-FGF2, Ad-SDF1α, and Ad-EGFP were synthesized and used for cell modification. UCB-MCs were isolated from UCB and transduced with adenoviral vectors. As part of our in vitro experiments, we evaluated the efficiency of transfection, the expression of recombinant genes, and the secretome profile. Later, we applied an in vivo Matrigel plug assay to assess engineered UCB-MC's angiogenic potential. We conclude that hUCB-MCs can be efficiently modified simultaneously with several adenoviral vectors. Modified UCB-MCs overexpress recombinant genes and proteins. Genetic modification of cells with recombinant adenoviruses does not affect the profile of secreted pro- and anti-inflammatory cytokines, chemokines, and growth factors, except for an increase in the synthesis of recombinant proteins. hUCB-MCs genetically modified with therapeutic genes induced the formation of new vessels. An increase in the expression of endothelial cells marker (CD31) was revealed, which correlated with the data of visual examination and histological analysis. The present study demonstrates that gene-engineered UCB-MC can be used to stimulate angiogenesis and possibly treat cardiovascular disease and diabetic cardiomyopathy.

Keywords: human umbilical cord blood mononuclear cells; angiogenesis; recombinant adenoviruses; gene modification; transgene expression; VEGF; FGF2; SDF1α; NUDE mice; Matrigel plugs; cytokine profile

1. Introduction

Angiogenesis is the growth of new blood vessels from pre-existing vessels, an essential process for development, wound healing, and the restoration of blood flow and oxygen supply to tissues after injury. One of the main tasks of modern medicine is the stimulation of the processes of angiogenesis in the treatment of vascular diseases. To date, many approaches have been proposed for the induction of therapeutic angiogenesis. Among the proposed methods are surgical methods [1], the use of inducer proteins [2], recombinant DNA molecules [3], inducer genes [4], and the use of various cell types [5,6], including ex vivo genetically modified cells [7]. In this aspect, human umbilical cord blood mononuclear cells (UCB-MC) seem to be a promising "tool" for stimulating angiogenesis through the delivery of genetic engineering systems, expression of recombinant proteins, and possibly direct participation in new vessel formation. The choice of UCB-MC in studies for cell and gene-cell therapy looks promising because of some advantages of this cellular source.

Umbilical cord blood contains many stem/progenitor cells and can be obtained easily [8]. The mononuclear fraction of UCB contains populations of different immature cells capable of differentiating into many cell types [9]. Cell populations that have been discovered in UCB are hematopoietic stem cells (HSCs), endothelial progenitor cells, mesenchymal stem cells (MSCs), unrestricted somatic stem cells (USSCs), and side population cells [10–13]. As cellular material for transplantation or carriers for genetic constructs, UCB-MCs have low immunogenicity because they do not express all the antigens on the cell membrane. This feature enhances the ability to cross donor-recipient HLA disparities. It allows for the usage of UCB-MC for transplantation in non-fully compatible HLA recipients with a much lower incidence of grade II-IV acute GVHD (graft versus host disease) cases after transplantation [14–17]. Furthermore, UCB-MCs can prevent the oncological transformation of recipient cells after transplantation [15]. Another appealing reason for using UCB cells for cell therapy is their ability to produce various biologically active molecules, such as proteins with antioxidant properties, angiogenic, neurotrophic, and growth factors [18–22], which make them suitable for effective stimulation of regenerative processes in non-compatible recipients for a short time before the immune system eliminates them. Overall, UCB cell transplantation can replace dead cells, prevent further death of surviving cells, and stimulate regeneration by secreting biologically active molecules. A genetic modification of UCB cells can enhance their ability to regenerate tissue [23,24]. This approach unites the advantages of cell- and gene therapy. Genetically modified UCB cells can provide targeted delivery of therapeutic genes and expression of recombinant molecules at the regeneration site. For example, our previous studies showed the positive effect of genetically modified umbilical cord blood mononuclear cells (UCB-MC) simultaneously produces three recombinant molecules (vascular endothelial growth factor (VEGF), glial cell-derived neurotrophic factor (GDNF), and neural cell adhesion molecule (NCAM) in animal models of amyotrophic lateral sclerosis [25], spinal cord injury [26] and stroke [7,27]. Many state-of-the-art methods and models for studying angiogenesis have been proposed, which are well analyzed in the review articles [28–30]. Among various models, the in vivo angiogenesis plug assay, which uses basement membrane extracts (BME) or Matrigel, is widely used for evaluating pro- or anti-angiogenic factors during in vivo angiogenesis. This assay is reliable, easy to perform without special equipment, reproducible, quantitative, and quick [31–33]. Matrigel predominantly contains laminin III, collagen IV, heparan sulfate proteoglycans, and various growth factors. The assay is performed by injecting the liquid Matrigel into the subcutaneous space of an animal at 4 °C, which solidifies to form a plug at body temperature. Over time, blood vessels sprout into the plug. The number of plug sites per animal can be several, allowing multiple test compounds or concentrations to be tested. Thus, drug screening can also be evaluated for effects on the activity of angiogenic or anti-angiogenic factors [34–37]. The drug can be placed in the plug with the test factor by mixing with the Matrigel matrix or given to the host animal. Cells or exosomes can also be examined when mixed into the gel to produce angiogenic factors. Furthermore, the assay is highly versatile. For example, the role of certain genes can be evaluated using genetically modified mice (overexpressing or ablating a protein gene) or animal models of diseases. This report aimed to study the effect of genetically modified umbilical cord blood mononuclear cells overexpressing recombinant proangiogenic proteins VEGF165, FGF2, and SDF1α on the induction of angiogenesis. Furthermore, we assessed the influence of all three factors on the tone of the secretory profile of modified UCB-MCs and tubule formation in the in vivo Matrigel plug assay. The present study shows that when combined with UCB, the three factors can enhance angiogenesis and be useful for developing new therapies.

2. Results

2.1. Characterization of Isolated Human UCB-MCs

Isolated cells demonstrated high viability (>97%) and included CD45+ lymphocytes (58.9%). CD45+CD3+ lymphocytes constituted 59.2%, while CD14+ macrophages constituted 7.3%. This ratio of the central populations of blood cells (lymphocytes, T-lymphocytes, and monocytes) is believed to be typical for human UCB-MCs. We also examined the percentage of CD34+ blood cells among isolated UCB-MCs. According to the obtained data, CD34+ cells constituted 0.4% of CD45+ cells. In addition, 91.8% of CD45+CD34+ cells expressed CD38. Furthermore, 90% of the CD45+CD34+ cells had the phenotype CD90+. The flow cytometry results are shown in Figure 1.

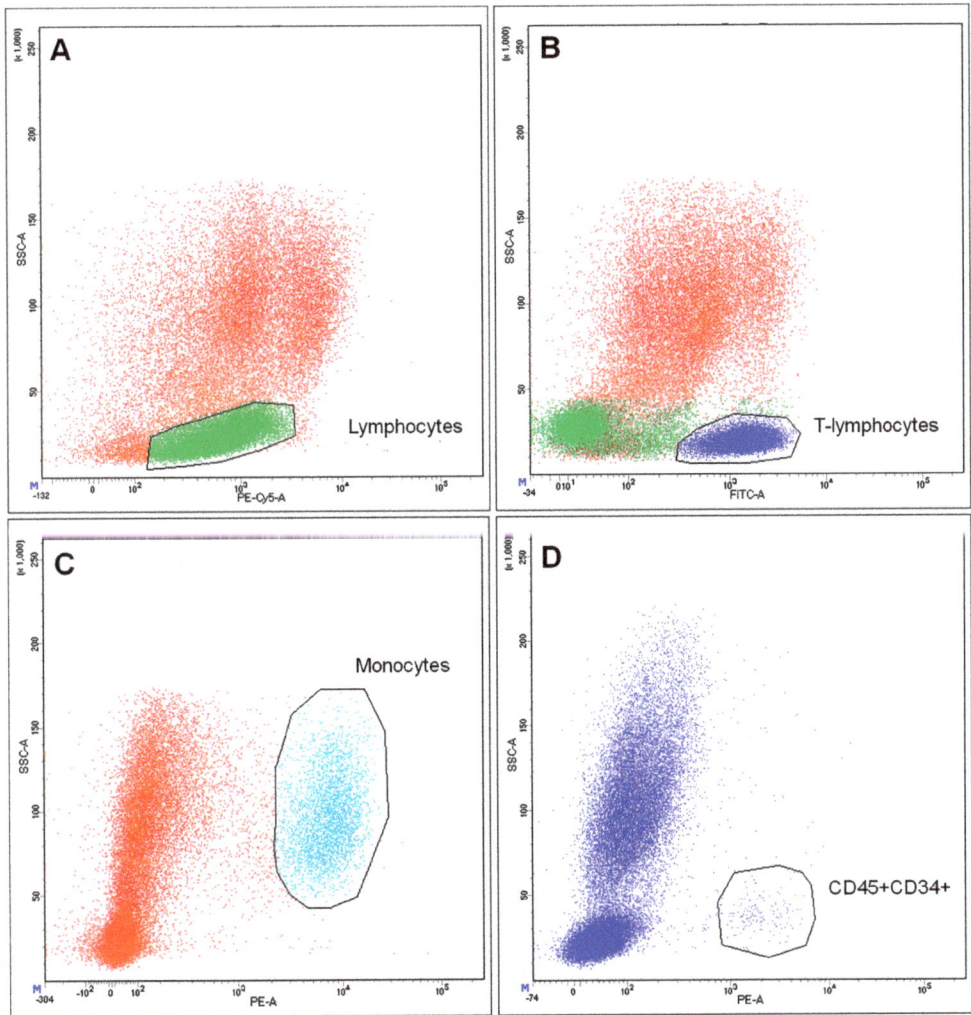

Figure 1. Determination of immune phenotype of umbilical cord blood mononuclear cells. (**A**) staining for CD45, selected green area–Lymphocytes; (**B**) staining for CD3, selected dark blue area–T-lymphocytes; (**C**) staining for CD14, selected blue area–Monocytes; (**D**) staining for CD34, selected events–CD45+ CD34+ cells.

Immunophenotyping of a pool of CD34-positive cells showed that genetic modification and expression of recombinant factors by cells did not affect the viability and endothelial cell markers (Figure 2).

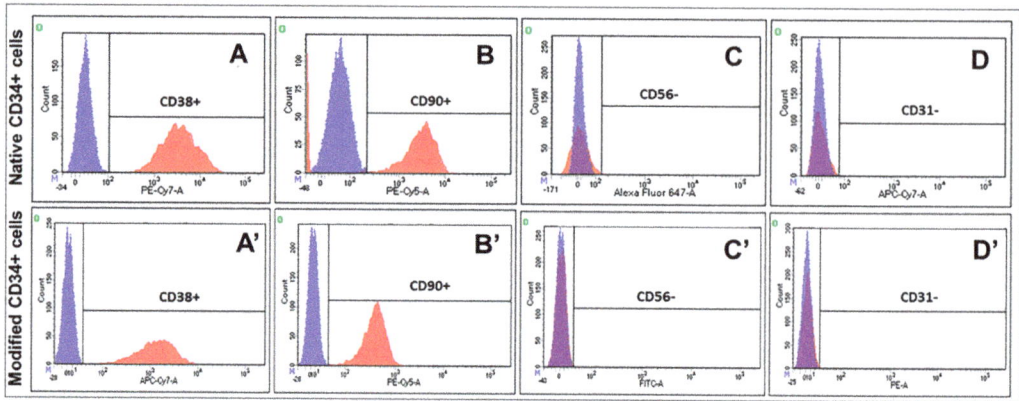

Figure 2. Determination of immune phenotype of CD34 positive cells. (**A,A′**) staining for CD38+; (**B,B′**) staining for CD90+; (**C,C′**) staining for CD56 (**D,D′**) staining for CD31. Native CD34+ cells—non-treated cells (UCB-MC-NTC). Modified CD34+ cells—cells modified with Ad-VEGF165, Ad-FGF2, and Ad-SDF1α (UCB-MC-VEGF-FGF2-SDF1α). Red peaks—staining with antibodies; blue peaks— negative control (isotype control).

2.2. Transduction of UC-MCs with Recombinant Adenoviruses Increased Transgene Expression

It has been demonstrated that genetic modification of the UCB-MCs with recombinant adenoviruses (Ad-VEGF, Ad-FGF2, Ad-SDF1α, or Ad-EGFP) did not affect cell viability. Moreover, it has been shown that UCB-MCs transduced with Ad-EGFP exhibited green fluorescence, confirming the efficiency of transduction (Figure 3A). Furthermore, EGFP expression was sustained for 30 days after a genetic modification of UCB-MCs. According to the flow cytometry results, EGFP+ cells constituted 28 ± 2.7% (Figure 3B). Analysis of the mRNA expression of VEGF165, FGF2, and SDF1α in genetically modified human UCB-MCs was carried out using qPCR. It has been established that genetic modification of hUCB-MCs with Ad-VEGF165 results in augmented VEGF expression (190.6 ± 8.9 fold). Simultaneous transduction with Ad-VEGF165, Ad-FGF2, and Ad-SDF1α resulted in the upregulation of VEGF, FGF2, and SDF1α expression (198.6 ± 0.45; 204.2 ± 0.36 and 140.9 ± 0.32 fold respectively) compared to non-transfected cells, and cells modified with Ad-EGFP (Figure 3C). The obtained results are evident for efficient modification of hUCB-MCs with developed genetic constructs which provide a synthesis of target genes in vitro.

2.3. Genetically Modified hUCB-MCs Produce a Broad Range of Cytokines, Chemokines, and Growth Factors

A complete analysis of all cytokines and chemokines measured in the Luminex assays demonstrated that gene modification and gene expression did not change levels of multiple anti and proinflammatory cytokines as well as chemokines. The results obtained from the eight donors in comparison to the untreated control are shown in Table 1 (Supplementary Table S1). We have not observed any statistically significant differences in cytokine and chemokine secretion between the groups of non-transfected cells and genetically modified ones except for upregulated levels of recombinant proteins in corresponding groups. Multiplex analysis revealed statistically significant ($p < 0.05$) upregulation of VEGF secretion (1087.12 ± 169.11 pg/mL) in UCB-MCs modified with Ad-VEGF compared to the UCB-MCs treated with Ad-EGFP (52.31 ± 10.36 pg/mL) and non-treated cells (51.75 ± 8.65 pg/mL). Simultaneous transduction with Ad-VEGF, Ad-FGF2, and Ad-SDF1 has resulted in the increased production of VEGF (701.94 ± 96.99 pg/mL), FGF2 (576.27 ± 57.83 pg/mL), and SDF1α (622.39 ± 113.07 pg/mL) (Figure 3D). Obtained results correlate with the data presented above of RT-qPCR and confirm the capacity of recombinant adenoviruses for infection of target cells.

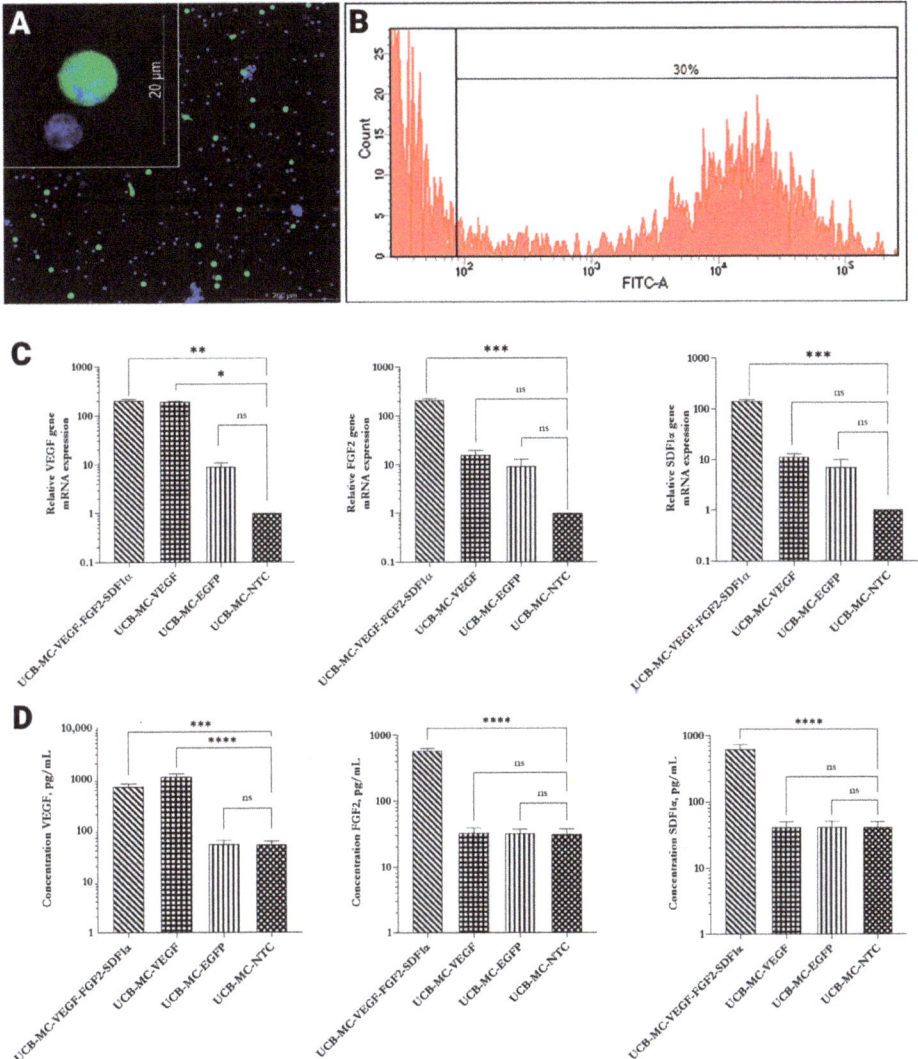

Figure 3. Analysis of transgenes expression in genetically modified UCB-MCs in vitro, 72 h post-transduction (MOI 10). (**A**) Fluorescent microscopy of EGFP+ cells (green fluorescence). Nuclei were stained with Hoechst 33342 (blue fluorescence). Scale bar 200 μm, 20 μm (inset). (**B**) Flow cytometry analysis of Ad-EGFP genetically modified UCB-MCs. Approximately 30% of genetically modified UCB-MCs exhibit green fluorescence. (**C**) Relative expression mRNA of therapeutic genes in genetically modified hUCB-MCs ($n = 3$). (**D**) The concentration of secreted proteins VEGF, FGF, and SDF1α in supernatants collected from genetically modified umbilical cord blood mononuclear cells ($n = 8$). UCB-MC-VEGF-FGF2-SDF1α—umbilical cord blood mononuclear cells modified with Ad-VEGF165, Ad-FGF2, and Ad-SDF1α. UCB-MC-VEGF—umbilical cord blood mononuclear cells modified with AdVEGF165. UCB-MC-EGFP—umbilical cord blood mononuclear cells modified with Ad-EGFP. UCB-MC-NTC—non-treated cells. Data presented as average ± standard error (SE); $p < 0.05$ * statistically significant differences compared to control (** $p < 0.01$; *** $p < 0.001$; **** $p < 0.0001$ ns—non-significant).

Table 1. Profile of cytokines, chemokines, and growth factors secreted by genetically modified human umbilical cord blood mononuclear cells (pg/mL).

Analytes	UCB-MC-VEGF-FGF2-SDF1α	±SE	UCB-MC-VEGF	±SE	UCB-MC-EGFP	±SE	UCB-MC-NTC	±SE
IL-1β	69.74	40.13	99.93	75.35	80.73	51.62	105.13	71.47
IL-1α	69.70	45.97	63.22	46.65	54.39	41.85	66.03	49.29
IL-2	59.47	33.06	42.08	22.04	25.98	10.33	25.04	11.27
IL-3	32.79	27.69	19.13	10.13	13.95	7.79	27.93	20.94
IL-4	6.66	3.57	6.38	4.01	5.83	3.66	5.80	3.52
IL-5	100.61	55.99	95.73	57.98	108.39	70.69	125.89	79.67
IL-6	5034.98	1779.01	4337.05	1813.94	4357.59	1815.39	4697.32	1818.03
IL-7	4.16	1.35	3.93	1.38	4.07	1.37	3.60	1.40
IL-8	1,250,075.63	1,235,541.48	912,305.50	859,150.84	818,724.14	803,905.26	1,245,513.75	1,232,516.25
IL-9	27.19	11.69	28.90	11.99	26.39	12.43	27.37	13.48
IL-10	45.47	20.64	42.49	21.28	43.65	18.37	42.49	21.09
IL-12p40	206.89	53.84	178.88	51.16	181.16	61.56	204.02	69.32
IL-12p70	12.46	4.89	17.68	7.54	17.25	9.49	11.73	3.85
IL-13	2.29	0.58	2.28	0.57	2.37	0.76	2.38	0.71
IL-15	270.89	74.22	268.08	82.28	269.57	83.57	271.49	94.11
IL-16	422.01	102.13	413.50	91.00	383.75	81.51	396.14	85.94
IL-17	31.08	13.63	26.312	10.39	18.20	5.07	16.37	5.79
IL-18	9.14	3.96	10.08	5.06	6.32	2.31	5.85	1.94
IL-1ra	2126.49	773.43	1643.84	749.23	1545.21	692.37	1432.40	787.78
IL-2Rα	42.72	13.75	65.02	30.70	33.75	14.06	40.06	15.24
G-CSF	1270.38	721.68	1169.91	793.85	1264.01	866.29	1310.72	883.87
M-CSF	15.36	5.21	17.41	5.92	14.55	4.82	20.66	9.79
GM-CSF	30.27	16.49	30.14	14.82	23.83	11.51	25.07	10.86
PDGF-bb	262.31	79.05	242.27	31.67	224.96	62.40	193.86	44.27
HGF	155.12	63.49	182.51	76.23	185.92	75.70	214.67	101.40
β-NGF	5.33	1.93	5.98	3.32	3.15	1.33	4.24	1.50
SCF	104.75	47.36	94.22	60.91	77.45	47.19	93.06	47.29
SCGF-β	13,765.94	9382.70	9029.33	4187.62	7519.49	3573.11	9416.14	4526.52
LIF	83.69	32.05	69.63	32.17	63.90	29.94	65.40	32.74
MIF	6675.05	4487.06	6253.82	3726.91	4901.44	2670.02	6055.06	3408.90
IFN-γ	110.16	33.54	103.33	36.70	96.52	33.54	120.61	50.39
IFN-α2	21.65	5.38	23.15	6.72	17.06	5.53	18.21	6.05
TNF-α	706.57	339.62	636.71	343.12	686.63	409.36	612.62	369.89
TNF-β	65.06	20.03	45.37	14.01	85.01	30.49	56.53	18.19
TRAIL	309.92	160.98	260.55	168.85	263.23	164.56	266.27	165.56
IP-10	2946.86	2203.50	2756.89	2121.07	461.92	218.19	494.25	238.09
MCP-1	5878.67	1279.98	5633.98	1331.15	5927.77	1267.73	5548.28	1370.69
MCP-3	759.58	650.42	752.86	651.29	748.83	652.17	786.16	648.42
MIP-1α	27,907.69	27,255.16	62,599.84	61,948.04	580.24	283.44	575.48	284.73
MIP-1β	2085.04	1401.47	2114.97	1399.79	2050.55	1407.29	1838.60	1411.79
RANTES	944.01	102.62	910.36	211.74	807.69	146.79	789.65	207.02
Eotaxin	10.89	3.84	8.24	2.94	7.34	1.93	7.35	1.95
CTACK	222.07	189.83	222.26	189.11	121.43	93.65	160.64	127.47
GROα	18,520.81	11,913.68	20,273.64	11,681.33	18,519.01	11,844.77	53,200.26	33,134.06
MIG	262.56	191.60	282.09	192.34	228.99	142.93	315.42	178.72
VEGF	701.94	96.99	1087.12	169.11	52.31	10.36	51.75	8.65
FGF-2	576.27	57.83	32.36	6.65	32.03	4.98	30.71	6.59
SDF-1α	622.39	113.07	40.59	9.05	41.07	9.66	40.41	9.10

It is also worth emphasizing that the UCB-MC-VEGF-FGF2-SDF1 and UCB-MC-VEGF did not differ from UCB-MC-EGFP and UCB-MC-NTC in vitro studies. What can be seen from the data of morphological, and phenotypic studies are the profiles of secreted factors. Therefore, UCB-MC-EGFP is the ideal control in our study in vivo.

2.4. Transplantation of Genetically Modified Cells Promotes Angiogenesis In Vivo

Matrigel mixtures were implanted into the subcutaneous space of the dorsal region in mice after seven days post-transplantation when implanted Matrigel samples containing genetically modified UCB-MCs were extracted from Balb/c nude mice. Embedded fragments represented discs with d = 10 mm and 2 mm height. The color of the implants correlated with vascularization density. The color of the implants varied from milky-white (Matrigel without cells and Matrigel with UCB-MC + Ad-EGFP) to red-brown (Matrigel with UCB-MC + Ad-VEGF165 + Ad-FGF2 + Ad-SDF1α) which is due to the vascular formation and presence of blood cells, particularly, erythrocytes (Figure 4A). Gross histological hematoxylin and eosin (H&E) staining of extracted plugs showed the absence of inflammatory sites. The skin and subcutaneous tissue in the area of implantation were not visually changed (Figure 4B). We have established that in isolated subcutaneous implants containing hUCB-MC, human-transduced Ad-VEGF165, or a combination of Ad-VEGF165, Ad-SDF1α, and Ad-FGF2, the hemoglobin concentration was significantly higher in comparison with Matrigel fragments without cell administration and implants with UCB-MCs transduced Ad-EGFP. Moreover, the significantly higher concentration of hemoglobin was determined in the samples containing UCB-MCs transduced with Ad-VEGF165, Ad-SDF1α, and Ad-FGF2 compared to the group with UCB-MCs transduced with single Ad-VEGF165 (Figure 4C). Moreover, we observed a two-fold increase of mCD31 mRNA expression in plugs containing hUCB-MC transduced Ad-VEGF165 or a combination of Ad-VEGF165, Ad-SDF1α, and Ad-FGF2 compared to controls. Moreover, we did not discover the difference between Ad-VEGF165 and the group with a mixture of Ad-VEGF165, Ad-SDF1α, and Ad-FGF2.

Analysis of the mRNA expression of VEGF165, FGF2, and SDF1α in genetically modified UCB-MCs in Matrigel implants was evaluated by RT-qPCR. Notably, obtained results confirmed the expression of target genes in genetically modified UCB-MCs implanted in Matrigel even at one-week post-transplantation. We discovered that the Matrigel complexes containing UCB-MC Ad-VEGF gave rise to more abundant VEGF mRNA than UCB-MC Ad-EGFP and PBS (Matrigel samples without UCB-MCs). Likewise, UCB-MCs contemporaneously transduced with Ad-VEGF, Ad-FGF2, and Ad-SDF1α exhibited upregulated levels of mRNA expression of VEGF, FGF2, and SDF1α. (Figure 5A). During histological analysis of implants, it has been shown that control—PBS (Matrigel samples without UCB-MCs) contained small amounts of migrated fibroblast-like cells. Visually, the implants were surrounded by a thin connective tissue capsule, which contained rare capillaries in a density of 1.5 ± 0.5 units/mm^2. In samples with implanted UCB-MCs transduced with a cocktail of adenoviruses (Ad-VEGF165, Ad-FGF2, and AdSDF1α), Matrigel mass contained a residual amount of VEGF+ cells. These vessels localized close to the capsule and migrated fibroblasts, some of which were positive for SDF1α and FGF2. In Matrigel samples with implanted UCB-MCs genetically modified with Ad-EGFP, we found single and small rounded clusters of EGFP-positive cells and rare migrated fibroblast-like cells. The implants were surrounded by a thin connective tissue capsule, from which strands of connective tissue grew into its depth with capillaries found in a density of 7.5 ± 3 units/mm^2. Vessels were located close to the capsule. Fibroblasts that migrated into Matrigel expressed SDF1α and FGF2. Expression of VEGF, FGF2, and SDF1α in the implanted UCB-MCs were not confirmed. In the group with UCB-MCs modified with Ad-VEGF165, implant samples presented Matrigel mass with single small, rounded clusters of VEGF-positive UCB-MCs cells and rare migrated fibroblast-like cells. The implants were surrounded by a thin connective tissue capsule, from which strands of connective tissue grew more profound into the central regions of the implant with capillaries' density of 16 ± 5 units/mm^2. In the group of UCB-MCs simultaneously transduced with a combination of Ad-VEGF165, Ad-SDF1α and Ad-FGF2, implant samples were represented by the mass of Matrigel with single and small rounded clusters of UCB-MCs, as well as rare migrated fibroblast-like cells. The implant was surrounded by a thin connective tissue capsule, from which the connective tissue and vessels of various calibers grew to the center of the implant with a capillary density of 23 ± 5 units/mm^2. Implanted UCB-MCs expressed VEGF, FGF2, and SDF1α (Figure 5B).

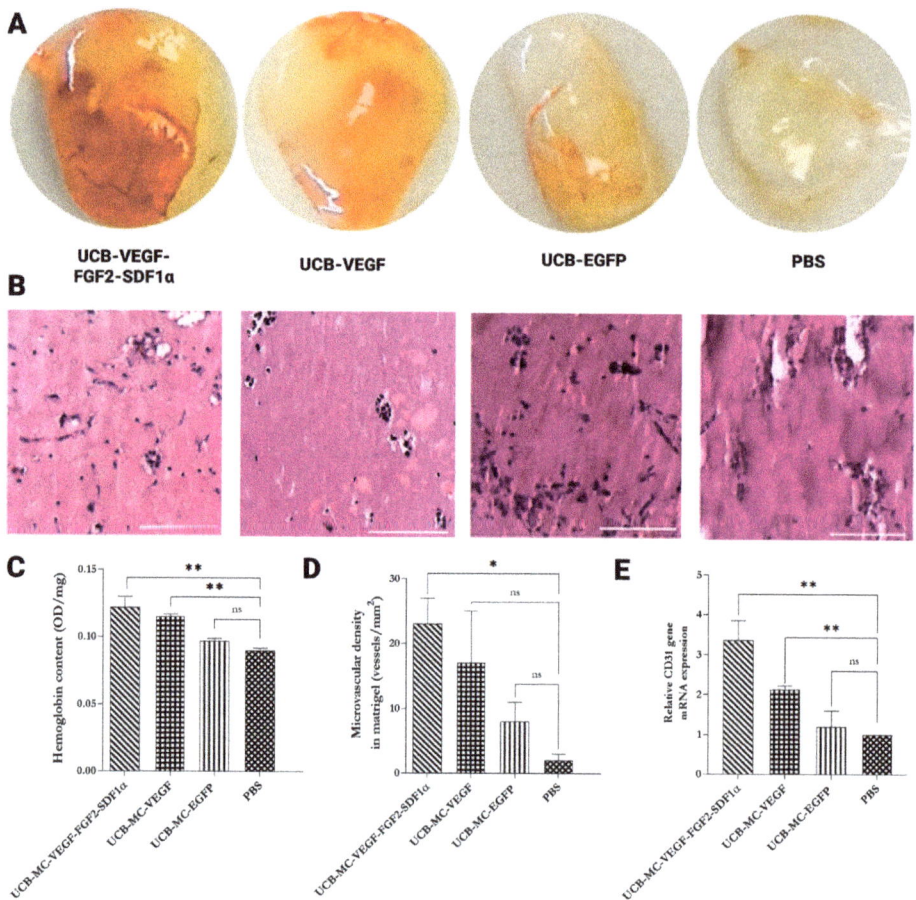

Figure 4. UCB-MC engineered with recombinant adenoviruses promotes angiogenesis in an in vivo Matrigel plug assay. Matrigel plugs collected at one-week post-implantation were appraised. (**A**) Representative gross morphological assessment and (**B**) microphotographs of harvested implants' hematoxylin & eosin (H&E) staining. Scale bar 100 μm. (**C**) Quantification of hemoglobin concentration in each Matrigel plug ($n = 4$). (**D**) Histomorphometric analysis of vascularization process, number of blood vessels in frozen sections of the Matrigel plugs. (**E**) RT-qPCR quantification of CD31 mRNA in Matrigel plugs. UCB-MC-VEGF-FGF2- SDF1α—umbilical cord blood mononuclear cells modified with Ad-VEGF165, Ad-FGF2, and Ad-SDF1α. UCB-MC-VEGF—umbilical cord blood mononuclear cells modified with Ad-VEGF165. UCB-MC-EGFP—umbilical cord blood mononuclear cells modified with Ad-EGFP. PBS—Matrigel containing PBS. Data presented as average ± standard error (SE); $p < 0.05$ * statistically significant differences compared to control ($n = 3$; ** $p < 0.01$; ns—non-significant).

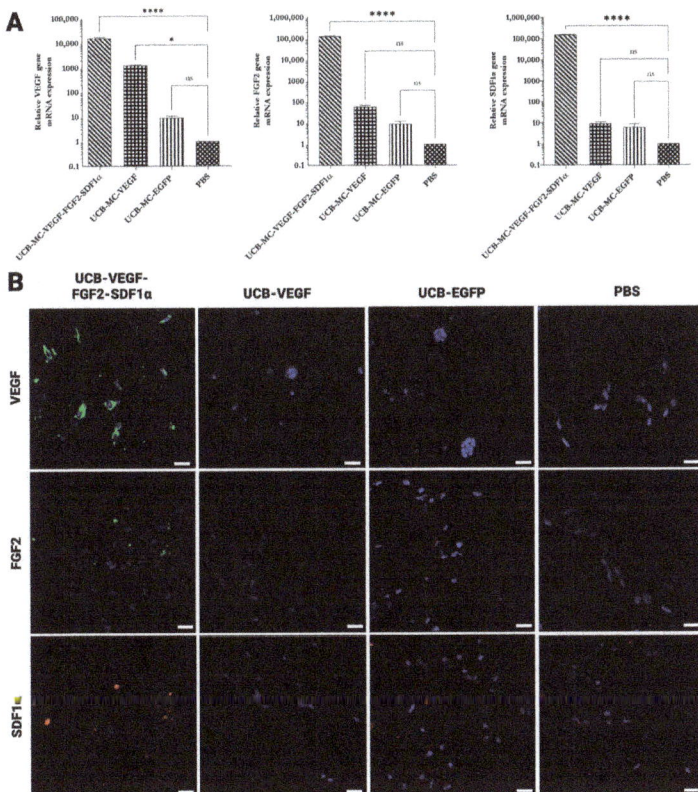

Figure 5. Analysis of transgenes expression in isolated Matrigel plugs one-week post-implantation. (**A**) Increasing concentrations of VEGF, FGF2, and SDF1α mRNA levels in Matrigel plugs. RT-qPCR analysis was performed on the total RNA extracted from gelled Matrigel samples. (**B**) Immunohistochemical analysis of Matrigel plugs. Staining for VEGF (green), FGF2 (green), and SDF1α (red). Nuclei were counterstained using a DAPI solution (4′,6-diamidino-2-phenylindole) (blue). Scale bars 50 µm. UCB-MC-VEGF-FGF2-SDF1—umbilical cord blood mononuclear cells modified with Ad-VEGF165, Ad-FGF2, and Ad-SDF1α. UCB-MC-VEGF—umbilical cord blood mononuclear cells modified with Ad-VEGF165. UCB-MC-EGFP—umbilical cord blood mononuclear cells modified with Ad-EGFP. PBS—Matrigel containing PBS. Data presented as average ± standard error (SE); $p < 0.05$ * statistically significant differences compared to control ($n = 3$; **** $p < 0.0001$; ns—non-significant.

3. Discussion

Adenoviruses mediate gene transfer into dividing and quiescent cells and can be produced with a significant titer. The high immunogenicity of adenoviruses as vehicles for the delivery of therapeutic genes represents one of the main disadvantages resulting in the activation of the immune response in immune-competent organisms and the absence of expression of the target therapeutic genes [38]. However, this negative effect is eliminated when using an ex vivo gene therapy approach. Moreover, adenoviral systems promote transient transgene expression due to their non-integration into the host cell genome [39]. However, this negative point might become beneficial for gene therapy based on growth factors: induction of angiogenesis does not require the prolonged expression of therapeutic proteins but, more importantly, their synergistic effect [40]. The absence of integration of adenoviruses eliminates the risk of insertional mutagenesis, which is a typical problem when using retroviral vectors [39]. Adenoviral vectors demonstrate comparatively low efficiency of genetic modification of hematopoietic cells, which might be increased with a higher concentration of virus [41] or its specific treatment, resulting in augmented tropism [42]. In the present study, we chose the adenovirus delivery vectors containing VEGF, FGF2, and SDF1α to investigate the angiogenic effect of UCB-MC in vitro and the Matrigel plug assay in Nude mice. In our investigation, cellular carriers expressed phenotype typical for UCB-MCs, and about 30% of the cells were efficiently transduced with an MOI of 10. The transduction efficiency correlated with previous results and other research groups' data [42,43]. After in vitro transduction, the UCB-MCs expressed the recombinant mRNA of proangiogenic factors in the cytoplasm and secreted those factors into the culture medium, which in our study we confirmed by RT-qPCR and immunological studies. The obtained data correlates with our previously published results [7].

Various approaches were proposed for stimulating therapeutic angiogenesis based on the delivery methods of genetically engineered systems expressing a broad range of proangiogenic factors. The therapeutic efficacy of proangiogenic factors has been proven in numerous experiments on animal models [44] and in several clinical studies [45,46]. The key inducers of angiogenesis, VEGF, FGF2, EGF, SDF1α, and PDGF-BB, are most often used as genetic components [2]. In particular, VEGF is perhaps the most characterized and frequently used mitogen in creating gene therapy systems and in the induction of therapeutic angiogenesis. VEGF is a crucial participant in forming new blood vessels and can induce the growth of pre-existing ones [47]. However, Zentilin et al. reported that overexpression of VEGF induced leaky neovessels that missed connecting correctly with existing vessels [48]. The FGF family includes vertebrates' most versatile growth factors that play critical roles in many biological processes, including angiogenesis [49]. FGF, similar to VEGF, is a pleiotropic molecule capable of acting on various cell types, including endodermal, mesenchymal, and neuroectodermal origin cells. It has been shown that FGF2 induces the expression of VEGF and several other factors by endothelial cells through autocrine and paracrine mechanisms [50,51]. SDF1α is a constitutively expressed and inducible chemokine, associated with various physiological and pathological processes, including embryonic development, homeostasis maintenance, and angiogenesis activation [52]. There is evidence that the administration of SDF1α increases blood flow and perfusion via the recruitment of endothelial progenitor cells (EPCs). SDF1α binds exclusively to CXCR4 and has CXCR4 as its only receptor [53]. Compared with the effects of other angiogenic growth factors, SDF1α has unique properties. The generation of hyperpermeable vessels, a significant characteristic of VEGF-stimulated angiogenesis, may not be observed after injection of SDF1α contributes to the stabilization of neovessel formation by recruiting CXCR4 + PDGFR+ cKit+ smooth muscle progenitor cells during recovery from vascular injury [54]. Extensive evidence suggests that SDF1α up-regulates VEGF synthesis in several cell types, whereas VEGF and basic FGF induce SDF1α and its receptor CXCR4 in endothelial cells [55]. However, it should also be noted that in a wide range of studies using various models, the mutual synergistic role of VEGF, FGF2, SDF1α, and countless other factors responsible for the formation of vessels has been shown [56–59]. It is generally known that an optional cellular source for allogenic transplantation should meet the following criteria: it must be less immunogenic and contain a sufficient amount of immature cells capable of differentiation in various directions; it should have a prolonged period of storage and potency for expansion. Most gene-cell-mediated therapy protocols intend genetic modification of target cells with different vectors, providing stable expression of target proteins. Human UCB-MCs might be easily isolated and characterized; these cells exhibit low immunogenicity and are composed of unique populations of progenitor cells capable of differentiation into endothelial, muscular, and neural cells, etc. Mononuclear cells from umbilical cord blood are a well-characterized group of cells that are extensively used in pre-clinical and clinical trials of therapy for various human diseases and the induction of therapeutic angiogenesis as well [60]. However, relatively small amounts of UCB-MCs for achieving sufficient therapeutic effect remain the main limitation for its extensive introduction in the clinic [61]. To increase its biological activity, it was proposed to mix different cell pools with further genetic modification [62]. Contemporary cell-mediated approaches to gene therapy suggest UCB-MC as a cell carrier for the delivery of various therapeutic genes. This concept assumes either the differentiation of transplanted cells into different cell types or the realization of therapeutic effects due to the secretion of a broad range of bioactive molecules [63]. Furthermore, our previous study has demonstrated that UCB-MCs are capable of transferring therapeutic genes and promoting evident therapeutic effects using different models, such as ALS [64], SCI [25,26], and stroke [27]. Similar results were obtained in investigations dedicated to therapies for hematologic and non-hematologic disorders [65–69]. At the same time, there is no current data about the influence of the simultaneous transduction of several recombinant adenoviruses on the secretome profile and angiogenic properties of modified hUCB-MCs. A sustained balance of proangiogenic factors and their synergetic effect is essential for functional vascular formation. In the present study, we developed the UCB-MC application to simultaneously deliver many genes (VEGF, FGF2, and SDF1α) to stimulate angiogenesis. Our previous studies also showed the approach of preventive gene therapy with many genes to positively affect stroke. Adenoviral vectors carrying genes encoding vascular VEGF, glial cell-derived neurotrophic factor (GDNF), and NCAM or gene-engineered umbilical cord blood mononuclear cells (UCB-MC) overexpressing recombinant proteins were intrathecally injected before distal occlusion of the middle cerebral artery in the rat. Morphometric and immunofluorescence analysis revealed a reduction in infarction volume and a lower number of apoptotic cells. It decreased the expression of Hsp70 in the peri-infarct region in gene-treated animals [7].

The heterogeneous cell population from the mononuclear fractions UCB-MCs secretes different anti-inflammatory, pro-inflammatory cytokines, chemokines end grow factors [70]. Previously, it was shown that the duration of cultivation, cultivation medium, and the additives used in the culture are the main factors influencing the production of cytokines by UCB-MCs. Our study describes the profile of cytokines and chemokines released by UCB-MC following their in vitro gene modification by adenoviruses. Five groups of secreted factors were investigated: pro-inflammatory cytokines (IL-6, IL-1β, and TNF), an anti-inflammatory cytokine (IL-4 and IL-10), TH1-type cytokines (IL-12 and IFN-γ), chemokines (IL-8, MIP-1α, MIP-1β, and MCP-1) and growth factors (VEGF, FGF2, and SDF1α). Interestingly, the range of cytokine, chemokine, and growth factor concentrations detected in the supernatants of UBC-MC varied between donors, indicating major individual heterogeneity, comparable with previously published data [71].

The highest secretion level by modified and unmodified cells was shown for IL-8 and MCP-1. These factors are known to be produced more intensively than any other chemokines in the human body and are seen as the first line of defense in inflammatory responses [72]. In addition, the cells also secreted high concentrations of GROα, IL-6, MIF, MIP-1α, MIP-1β, and SCGF-β. Unfortunately, adenoviruses are potent activators of the innate and adaptive immune systems. The administration of high doses of Ad-based vectors to animals or patients, primarily through the intravascular pathway, leads to severe immunopathology manifested by cytokine storm syndrome, disseminated intravascular coagulation, thrombocytopenia, and hepatotoxicity, which can lead to morbidity and also death [69]. Research by Teigler et al. on peripheral blood mononuclear cells (PBMCs) showed that their stimulation with the Ad vector increases the secretion of IFN-γ, INF2α, IL-15, G-CSF, MIG, and IP-10. Supporting this perspective, it is worth emphasizing that the study's authors used 10^3 vp/cell [73]. Previous studies have shown that treatment of myeloid dendritic cells and plasmacytoid dendritic cells with Ad5 does not lead to an increase in IFN production by them, even at the highest exposed rAd (100 vp/cell) [43]. Our previous examination has shown that genetic modification UCB-MC and expression of transgenes VEGF or EGFP did not influence the global transcriptome landscape [74]. In this study, we demonstrate that a gene-cell system with simultaneous delivery of genes based on UCB-MC can generate the expression of several transgenes both in vitro and in vivo. Furthermore, the UCB-MC-VEGF165 and UCB-MC-VEGF-FGF2-SDF1α Matrigel plugs in mice were filled with red blood cells and showed vessel-like structure formation. We did not find significant differences between the UCB-MC-VEGF and UCB-MC-VEGF-FGF2-SDF1α groups in the present study. Although in line with the RT-qPCR data and immunology tests, levels of expression of VEGF, SDF1α, and FGF varied. Perhaps this is because we used a small amount of cellular material and a short exposure period to Matrigel fragments.

Furthermore, the UCB-MC-VEGF165 and UCB-MC-VEGF-FGF2-SDF1α Matrigel plugs in mice were filled with red blood cells and showed vessel-like structure formation. We did not find significant differences between the UCB-MC-VEGF165 and UCB-MC-VEGF-FGF2-SDF1α groups in the present study. Although in line with the RT-qPCR data and immunology tests, levels of expression of VEGF, SDF1α, and FGF varied. Perhaps this is because we used a small amount of cellular material and a short exposure period to Matrigel fragments.

4. Materials and Methods

4.1. Obtaining Recombinant Adenovirus Ad-SDF1α

The creation of expression constructs based on adenovirus was carried out by using molecular cloning methods of Gateway-cloning technology (Invitrogen), as described previously [75]. Briefly, subcloning of SDF1α from the plasmid vector pBud-VEGF-SDF1α into the intermediate vector pDONR221 was performed [76].

4.2. The Production of Recombinant Adenoviruses

The HEK293A cells were infected with a coarse viral runoff to prepare the necessary amounts of Ad-VEGF, Ad-FGF2, Ad-SDF1α, and Ad-EGFP adenoviruses. To purify viral particles from cell debris, supernatants were filtered through 0.22 μm filters and centrifuged in a gradient of cesium chloride. Virus dialysis was performed using a membrane with a pore throughput of 3.5 kDa in two stages. After purification and concentration, the resulting recombinant adenoviruses were titrated by optical density, as well as by plaque formation. The titer of the recombinant adenoviruses we obtained was from 1 to 3.8×10^9 PFU/mL. The viral titer values were guided by the genetic modification of human UCB-MC.

4.3. UCB-MC Isolation and Characterization

All UCB-MC units were collected from healthy donors with a gestation period of 37–40 weeks in maternity public hospitals in Kazan. Blood collections were carried out into single blood bags of 250 mL, with the blood preservative CPDA-1 (GCMS, Republic of Korea). Exclusion criteria were maternal infections or viral diseases. Isolations of mononuclear cells were conducted within 16–18 h after blood collection. Nucleated blood cells were isolated using SepMate ™-50 tubes according to the manufacturer's protocol (STEMCELL Technologies Inc., Vancouver, BC, Canada). The viability of the isolated cells was determined in a hemocytometer with a 0.4% trypan blue solution. Cell viability, as measured by trypan blue exclusion, was >97%. The immune phenotype of isolated cells was analyzed by staining with monoclonal antibodies CD45—PerCP (BioLegend, San Diego, CA, USA), CD3-FITC (BioLegend, San Diego, CA, USA) CD14-APC/Cy7 (BioLegend, San Diego, CA, USA), CD38-APC/Cy7 (BioLegend, San Diego, CA, USA) CD34-FITC (BioLegend, San Diego, CA, USA), CD90-PE/Cy5 (BioLegend, San Diego, CA, USA). Expression of CD markers were analyzed by flow cytometry using BD FACS Aria III (BD bioscience, San Jose, CA, USA)

4.4. Analysis of Adenoviral Transduction of hUCB-MCs

Genetic modification of hUCB-MCs with recombinant adenoviruses (MOI 10 for each virus) was performed according to a previously developed protocol [77].

The efficiency of genetic modification was assessed after 72 h by means of fluorescent microscopy on AxioObserverZ1 (Carl Zeiss, Oberkochen, Germany) and flow cytometry using BD FACS Aria III (BD Bioscience, San Jose, CA, USA).

4.5. Total RNA Extraction and RT-qPCR

Analysis of the mRNA expression of VEGF165, FGF2, and SDF1α in genetically modified cells and isolated Matrigel implants was carried out by qPCR with further statistical analysis. Isolation of total RNA was performed by using the TRIzol (Thermo Fisher Scientific, Waltham, MA, USA) reagent according to the manufacturer's recommendations with further cDNA synthesis. Real-Time PCR was set up on the Real-Time CFX96 Touch (BioRad Laboratories, CA, USA). The nucleotide sequences of the primers and probes used in RT-qPCR are mentioned in Table 2. All reactions for each sample were performed in triplicate with a further calculation of ΔΔCt values and normalization to the housekeeping gene of β-actin rRNA. Standard curves for quantitative analysis were created using serial dilutions of plasmid DNA with corresponding inserts (VEGF, FGF2, and SDF1α). Expression of target genes in non-transduced UCB-MCs was considered 100%. The level of the murine target gene mCD31 was normalized to the mouse housekeeping gene of mGAPDH. The statistical analysis of the obtained results was carried out in MS Excel 2007 with further calculation using U criteria (Mann-Whitney).

Table 2. Nucleotide sequences of primers used for RT-qPCR.

Name	Nucleotide Sequence
β-actin-TM-F (human)	GCGAGAAGATGACCCAGGATC
β-actin-TM-R (human)	CCAGTGGTACGGCCAGAGG
β-actin-TMprobe (human)	CCAGCCATGTACGTTGCTATCCAGGC
hVEGF-TM49F	TACCTCCACCATGCCAAGTG
hVEGF-TM110R	TGATTCTGCCCTCCTCCTTCT
hVEGF-TMProbe	TCCCAGGCTGCACCCATGG
hFGF2-TM134F	CCGACGGCCGAGTTGAC
hFGF2-TM203R	TCTTCTGCTTGAAGTTGTAGCTTGA
hFGF2-TMprobe	CCGGGAGAAGAGCGACCCTCAC
hSDF1-TM-F	TGACCGCTAAAGTGGTCGTC
hSDF1-TM-R	ACGTGGCTCTCAAAGAACCT
hSDF1-TMprobe	CCCTGTGCCTGTCCGATGGA
mCD31-F	CGGTTATGATGATGTTTCTGGA
mCD31-R	AAGGGAGGACACTTCCACTTCT
mGAPDH-F	GAAGGTCGGTGTGAACGGATT
mGAPDH-R	TGACTGTGCCGTTGAATTTG

4.6. Analysis of Cytokines and Chemokines

Supernatant cytokine profiles were analyzed using Bio-Plex Pro Human Cytokine 27-plex Panel and Bio-Plex Human Cytokine 21-plex Panel (Bio-Rad, Hercules, CA, USA) multiplex magnetic bead-based antibody detection kits, following the manufacturer's instructions. Supernatant aliquots (50 µL) were used for analysis, with a minimum of 50 beads per analyte acquired. Median fluorescence intensities were measured using a Bioplex 200 (Bio-Rad, Hercules, CA, USA) analyzer. Data collected were analyzed with MasterPlex CT control software and MasterPlex QT analysis software (Hitachi Software, San Bruno, CA, USA). Standard curves for each analyte were generated using standards provided by the manufacturer.

4.7. In Vivo Experiments

In vivo experiments were performed using immune-deficient mice of Balb/c nude lineage of both sexes for 7–8 weeks. The animals were bred using the animal facilities in

Puschino's laboratory. After quarantine, animals were held in an SPF vivarium with HEPA filters according to GLP standards. In the area of the withers, mice were subcutaneously injected with 2 million human transduced or native UCB-MCs mixed with 300 µL of Matrigel matrix. Female and male Balb/c nude mice were randomly assigned to a few groups: 1. Matrigel without UCB-MCs; 2. UCB-MCs transduced with Ad-EGFP in Matrigel; 3. UCB-MCs transduced with Ad-VEGF165 in Matrigel; 4. UCB-MCs transduced with a combination of adenoviruses Ad-VEGF165; Ad-SDF1α and Ad-FGF2 in Matrigel. All experiments were performed in quadruplicates. After seven days post-transplantation mice were taken from the experiment. The status of subcutaneous Matrigel implants was evaluated visually, and concentrations of hemoglobin were evaluated. Levels of the expression of therapeutic genes were analyzed by RT-qPCR. Production of therapeutic proteins was assessed via immunohistochemistry.

4.8. Analysis of Hemoglobin Concentration

The analysis of hemoglobin concentration in subcutaneous implants was evaluated colorimetrically. Implants were balanced by weight and homogenized in DPBS using the Mini-Bead Beater-16 (BioSpec, Bartlesville, OK, USA) with zirconium beads (d = 2 mm for 100 mg), during 2 cycles for 20 sec each. The obtained homogenates were centrifuged at 15,000× g for 15 min. Supernatants containing hemoglobin were examined on a microplate reader Tecan Infinite Pro 2000 with an OD of 540 nm (Tecan Austria GmbH, Grödig, Austria).

4.9. Histological Analysis

For histological analysis, Matrigel implants were fixed in a 10% buffered formalin solution for 48 h. After fixation, implants were dehydrated in increasing concentrations of ethanol and embedded in paraffin (Histomix, Biovitrum, Saint Petersburg, Russia). Paraffin slides with 5 µm thickness were prepared at the rotary microtome HM 355S (Thermo Fisher Scientific, Waltham, MA, USA). For general morphological characterization, slides were deparaffinized and stained with hematoxylin and eosin according to the standard protocol. For immunological studies, serial sections were deparaffinized and incubated in a citric buffer for 30 min to unmask epitopes. Cell membranes were permeabilized in a 0.1% solution of Tween-20 in PBS. Non-specific binding was blocked by incubation in a 10% solution of donkey serum for 30 min. Sections were stained with the antibodies to VEGF (mab293), FGF2 (sc-1390), and SFD1α (sc-28876), diluted 1:100 for 1 h at room temperature. After washing sections were stained for 1 h with secondary antibodies at room temperature followed by washing and DAPI staining (1:50,000 dilution) for 10 min. Primary and secondary antibodies are shown in Table 3. The microvessel density (MVD) was examined by counting the vessels in each implant, reported as the vessel number per square millimeter (vessel/mm^2). Visualization of the results was performed on a scanning laser microscope LSM780 (Carl Zeiss, Oberkochen, Germany). Image analysis was performed using Image J software (https://fiji.sc/ (accessed on 18 December 2021)).

Table 3. Primary and secondary antibodies used in immunofluorescent staining.

Antibody	Host	Dilution	Source
SFD1-α (sc-28876)	Rabbit	1:100	Abcam
FGF2 (SC-1390)	Goat	1:100	Santa Cruz
VEGF (mab293)	Mouse	1:100	R&D Systems
Anti- Mouse IgG Alexa Fluor 488	Goat	1:200	Invitrogen
Anti-Rabbit IgG H&L Alexa Fluor 647	Goat	1:200	Invitrogen
Anti-Rabbit IgG H&L Alexa Fluor 488	Goat	1:200	Abcam

4.10. Statistical Analysis

GraphPad Prism® 7 software was used to show all data reports (GraphPad, Inc., La Jolla, CA, USA). The data are presented as the mean ± standard error (SE). p-values were analyzed using a one-way analysis of variance (ANOVA) followed by Tukey's test. Statistical significance is denoted by $p < 0.05$. All tests with animals, morphometric and statistical analyses were performed in a "blinded" manner with respect to the experimental groups.

5. Conclusions

The study suggests that UCB-MCs continue to be an essential source of stem cells for therapy and various gene-cell strategies for tissue regeneration. It is important to emphasize that the transplantation of genetically modified UCB-MCs is safer and more effective than direct gene therapy. Human UCB-MCs can be efficiently simultaneously modified with adenoviral vectors encoding VEGF, FGF2, and SDF1α. Modified UCB-MCs overexpress recombinant genes. Genetic modification of cells with recombinant adenoviruses (MOI 10) does not affect the profile of secreted pro- and anti-inflammatory cytokines, chemokines, or growth factors, except for an increase in the synthesis of recombinant proteins by cells. Modified cells can induce the formation of new vessels. Although many unresolved problems remain before modified UCB-MCs can be applied in the clinic, our results remain promising in terms of the induction of therapeutic angiogenesis in future clinical trials, including the treatment of decompensated forms.

Supplementary Materials: The following supporting information can be downloaded at: https://www.mdpi.com/article/10.3390/ijms24054396/s1.

Author Contributions: Conceptualization, R.R.I., A.A.R. and I.I.S.; investigation, D.Z.G., I.M.G., E.E.G., M.N.Z., S.S.A., M.A.G., M.O.G. and I.I.S.; writing—original draft preparation, D.Z.G., I.I.S. and M.N.Z.; writing—review and editing, A.A.R., E.E.G. and I.I.S.; visualization, D.Z.G., I.M.G., S.S.A. and M.N.Z.; methodology, I.M.G., S.S.A., E.E.G. and M.N.Z.; supervision, R.R.I., A.A.R. and I.I.S.; funding acquisition, I.I.S. and A.A.R. All authors have read and agreed to the published version of the manuscript.

Funding: The reported study was funded by RSF according to the research project №21-75-10035. Also, this paper has been supported by the Kazan Federal University Strategic Academic Leadership Program (PRIORITY-2030).

Institutional Review Board Statement: The study was conducted according to the guidelines of the Declaration of Helsinki, and approved by the Local Ethics Committee Kazan Federal University. All procedures using animals were approved by the Kazan Federal University Animal Care and Use Committee (protocol 23 dated 30 June 2020).

Informed Consent Statement: Informed consent was obtained from each study subject according to the guidelines approved under this protocol (article 20, Federal Law "Protection of Health Rights of Citizens of Russian Federation" N323-FZ, 11.21.2011).

Data Availability Statement: The data presented in this study are available on request from the corresponding author.

Conflicts of Interest: The authors declare no conflict of interest.

References

1. Rodriguez, A.E.; Pavlovsky, H.; Del Pozo, J.F. Understanding the Outcome of Randomized Trials with Drug-Eluting Stents and Coronary Artery Bypass Graft in Patients with Multivessel Disease: A Review of a 25-Year Journey. *Clin. Med. Insights Cardiol.* **2016**, *10*, 195–199. [CrossRef] [PubMed]
2. Kuwano, M.; Fukushi, J.; Okamoto, M.; Nishie, A.; Goto, H.; Ishibashi, T.; Ono, M. Angiogenesis factors. *Intern. Med.* **2001**, *40*, 565–572. [CrossRef] [PubMed]
3. Ripa, R.S.; Wang, Y.; Jørgensen, E.; Johnsen, H.E.; Hesse, B.; Kastrup, J. Intramyocardial injection of vascular endothelial growth factor-A165 plasmid followed by granulocyte-colony stimulating factor to induce angiogenesis in patients with severe chronic ischaemic heart disease. *Eur. Heart J.* **2006**, *27*, 1785–1792. [CrossRef]

4. Rissanen, T.T.; Ylä-Herttuala, S. Current status of cardiovascular gene therapy. *Mol. Ther.* **2007**, *15*, 1233–1247. [CrossRef]
5. Lee, E.J.; Park, H.W.; Jeon, H.J.; Kim, H.S.; Chang, M.S. *Potentiated Therapeutic Angiogenesis by Primed Human Mesenchymal Stem Cells in a Mouse Model of Hindlimb Ischemia*; Future Medicine Ltd.: London, UK, 2013; Volume 8. [CrossRef]
6. Boyle, A.J.; McNiece, I.K.; Hare, J.M. Mesenchymal stem cell therapy for cardiac repair. *Stem Cells Myocard. Regen.* **2010**, *660*, 65–84.
7. Markosyan, V.; Safiullov, Z.; Izmailov, A.; Fadeev, F.; Sokolov, M.; Kuznetsov, M.; Trofimov, D.; Kim, E.; Kundakchyan, G.; Gibadullin, A.; et al. Preventive Triple Gene Therapy Reduces the Negative Consequences of Ischemia-Induced Brain Injury after Modelling Stroke in a Rat. *Int. J. Mol. Sci.* **2020**, *21*, 6858. [CrossRef] [PubMed]
8. Urashima, M.; Hoshi, Y.; Shishikura, A.; Kamijo, M.; Kato, Y.; Akatsuka, J.; Maekawa, K. Umbilical cord blood as a rich source of immature hematopoietic stem cells. *Pediatr. Int.* **1994**, *36*, 649–655. [CrossRef]
9. Harris, D.T.; Rogers, I. Umbilical cord blood: A unique source of pluripotent stem cells for regenerative medicine. *Curr. Stem Cell Res. Ther.* **2007**, *2*, 301–309. [CrossRef]
10. Divya, M.S.; Roshin, G.E.; Divya, T.S.; Rasheed, V.A.; Santhoshkumar, T.R.; Elizabeth, K.E.; James, J.; Pillai, R.M. Umbilical cord blood-derived mesenchymal stem cells consist of a unique population of progenitors co-expressing mesenchymal stem cell and neuronal markers capable of instantaneous neuronal differentiation. *Stem Cell Res. Ther.* **2012**, *3*, 57. [CrossRef]
11. Kögler, G.; Sensken, S.; Airey, J.A.; Trapp, T.; Müschen, M.; Feldhahn, N.; Liedtke, S.; Sorg, R.V.; Fischer, J.; Rosenbaum, C.; et al. A new human somatic stem cell from placental cord blood with intrinsic pluripotent differentiation potential. *J. Exp. Med.* **2004**, *200*, 123–135. [CrossRef]
12. Gluckman, E. Ten years of cord blood transplantation: From bench to bedside. *Br. J. Haematol.* **2009**, *147*, 192–199. [CrossRef] [PubMed]
13. Pimentel-Coelho, P.M.; Rosado-de Castro, P.H.; Da Fonseca, L.M.B.; Mendez-Otero, R. Umbilical cord blood mononuclear cell transplantation for neonatal hypoxic-ischemic encephalopathy. *Pediatr. Res.* **2012**, *71*, 464–473. [CrossRef] [PubMed]
14. Goldstein, G.; Toren, A.; Nagler, A. Transplantation and other uses of human umbilical cord blood and stem cells. *Curr. Pharm. Des.* **2007**, *13*, 1363–1373. [CrossRef]
15. Balassa, K.; Rocha, V. Anticancer cellular immunotherapies derived from umbilical cord blood. *Expert Opin. Biol. Ther.* **2018**, *18*, 121–134. [CrossRef]
16. Rocha, V.; Labopin, M.; Sanz, G.; Arcese, W.; Schwerdtfeger, R.; Bosi, A.; Jacobsen, N.; Ruutu, T.; de Lima, M.; Finke, J.; et al. Transplants of umbilical-cord blood or bone marrow from unrelated donors in adults with acute leukemia. *N. Engl. J. Med.* **2004**, *351*, 2276–2285. [CrossRef]
17. Wagner, J.E.; Rosenthal, J.; Sweetman, R.; Shu, X.O.; Davies, S.M.; Ramsay, N.K.; McGlave, P.B.; Sender, L.; Cairo, M.S. Successful transplantation of HLA-matched and HLA-mismatched umbilical cord blood from unrelated donors: Analysis of engraftment and acute graft-versus-host disease. *Blood* **1996**, *88*, 795–802. [CrossRef] [PubMed]
18. Arien-Zakay, H.; Lecht, S.; Bercu, M.M.; Tabakman, R.; Kohen, R.; Galski, H.; Nagler, A.; Lazarovici, P. Neuroprotection by cord blood neural progenitors involves antioxidants, neurotrophic and angiogenic factors. *Exp. Neurol.* **2009**, *216*, 83–94. [CrossRef]
19. Bachstetter, A.D.; Pabon, M.M.; Cole, M.J.; Hudson, C.E.; Sanberg, P.R.; Willing, A.E.; Bickford, P.C.; Gemma, C. Peripheral injection of human umbilical cord blood stimulates neurogenesis in the aged rat brain. *BMC Neurosci.* **2008**, *9*, 22. [CrossRef]
20. Dasari, V.R.; Spomar, D.G.; Li, L.; Gujrati, M.; Rao, J.S.; Dinh, D.H. Umbilical cord blood stem cell mediated downregulation of fas improves functional recovery of rats after spinal cord injury. *Neurochem. Res.* **2008**, *33*, 134–149. [CrossRef]
21. Schira, J.; Gasis, M.; Estrada, V.; Hendricks, M.; Schmitz, C.; Trapp, T.; Kruse, F.; Kögler, G.; Wernet, P.; Hartung, H.P.; et al. Significant clinical, neuropathological and behavioural recovery from acute spinal cord trauma by transplantation of a welldefined somatic stem cell from human umbilical cord blood. *Brain* **2012**, *135*, 431–446. [CrossRef]
22. Xiao, J.; Nan, Z.; Motooka, Y.; Low, W.C. Transplantation of a novel cell line population of umbilical cord blood stem cells ameliorates neurological deficits associated with ischemic brain injury. *Stem Cells Dev.* **2005**, *14*, 722–733. [CrossRef]
23. Ikeda, Y.; Fukuda, N.; Wada, M.; Matsumoto, T.; Satomi, A.; Yokoyama, S.I.; Saito, S.; Matsumoto, K.; Kanmatsuse, K.; Mugishima, H. Development of angiogenic cell and gene therapy by transplantation of umbilical cord blood with vascular endothelial growth factor gene. *Hypertens. Res.* **2004**, *27*, 119–128. [CrossRef]
24. Chen, H.K.; Hung, H.F.; Shyu, K.G.; Wang, B.W.; Sheu, J.R.; Liang, Y.J.; Chang, C.C.; Kuan, P. Combined cord blood stem cells and gene therapy enhances angiogenesis and improves cardiac performance in mouse after acute myocardial infarction. *Eur. J. Clin. Investig.* **2005**, *35*, 677–686. [CrossRef]
25. Islamov, R.R.; Sokolov, M.E.; Bashirov, F.V.; Fadeev, F.O.; Shmarov, M.M.; Naroditskiy, B.S.; Povysheva, T.V.; Shaymardanova, G.F.; Yakupov, R.A.; Chelyshev, Y.A.; et al. A pilot study of cell-mediated gene therapy for spinal cord injury in mini pigs. *Neurosci. Lett.* **2017**, *644*, 67–75. [CrossRef]
26. Izmailov, A.A.; Povysheva, T.V.; Bashirov, F.V.; Sokolov, M.E.; Fadeev, F.O.; Garifulin, R.R.; Naroditsky, B.S.; Logunov, D.Y.; Salafutdinov, I.I.; Chelyshev, Y.A.; et al. Spinal Cord Molecular and Cellular Changes Induced by Adenoviral Vector- and Cell-Mediated Triple Gene Therapy after Severe Contusion. *Front. Pharmacol.* **2017**, *8*, 813. [CrossRef]
27. Sokolov, M.E.; Bashirov, F.V.; Markosyan, V.A.; Povysheva, T.V.; Fadeev, F.O.; Izmailov, A.A.; Kuztetsov, M.S.; Safiullov, Z.Z.; Shmarov, M.M.; Naroditskyi, B.S.; et al. Triple-Gene Therapy for Stroke: A Proof-of-Concept in Vivo Study in Rats. *Front. Pharmacol.* **2018**, *9*, 111. [CrossRef]

28. Nowak-Sliwinska, P.; Alitalo, K.; Allen, E.; Anisimov, A.; Aplin, A.C.; Auerbach, R.; Augustin, H.G.; Bates, D.O.; van Beijnum, J.R.; Bender, R.H.F.; et al. Consensus guidelines for the use and interpretation of angiogenesis assays. *Angiogenesis* **2018**, *21*, 425–532. [CrossRef]
29. Stryker, Z.I.; Rajabi, M.; Davis, P.J.; Mousa, S.A. Evaluation of angiogenesis assays. *Biomedicines* **2019**, *7*, 37. [CrossRef]
30. Simons, M.; Alitalo, K.; Annex, B.H.; Augustin, H.G.; Beam, C.; Berk, B.C.; Dimmeler, S. State-of-the-art methods for evaluation of angiogenesis and tissue vascularization: A scientific statement from the American Heart Association. *Circ. Res.* **2015**, *116*, e99–e132. [CrossRef]
31. Passaniti, A.; Taylor, R.; Pili, R.; Guo, Y.; Long, P.; Haney, J.; Pauly, R.; Grant, D.; Martin, G. A simple, quantitative method for assessing angiogenesis and antiangiogenic agents using reconstituted basement membrane, heparin, and fibroblast growth factor. *Lab. Investig.* **1992**, *67*, 519–528.
32. Martin, S.; Murray, J.C. (Eds.) *Angiogenesis Protocols*, 2nd ed.; Methods in Molecular Biology; Springer: New York, NY, USA, 2009; Volume 467. [CrossRef]
33. Norrby, K. In vivo models of angiogenesis. *J. Cell. Mol. Med.* **2006**, *10*, 588–612. [CrossRef] [PubMed]
34. Kibbey, M.C.; Corcoran, M.L.; Wahl, L.M.; Kleinman, H.K. Laminin SIKVAV peptide-induced angiogenesis in vivo is potentiated by neutrophils. *J. Cell. Physiol.* **1994**, *160*, 185–193. [CrossRef] [PubMed]
35. Salcedo, R.; Ponce, M.L.; Young, H.A.; Wasserman, K.; Ward, J.M.; Kleinman, H.K.; Oppenheim, J.J.; Murphy, W.J. Human endothelial cells express CCR2 and respond to MCP-1: Direct role of MCP-1 in angiogenesis and tumor progression. *Blood* **2000**, *96*, 34–40. [CrossRef] [PubMed]
36. Johns, A. Disruption of estrogen receptor gene prevents 17 beta estradiol-induced angiogenesis in transgenic mice. *Endocrinology* **1996**, *137*, 4511–4513. [CrossRef] [PubMed]
37. Ferber, S.; Tiram, G.; Satchi-Fainaro, R. Monitoring functionality and morphology of vasculature recruited by factors secreted by fast-growing tumor-generating cells. *J. Vis. Exp.* **2014**, *93*, 1–6. [CrossRef]
38. Barkats, M.; Bilang-Bleuel, A.; Buc-Caron, M.H.; Castel-Barthe, M.N.; Corti, O.; Finiels, F.; Horellou, P.; Revah, F.; Sabate, O.; Mallet, J. Adenovirus in the brain: Recent advances of gene therapy for neurodegenerative diseases. *Prog. Neurobiol.* **1998**, *55*, 333–341. [CrossRef]
39. Lee, C.S.; Bishop, E.S.; Zhang, R.; Yu, X.; Farina, E.M.; Yan, S.; Zhao, C.; Zheng, Z.; Shu, Y.; Wu, X.; et al. Adenovirus-Mediated Gene Delivery: Potential Applications for Gene and Cell-Based Therapies in the New Era of Personalized Medicine. *Genes Dis.* **2017**, *4*, 43–63. [CrossRef]
40. Seggern, D.J.; Nemerow, G.R. Adenoviral Vectors For Protein Expression. In *Gene Expr. Systems*; Fernandez, J., Hoeffler, J.P., Eds.; Academic: London, UK, 1998; pp. 111–156. [CrossRef]
41. MacKenzie, K.L.; Hackett, N.R.; Crystal, R.G.; Moore, M.A.S. Adenoviral vector–mediated gene transfer to primitive human hematopoietic progenitor cells: Assessment of transduction and toxicity in long-term culture. *Blood* **2000**, *96*, 100–108. [CrossRef]
42. Adams, W.C.; Gujer, C.; McInerney, G.; Gall, J.G.D.; Petrovas, C.; Karlsson Hedestam, G.B.; Koup, R.A.; Loré, K. Adenovirus type-35 vectors block human CD4+ T-cell activation via CD46 ligation. *Proc. Natl. Acad. Sci. USA* **2011**, *108*, 7499–7504. [CrossRef]
43. Loré, K.; Adams, W.C.; Havenga, M.J.E.; Precopio, M.L.; Holterman, L.; Goudsmit, J.; Koup, R.A. Myeloid and plasmacytoid dendritic cells are susceptible to recombinant adenovirus vectors and stimulate polyfunctional memory T cell responses. *J. Immunol.* **2007**, *179*, 1721–1729. [CrossRef]
44. Liu, M.; Xie, S.; Zhou, J. Use of animal models for the imaging and quantification of angiogenesis. *Exp. Anim.* **2018**, *67*, 1–6. [CrossRef]
45. Iyer, S.R.; Annex, B.H. Therapeutic Angiogenesis for Peripheral Artery Disease: Lessons Learned in Translational Science. *JACC Basic Transl. Sci.* **2017**, *2*, 503–512. [CrossRef]
46. Gupta, R.; Tongers, J.; Losordo, D.W. Human studies of angiogenic gene therapy. *Circ. Res.* **2009**, *105*, 724–736. [CrossRef]
47. Duffy, A.M.; Bouchier-Hayes, D.J.; Harmey, J.H. Vascular Endothelial Growth Factor (VEGF) and Its Role in Non-Endothelial Cells: Autocrine Signalling by VEGF. *Madame Curie Bioscience Database* [Internet]. Landes Bioscience. 2013. Available online: https://www.ncbi.nlm.nih.gov/books/NBK6482/ (accessed on 12 September 2022).
48. Zentilin, L.; Tafuro, S.; Zacchigna, S.; Arsic, N.; Pattarini, L.; Sinigaglia, M.; Giacca, M. Bone marrow mononuclear cells are recruited to the sites of VEGF-induced neovascularization but are not incorporated into the newly formed vessels. *Blood* **2006**, *107*, 3546–3554. [CrossRef]
49. Itoh, N. The Fgf families in humans, mice, and zebrafish: Their evolutional processes and roles in development, metabolism, and disease. *Biol. Pharm. Bull.* **2007**, *30*, 1819–1825. [CrossRef]
50. Seghezzi, G.; Patel, S.; Ren, C.J.; Gualandris, A.; Pintucci, G.; Robbins, E.S.; Shapiro, R.L.; Galloway, A.C.; Rifkin, D.B.; Mignatti, P. Fibroblast growth factor-2 (FGF-2) induces vascular endothelial growth factor (VEGF) expression in the endothelial cells of forming capillaries: An autocrine mechanism contributing to angiogenesis. *J. Cell Biol.* **1998**, *141*, 1659–1673. [CrossRef]
51. Jin, S.; Yang, C.; Huang, J.; Liu, L.; Zhang, Y.; Li, S.; Zhang, L.; Sun, Q.; Yang, P. Conditioned medium derived from FGF-2-modified GMSCs enhances migration and angiogenesis of human umbilical vein endothelial cells. *Stem Cell Res. Ther.* **2020**, *11*, 68. [CrossRef]
52. Ho, T.K.; Shiwen, X.; Abraham, D.; Tsui, J.; Baker, D. Stromal-Cell-Derived Factor-1 (SDF-1)/CXCL12 as Potential Target of Therapeutic Angiogenesis in Critical Leg Ischaemia. *Cardiol. Res. Pract.* **2012**, *2012*, 143209. [CrossRef]

53. Horuk, R. Chemokine receptors. *Cytokine Growth Factor Rev.* **2001**, *12*, 313–335. [CrossRef]
54. Zernecke, A.; Schober, A.; Bot, I.; von Hundelshausen, P.; Liehn, E.A.; Möpps, B.; Mericskay, M.; Gierschik, P.; Biessen, E.A.; Weber, C. SDF-1alpha/CXCR4 axis is instrumental in neointimal hyperplasia and recruitment of smooth muscle progenitor cells. *Circ. Res.* **2005**, *96*, 784–791. [CrossRef]
55. Salcedo, R.; Oppenheim, J.J. Role of chemokines in angiogenesis: CXCL12/SDF-1 and CXCR4 interaction, a key regulator of endothelial cell responses. *Microcirculation* **2003**, *10*, 359–370. [CrossRef] [PubMed]
56. Kano, M.R.; Morishita, Y.; Iwata, C.; Iwasaka, S.; Watabe, T.; Ouchi, Y.; Miyazono, K.; Miyazawa, K. VEGF-A and FGF-2 synergistically promote neoangiogenesis through enhancement of endogenous PDGF-B-PDGFRbeta signaling. *J. Cell Sci.* **2005**, *118*, 3759–3768. [CrossRef] [PubMed]
57. Hu, G.j.; Feng, Y.g.; Lu, W.p.; Li, H.t.; Xie, H.w.; Li, S.f. Effect of combined VEGF165/ SDF-1 gene therapy on vascular remodeling and blood perfusion in cerebral ischemia. *J. Neurosurg.* **2017**, *127*, 670–678. [CrossRef] [PubMed]
58. Ljubimov, A.V. Growth Factor Synergy in Angiogenesis. In *Retinal and Choroidal Angiogenesis*; Penn, J.S., Ed.; Springer: Berlin/Heidelberg, Germany, 2008; pp. 289–310. [CrossRef]
59. de Paula, E.V.; Flores-Nascimento, M.C.; Arruda, V.R.; Garcia, R.A.; Ramos, C.D.; Guillaumon, A.T.; Annichino-Bizzacchi, J.M. Dual gene transfer of fibroblast growth factor-2 and platelet derived growth factor-BB using plasmid deoxyribonucleic acid promotes effective angiogenesis and arteriogenesis in a rodent model of hindlimb ischemia. *Transl. Res.* **2009**, *153*, 232–239. [CrossRef] [PubMed]
60. Hwang, S.; Choi, J.; Kim, M. Combining Human Umbilical Cord Blood Cells With Erythropoietin Enhances Angiogenesis/Neurogenesis and Behavioral Recovery After Stroke. *Front. Neurol.* **2019**, *10*, 357. [CrossRef] [PubMed]
61. Ballen, K.K.; Gluckman, E.; Broxmeyer, H.E. Umbilical cord blood transplantation: The first 25 years and beyond. *Blood* **2013**, *122*, 491–498. [CrossRef]
62. Garbuzova-Davis, S.; Ehrhart, J.; Sanberg, P.R. Cord blood as a potential therapeutic for amyotrophic lateral sclerosis. *Expert Opin. Biol. Ther.* **2017**, *17*, 837–851. [CrossRef]
63. Rizvanov, A.A.; Guseva, D.S.; Salafutdinov, I.I.; Kudryashova, N.V.; Bashirov, F.V.; Kiyasov, A.P.; Yalvaç, M.E.; Gazizov, I.M.; Kaligin, M.S.; Sahin, F.; et al. Genetically modified human umbilical cord blood cells expressing vascular endothelial growth factor and fibroblast growth factor 2 differentiate into glial cells after transplantation into amyotrophic lateral sclerosis transgenic mice. *Exp. Biol. Med.* **2011**, *236*, 91–98. [CrossRef]
64. Islamov, R.R.; Rizvanov, A.A.; Fedotova, V.Y.; Izmailov, A.A.; Safiullov, Z.Z.; Garanina, E.E.; Salafutdinov, I.I.; Sokolov, M.E.; Mukhamedyarov, M.A.; Palotás, A. Tandem Delivery of Multiple Therapeutic Genes Using Umbilical Cord Blood Cells Improves Symptomatic Outcomes in ALS. *Mol. Neurobiol.* **2017**, *54*, 4756–4763. [CrossRef]
65. Zhu, H.; Poon, W.; Liu, Y.; Leung, G.K.K.; Wong, Y.; Feng, Y.; Ng, S.C.P.; Tsang, K.S.; Sun, D.T.F.; Yeung, D.K.; et al. Phase I-II Clinical Trial Assessing Safety and Efficacy of Umbilical Cord Blood Mononuclear Cell Transplant Therapy of Chronic Complete Spinal Cord Injury. *Cell Transplant.* **2016**, *25*, 1925–1943. [CrossRef]
66. Liu, J.; Han, D.; Wang, Z.; Xue, M.; Zhu, L.; Yan, H.; Zheng, X.; Guo, Z.; Wang, H. Clinical analysis of the treatment of spinal cord injury with umbilical cord mesenchymal stem cells. *Cytotherapy* **2013**, *15*, 185–191. [CrossRef]
67. Ichim, T.E.; Solano, F.; Lara, F.; Paris, E.; Ugalde, F.; Rodriguez, J.P.; Minev, B.; Bogin, V.; Ramos, F.; Woods, E.J.; et al. Feasibility of combination allogeneic stem cell therapy for spinal cord injury: A case report. *Int. Arch. Med.* **2010**, *3*, 30. [CrossRef]
68. Yang, W.Z.; Zhang, Y.; Wu, F.; Min, W.P.; Minev, B.; Zhang, M.; Luo, X.L.; Ramos, F.; Ichim, T.E.; Riordan, N.H.; et al. Safety evaluation of allogeneic umbilical cord blood mononuclear cell therapy for degenerative conditions. *J. Transl. Med.* **2010**, *8*, 75. [CrossRef] [PubMed]
69. Atasheva, S.; Yao, J.; Shayakhmetov, D.M. Innate immunity to adenovirus: Lessons from mice. *FEBS Lett.* **2019**, *593*, 3461–3483. [CrossRef] [PubMed]
70. Garanina, E.E.; Gatina, D.; Martynova, E.V.; Rizvanov, A.; Khaiboullina, S.; Salafutdinov, I. Cytokine Profiling of Human Umbilical Cord Plasma and Human Umbilical Cord Blood Mononuclear Cells. *Blood* **2017**, *130*, 4814. [CrossRef]
71. Reuschel, E.; Toelge, M.; Entleutner, K.; Deml, L.; Seelbach-Goebel, B. Cytokine profiles of umbilical cord blood mononuclear cells upon in vitro stimulation with lipopolysaccharides of different vaginal gram-negative bacteria. *PLoS ONE* **2019**, *14*, e0222465. [CrossRef]
72. Newman, M.B.; Willing, A.E.; Manresa, J.J.; Sanberg, C.D.; Sanberg, P.R. Cytokines produced by cultured human umbilical cord blood (HUCB) cells: Implications for brain repair. *Exp. Neurol.* **2006**, *199*, 201–208. [CrossRef]
73. Teigler, J.E.; Iampietro, M.J.; Barouch, D.H. Vaccination with adenovirus serotypes 35, 26, and 48 elicits higher levels of innate cytokine responses than adenovirus serotype 5 in rhesus monkeys. *J. Virol.* **2012**, *86*, 9590–9598. [CrossRef]
74. Salafutdinov, I.I.; Gatina, D.Z.; Markelova, M.I.; Garanina, E.E.; Malanin, S.Y.; Gazizov, I.M.; Khaiboullina, S.; Rizvanov, A.A.; Islamov, R.I. Transcriptomic Landscape of Umbilical Cord Blood Mononuclear Cells after Genetic Modification. *Blood* **2020**, *136*, 33–34. [CrossRef]
75. Cherenkova, E.; Fedotova, V.; Borisov, M.; Islamov, R.; Rizvanov, A. Generation of recombinant adenoviruses and lentiviruses expressing angiogenic and neuroprotective factors using Gateway cloning technology. *Cell. Transplant. Tissue Eng.* **2012**, *7*, 164–168.

76. Solovyeva, V.V.; Chulpanova, D.S.; Tazetdinova, L.G.; Salafutdinov, I.I.; Bozo, I.Y.; Isaev, A.A.; Deev, R.V.; Rizvanov, A.A. In Vitro Angiogenic Properties of Plasmid DNA Encoding SDF-1α and VEGF165 Genes. *Appl. Biochem. Biotechnol.* **2020**, *190*, 773–788. [CrossRef] [PubMed]
77. Islamov, R.; Rizvanov, A.; Mukhamedyarov, M.; Salafutdinov, I.; Garanina, E.; Fedotova, V.; Solovyeva, V.; Mukhamedshina, Y.; Safiullov, Z.; Izmailov, A.; et al. Symptomatic improvement, increased life-span and sustained cell homing in amyotrophic lateral sclerosis after transplantation of human umbilical cord blood cells genetically modified with adeno-viral vectors expressing a neuro-protective factor and a neural cell adhesion molecule. *Curr. Gene Ther.* **2015**, *15*, 266–276. [PubMed]

Disclaimer/Publisher's Note: The statements, opinions and data contained in all publications are solely those of the individual author(s) and contributor(s) and not of MDPI and/or the editor(s). MDPI and/or the editor(s) disclaim responsibility for any injury to people or property resulting from any ideas, methods, instructions or products referred to in the content.

Article

Non-Muscle MLCK Contributes to Endothelial Cell Hyper-Proliferation through the ERK Pathway as a Mechanism for Vascular Remodeling in Pulmonary Hypertension

Mariam Anis [1], Janae Gonzales [2], Rachel Halstrom [2], Noman Baig [2], Cat Humpal [2], Regaina Demeritte [2], Yulia Epshtein [2], Jeffrey R. Jacobson [2] and Dustin R. Fraidenburg [2,*]

[1] Northwestern Medical Group, Lake Forest, IL 60045, USA
[2] Department of Medicine, University of Illinois at Chicago, Chicago, IL 60612, USA
* Correspondence: dfraiden@uic.edu; Tel.: +1-312-355-5918

Citation: Anis, M.; Gonzales, J.; Halstrom, R.; Baig, N.; Humpal, C.; Demeritte, R.; Epshtein, Y.; Jacobson, J.R.; Fraidenburg, D.R. Non-Muscle MLCK Contributes to Endothelial Cell Hyper-Proliferation through the ERK Pathway as a Mechanism for Vascular Remodeling in Pulmonary Hypertension. *Int. J. Mol. Sci.* **2022**, *23*, 13641. https://doi.org/10.3390/ijms232113641

Academic Editors: Elisabeth Deindl and Paul Quax

Received: 30 September 2022
Accepted: 1 November 2022
Published: 7 November 2022

Publisher's Note: MDPI stays neutral with regard to jurisdictional claims in published maps and institutional affiliations.

Copyright: © 2022 by the authors. Licensee MDPI, Basel, Switzerland. This article is an open access article distributed under the terms and conditions of the Creative Commons Attribution (CC BY) license (https://creativecommons.org/licenses/by/4.0/).

Abstract: Pulmonary arterial hypertension (PAH) is characterized by endothelial dysfunction, uncontrolled proliferation and migration of pulmonary arterial endothelial cells leading to increased pulmonary vascular resistance resulting in great morbidity and poor survival. Bone morphogenetic protein receptor II (BMPR2) plays an important role in the pathogenesis of PAH as the most common genetic mutation. Non-muscle myosin light chain kinase (nmMLCK) is an essential component of the cellular cytoskeleton and recent studies have shown that increased nmMLCK activity regulates biological processes in various pulmonary diseases such as asthma and acute lung injury. In this study, we aimed to discover the role of nmMLCK in the proliferation and migration of pulmonary arterial endothelial cells (HPAECs) in the pathogenesis of PAH. We used two cellular models relevant to the pathobiology of PAH including BMPR2 silenced and vascular endothelial growth factor (VEGF) stimulated HPAECs. Both models demonstrated an increase in nmMLCK activity along with a robust increase in cellular proliferation, inflammation, and cellular migration. The upregulated nmMLCK activity was also associated with increased ERK expression pointing towards a potential integral cytoplasmic interaction. Mechanistically, we confirmed that when nmMLCK is inhibited by MLCK selective inhibitor (ML-7), proliferation and migration are attenuated. In conclusion, our results demonstrate that nmMLCK upregulation in association with increased ERK expression may contribute to the pathogenesis of PAH by stimulating cellular proliferation and migration.

Keywords: pulmonary hypertension; endothelial cells; myosin light chain kinase; cytoskeleton; vascular remodeling

1. Introduction

Pulmonary arterial hypertension (PAH) is a severe and progressive disease characterized by obstruction of small pulmonary arteries leading to increased pulmonary vascular resistance and right heart failure [1,2]. Despite several new advancements and therapies for PAH in the recent years, mortality remains unacceptably high with a 3-year rate of 22% [2]. Genetic studies have demonstrated that mutations in bone morphogenic protein receptor type 2 (BMPR2) are present in 80% of hereditable PAH leading to loss of function and reduced downstream signaling [3–6]. Decreased BMPR2 expression is also identified in many other forms of PAH not associated with clear BMPR2 mutations [7]. However, the mechanism by which BMPR2 deficiency causes PAH is under ongoing examination with several pathways and cell types being explored [4,8–11]. In the lung, BMPR2 is highly expressed on the vascular endothelium of pulmonary arteries [9]. BMPR2 is known to interact with the cytoskeleton as it directly binds and modulates proteins related to cytoskeletal organization, including LIM domain kinase (LIMK), light chain of cytoplasmic dynein (TCTEX), and non-receptor tyrosine kinase family (SRC) [12–14], and has been shown to regulate cytoskeletal functions including adhesion [15] and migration [16]. BMPR2 loss has been shown to induce increased endothelial cell permeability leading to increased inflammation, in turn resulting in development of PAH [15]. Cytoskeletal defects are thus broadly seen in PAH patients and could be mechanistically linked to BMPR2 dysfunction. Endothelial cell hyper-proliferation, inhibited apoptosis, and alterations of biochemical-metabolic pathways are the unifying pathobiology of the disease [17]. Several clinical studies have provided evidence that the pulmonary endothelium is essential for the production of various mediators of vascular remodeling including vasoactive peptides, nitric oxide (NO), prostaglandin -I2 (PGI2), endothelin-1 (ET-1), fibroblast growth factor (FGF), angiotensin II (Ang II), cytokines (IL-1, IL-6), and cross talk between endothelial and smooth muscle cells; which are crucial to the development and progression of pulmonary hypertension [11,18].

BMPR2 has been a significant focus of experimental and human PAH over the past several years; however, the signaling of vascular endothelial growth factor (VEGF) and its effects on abnormal angiogenesis is also gaining increasing recognition. VEGF is abundant in the lungs with several functions including maintenance of the pulmonary endothelium [19,20]. VEGF also influences the pulmonary vasculature and aids in the production of nitric oxide (NO) and prostacyclin, which regulates vasoconstriction and dilatation, an important stimulus for PAH development and therapy [19,20]. Abnormal VEGF signaling has been identified in several disease processes including tumor angiogenesis [21] and roles in pulmonary pathologies including obstructive lung diseases [22,23] and lung injury [24]. Patients with PAH have elevated plasma levels of VEGF [25–27] and plexiform lesions of explanted lungs demonstrate increased levels of VEGF [28,29]. Experimental animal models chronically exposed to hypoxia express increased levels of VEGF and develop PAH [30–32]. Both human and experimental studies implicate a role for VEGF in the hyperproliferation of pulmonary arterial cells and hypoxia-induced vascular remodeling; however, the extensive physiologic involvement of VEGF on endothelial cells makes exploration of specific cell signaling in the pathogenesis of PAH challenging. At the molecular level, the proangiogenic effects of VEGF include cellular proliferation, migration, and reorganization of the actin cytoskeleton and some effects have been associated with MAPK/ERK signaling [33,34]. Similarly, the pro-proliferative and apoptosis resistant phenotype of pulmonary artery endothelial cells (HPAEC) in the absence of BMPR2 has been linked to activation of ERK1/2 and p38MAPK [35], creating overlap in the signaling of two important stimuli in the pathobiology of PAH.

The role of endothelial cells has been increasingly recognized in the pathogenesis of PAH, but there is still much left to be explored. Additionally, although increased expression of myosin light chain kinase (MLCK) in pulmonary arterial smooth muscle cells (PASMCs) from patients with PAH compared with controls has been established [36], the same does not hold true for HPAEC. MLCK, a central cytoskeletal regulator is a Ca^{2+}/Calmodulin dependent enzyme encoded by myosin light chain kinase encoding gene (MYLK)that phosphorylates myosin light chain (MLC) and plays a key pathophysiological role in complex diseases including acute lung injury (ALI) and asthma [37]. Garcia et al. first identified a non-muscle MLCK (nmMLCK) that encodes four high molecular weight MLCK isoforms (MLCK1-4) in the pulmonary endothelium [38]. A major distinguishing feature of the two isoforms is that nmMLCK contains an additional 922-amino-acid stretch at the N-terminus that is not present in smooth muscle MLCK (smMLCK). This amino-acid stretch is involved in distinct cellular functions through unique interactions with other contractile proteins and is shown to influence junctional disruption and paracellular gap formation, cell division, proliferation and cell shape [39].

Therefore, the aim of this study is to examine the downstream effects of nmMLCK activation on cellular proliferation and migration in HPAECs in the presence of two pathologically relevant stimuli in the pathogenesis of PAH, namely BMPR2 deficiency and VEGF stimulation. Utilizing cell proliferation and migration, we hypothesize that increased nmMLCK activity leads to increased ERK/MAPK activity, and further suggesting that nmMLCK activity is important in the development and progression of PAH.

2. Results

2.1. Upregulated Protein Expression of nmMLCK and ERK in BMPR2 Silenced Human Pulmonary Artery Endothelial Cells

BMPR2 mutations and the downregulation of BMPR2 expression is known to contribute to the pathogenesis of PAH. As BMPR2 is highly expressed in the endothelium of pulmonary arteries, we chose to conduct our experiments on normal HPAECs. To examine a potential mechanistic relationship between BMPR2 and the cytoskeleton, we silenced BMPR2 in HPAECs and used Western blot analysis to measure protein levels of cytoskeletal regulating pathways. Myosin light chain when phosphorylated leads to activation of the quiescent endothelium and cytoskeletal reorganization defined by stress fiber formation and contractility. ERK is a known upstream regulator of MLCK and phosphorylated-ERK (p-ERK) is the activated form. After 48 h of targeted knockdown, BMPR2 protein levels were reduced by >90% (Figure 1A,B; $p \leq 0.01$). BMPR2 silenced HPAECs demonstrated an approximately 10-fold increase in phosphorylated-myosin light chain (p-MLC) protein expression when compared to control (Figure 1A,C; $p \leq 0.01$). The upregulated expression of nmMLCK was associated with a nearly five-fold increase in expression of p-ERK (Figure 1A,D; $p \leq 0.01$). These results indicate that repression of BMPR2 in HPAECs leads to activation of both nmMLCK and ERK pathways, which are known to play an important role in cytoskeletal regulation.

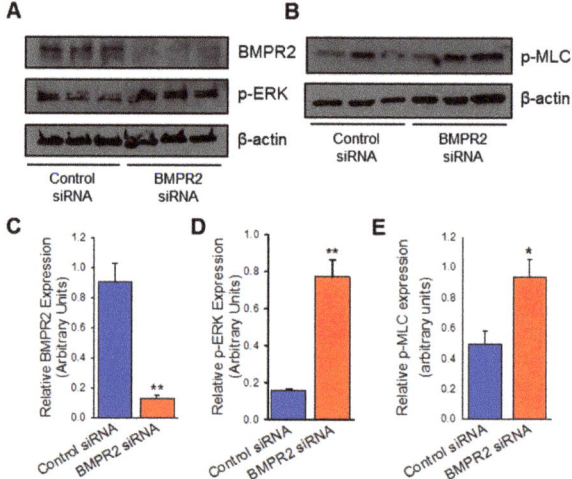

Figure 1. BMPR2 silencing in human pulmonary artery endothelial cells (HPAECs) is associated with ERK/MAPK activation. Representative Western blot images denoting BMPR2 and p-ERK (**A**) as well as p-MLC (**B**) protein expression between HPAECs transfected with BMPR2 siRNA or control siRNA for 48 h. Bar graphs summarizing relative BMPR2 (**C**), phosphorylated ERK (**D**), and phosphorylated MLC (**E**) protein expression at 48 h in BMPR2 silenced HPAECs compared to control (n = 3). * indicates $p < 0.05$; **, $p \leq 0.01$; BMPR2, Bone morphogenetic protein receptor type II; p-ERK, phosphorylated extracellular signal-regulated kinase; p-MLC, phosphorylated myosin light-chain; siRNA, small interfering (silencing) RNA.

2.2. BMPR2 Silencing Induces Increased Endothelial Cell Viability, Proliferation, and Cytokine Release Which Are Attenuated by MLCK Inhibition

Since BMPR2 silencing demonstrated upregulation of cytoskeletal proteins, we next sought to evaluate for the endothelial dysfunction that is characteristic of vascular remodeling. Endothelial dysfunction in PAH leads to a hyperproliferative and apoptosis-resistant phenotype; to evaluate for this, BMPR2 silenced HPAECs were subjected to established measurements of cellular proliferation including Western blot analysis of proliferating cell nuclear antigen (PCNA) protein expression and water-soluble tetrazolium salt-1 (WST-1) assay. PCNA is a component of cell replication machinery and WST-1 measures the activity of cellular mitochondrial dehydrogenases. BMPR2 silenced HPAECs demonstrated over 200% increase in PCNA protein expression (Figure 2A,B; $p \leq 0.05$) and approximately 50% increase in endothelial cell proliferation measured by WST-1 assay when compared to control (Figure 2C; $p \leq 0.01$). To evaluate the potential role of the cytoskeleton in the hyperproliferation of BMPR2 silenced HPAECs, an MLCK specific inhibitor was used to prevent phosphorylation of MLC. Pre-treatment with ML-7 in conjunction with BMPR2 silencing led to a 50% decrease in cell viability and proliferation when compared to BMPR2 silencing alone (Figure 2C; $p \leq 0.01$).

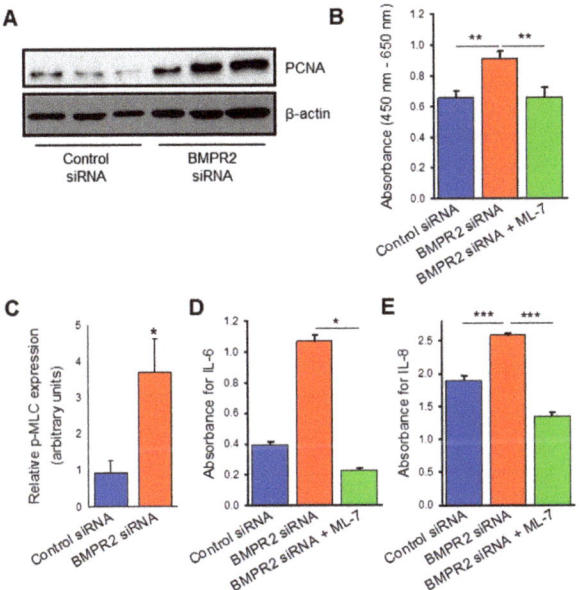

Figure 2. BMPR2 silencing increases HPAEC proliferation and cytokine release which is attenuated by inhibition of myosin light chain kinase (MLCK). Representative Western blot (**A**) and accompanying bar graph (**B**) depicting PCNA protein expression at 48 h between HPAECs transfected with BMPR2 or control siRNA (n = 3). (**C**) Bar graph representing changes in proliferation and viability as measured by WST-1 assay in HPAECs transfected with BMPR2 siRNA, control siRNA, and BMPR2 siRNA with ML-7 (10 µM) pre-treatment for 48 h (n = 9). Bar graph measuring IL-6 (**D**) and IL-8 (**E**) concentrations in the media of HPAECs transfected with BMPR2 siRNA, control siRNA, and BMPR2 siRNA with ML-7 (10 µM) pre-treatment for 48 h (n = 3). * indicates $p < 0.05$; **, $p \leq 0.01$; ***, $p \leq 0.001$; PCNA, proliferating cell nuclear antigen; BMPR2, Bone morphogenetic protein receptor type II; siRNA, small interfering (silencing) RNA; ML-7, myosin light chain kinase specific inhibitor; IL-6, Interleukin-6; IL-8, Interleukin-8.

Endothelial dysfunction was also evaluated in the form of inflammation by measurement of cytokine release. Interleukins 6 and 8 are known to be upregulated in PAH and were measured in our BMPR2 silenced HPAECs by ELISA technique. In Figure 2D,E, IL-6 ($p \leq 0.05$) and IL-8 ($p \leq 0.001$) secretion were both increased in the basal media of BMPR2 silenced HPAECs when compared to control. MLCK specific inhibition with ML-7 pre-treatment prevented the increased cytokine release induced by BMPR2 known down (IL-6 $p < 0.05$; IL-8 $p \leq 0.001$). Taken together, BMPR2 silencing in HPAECs leads to endothelial dysfunction as measured by increased cellular proliferation, viability, and cytokine release, which are all attenuated by inhibition of the nmMLCK pathway.

2.3. BMPR2 Silencing Significantly Increases nmMLCK Dependant HPAEC Migration

Endothelial dysfunction in PAH is also represented by increased and disorganized cellular migration. We next sought to evaluate migration in our BMPR2 silenced HPAECs and how cellular motility is affected by MLCK inhibition. To measure migration, the cell monolayer is disrupted, and the rate of recovery is measured; electrical cell impedance sensing (ECIS) wound assay and scratch assay are two established modalities for measuring isolated cell migration. BMPR2 transfected HPAECs were grown to confluence on electrodes, subjected to ECIS-based wounding, and transendothelial resistance (TER) was measured over time, representing the rate of recovery. In Figure 3A, BMPR2 silenced HPAECs demonstrate a trend toward increased rate of recovery after wounding and increased area under the curve when compared to control. Pre-treatment with MLCK inhibitor in conjunction with BMPR2 silencing reduces HPAEC migration ($p \leq 0.01$). Cellular migration was also evaluated by wound healing scratch assay. By this method, a scratch is created in the monolayer of HPAECs with respective treatments, the size of the created wound is measured over time, and the rate of wound closure is representative of cellular migration. BMPR2 transfected HPAECs demonstrated a nonsignificant trend towards increased wound closure at 24 h (Figure 3C,D). Pre-treatment with ML-7 in BMPR2 transfected HPAECs had a significant reduction in endothelial cell motility and migration when compared to BMPR2 transfected HPAECs alone (Figure 3C,D; $p < 0.05$). Taken together, MLCK inhibition has a significant effect on migration in BMPR2 deficient HPAECs.

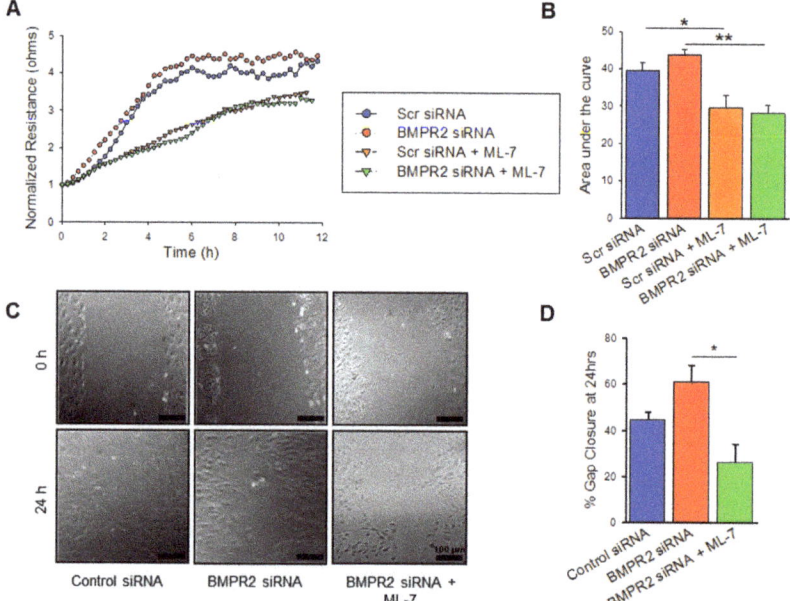

Figure 3. BMPR2 silencing increases HPAEC migration which is attenuated by MLCK inhibition. (**A**) Plot demonstrating the transendothelial resistance by ECIS-based wounding over time in HPAECs transfected with BMPR2 siRNA, control siRNA, BMPR2 siRNA with ML-7 (10 μM), and control siRNA with ML-7 (10 μM) ($n = 2$). (**B**) Bar graph denoting the area under the curve measurements at 12 h for the ECIS-based wounding experiments ($n = 2$). (**C**) Representative images of wound healing assay depicting scratches created in confluent cultures of HPAECs transfected with BMPR2 siRNA, control siRNA, and BMPR2 siRNA with ML-7 (10 μM) at 0 h and 24 h time points; scale bar = 100 μm. (**D**) Bar graph summarizing percent gap closure at 24 h for the wound healing assay experiments ($n = 3$). * indicates $p < 0.05$; **, $p \leq 0.01$; BMPR2, Bone morphogenetic protein receptor type II; siRNA, small interfering (silencing) RNA; ML-7, myosin light chain kinase specific inhibitor.

2.4. VEGF Treatment Leads to Increased nmMLCK Activity

BMPR2 deficiency in HPAECs led to endothelial dysfunction seen in PAH characterized by hyperproliferation and cytokine release; these processes were dependent on the nmMLCK pathway, as specific inhibition negated these adverse effects. To further validate the importance of nmMLCK activation on the early cellular mechanisms that lead to the development of PAH, VEGF stimulation was also explored in a similar context. HPAECs were treated with VEGF for 72 h and demonstrated increased expression of p-MLC when compared to vehicle control, consistent with increased nmMLCK activity (Figure 4A,B; $p \leq 0.01$). Pre-treatment with MLCK specific inhibitor, ML-7 on VEGF-treated HPAECs decreased expression of p-MLC (Figure 4A,B; $p \leq 0.001$). Similarly, VEGF treated HPAECs demonstrated upregulated p-ERK (Figure 4A,C) with decreased expression after ML-7 pretreatment (Figure 4A,C; $p \leq 0.001$).

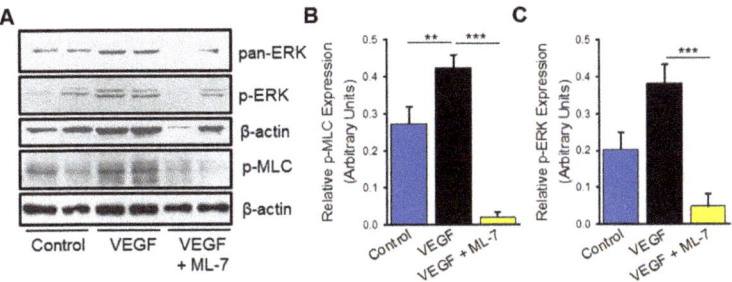

Figure 4. VEGF stimulation in HPAECs leads to upregulation of ERK/MAPK, which is abrogated by MLCK inhibition. (**A**) Representative Western blot images denoting p-ERK and p-MLC protein expression in HPAECs treated with VEGF (100 ng/mL), control (PBS vehicle), and VEGF with ML-7 (10 μM) pre-treatment for 72 h. Bar graphs summarizing relative phosphorylated ERK (**B**) and phosphorylated MLC (**C**) protein expression at 72 h in HPAECs treated with VEGF, control, and VEGF with ML-7 (n = at least 4 for p-ERK, n = at least 7 for p-MLC). ** indicates $p \leq 0.01$; ***, $p \leq 0.001$; VEGF, vascular endothelial growth factor; p-ERK, phosphorylated extracellular signal-regulated kinase; p-MLC, phosphorylated myosin light-chain; ML-7, myosin light chain kinase specific inhibitor.

2.5. VEGF Treatment Increases Endothelial Cell Proliferation and Cytokine Release Which Is Negated by MLCK Inhibition

Endothelial dysfunction measured by hyperproliferation and inflammation was also measured in VEGF stimulated HPAECs. After 72 h, VEGF-treated HPAECs demonstrated a non-significant trend toward increased cellular proliferation compared to vehicle control measured by PCNA expression (Figure 5A,B). Significantly increased cellular viability and proliferation was measured by WST assay in VEGF-treated HPAECs compared to control (Figure 5C; $p \leq 0.001$). ML-7 pre-treatment reduced hyperproliferation and viability in VEGF-treated cells when compared to VEGF treatment alone both by PCNA expression (Figure 5A,B; $p \leq 0.01$) and cellular proliferation (WST-1) assay (Figure 5C; $p \leq 0.001$). VEGF stimulation leads to a hyperproliferative phenotype in HPAECs which is mediated by the nmMLCK pathway.

2.6. VEGF Treatment Significantly Increases nmMLCK Dependant HPAEC Migration

Lastly, VEGF-treated cells were also measured for hypermigration measured by both ECIS-based wounding and traditional scratch assay. VEGF treatment enhanced HPAEC migration compared to control as shown in Figure 6A, and pre-treatment with ML-7 abrogated this increased migration. Area under the curve analysis showed a significant increase in cellular motility in VEGF treated HPAECs compared to control (Figure 6B; $p \leq 0.01$), and pre-treatment with ML-7 decreased the HPAEC migration noted in VEGF treatment alone ($p < 0.05$). These findings were confirmed with scratch assay, which demonstrated increased wound gap closure after 24 h in VEGF treatment compared to control (Figure 6C,D; $p < 0.05$), and similar attenuation of migration with MLCK specific inhibition (Figure 6C,D; $p \leq 0.001$). Overall, these results confirm that in addition to BMPR2 silencing, HPAECs treated with VEGF, another known PAH-inducing cellular mechanism, leads to endothelial dysfunction and hyper-migration mediated by the nmMLCK pathway.

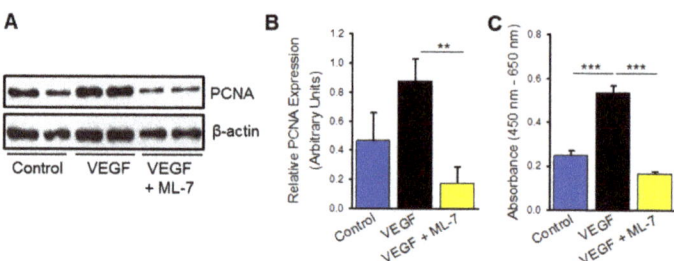

Figure 5. VEGF treatment increases HPAEC proliferation and cytokine release which is attenuated by inhibition of myosin light chain kinase (MLCK). Representative Western blot (**A**) and accompanying bar graph (**B**) depicting PCNA protein expression at 72 h between HPAECs treated with VEGF (100 ng/mL), control (PBS vehicle), and VEGF with ML-7 (10 µM) pre-treatment (n = at least 2). β-actin loading control is identical to Figure 4A as the representative blot was derived from the same experiment. (**C**) Bar graph denoting changes in proliferation and viability as measured by WST-1 assay in HPAECs treated with VEGF, control, and VEGF with ML-7 for 72 h (n = 4). ** indicated $p \leq 0.01$; *** indicates $p \leq 0.001$; PCNA, proliferating cell nuclear antigen; VEGF, vascular endothelial growth factor; ML-7, myosin light chain kinase specific inhibitor; IL-6, Interleukin-6; IL-8, Interleukin-8.

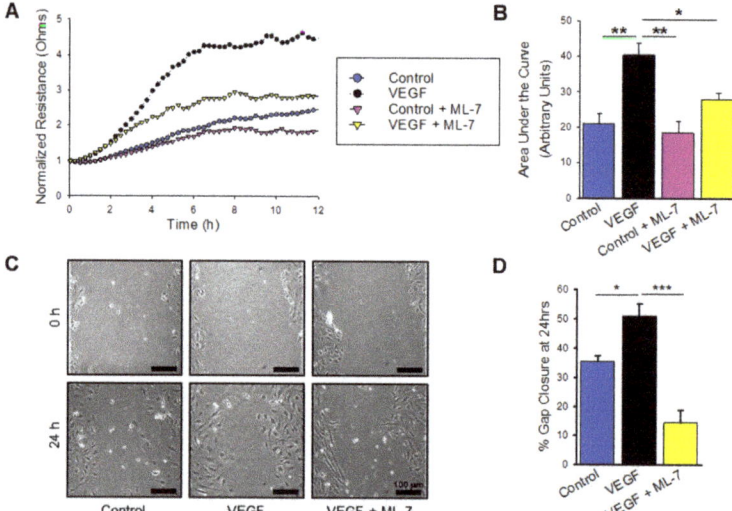

Figure 6. VEGF treatment increases HPAEC migration which is attenuated by MLCK inhibition. (**A**) Plot demonstrating the transendothelial resistance by ECIS-based wounding over time in HPAECs treated with VEGF (100 ng/mL), control (PBS vehicle), VEGF with ML-7 (10 µM) pre-treatment, and control with ML-7 (n = at least 3). (**B**) Bar graph denoting the area under the curve measurements at 12 h for the ECIS-based wounding experiments (n = at least 3). (**C**) Representative images of wound healing assays depicting scratches created in confluent cultures of HPAECs treated with VEGF (100 ng/mL), control (PBS vehicle), VEGF with ML-7 (10 µM) pre-treatment at 0 h and 24 h time points; scale bar = 100 µm. (**D**) Bar graph summarizing percent gap closure at 24 h for the wound healing assay experiments (n = 2). * indicates $p \leq 0.05$; **, $p \leq 0.01$; ***, $p \leq 0.001$ VEGF, vascular endothelial growth factor; ML-7, myosin light chain kinase specific inhibitor.

3. Discussion

Human pulmonary endothelial cells from patients with idiopathic pulmonary arterial hypertension are known to grow faster in culture due to both increased proliferation and resistance to apoptosis [40]. These cells also demonstrate increased cellular migration due to endothelial dysfunction [41]. BMPR2 silencing in pulmonary endothelial cells has reiterated the cellular dysfunction noted in PAH pathogenesis [10,42–46]. BMP/BMPR2 effects mediate SMAD 1/5/8 phosphorylation leading to characteristic cellular phenotype observed in PAH [47]. There are SMAD independent pathways such as Wnt/catenin and PPAR which also regulate cell signaling [48,49].

Pulmonary arterial endothelial cell proliferation and migration play a pivotal role in PAH pathogenesis. In this study, we identified that nmMLCK upregulation in human pulmonary endothelial cells is associated with many of the abnormalities and pathogenic mechanisms observed in PAH. The effect of nmMLCK in driving PAH pathogenesis is further strengthened by the loss of this effect when MLCK is inhibited which we have also shown to have effect on the ERK pathway. Human PAEC transfected with BMPR2 siRNA or treated with VEGF showed increased proliferation both by WST assay and protein expression when compared to scrambled siRNA or control conditions. Increased migration was observed after VEGF treatment in both cell scratch and ECIS wound healing assays. Treatment of BMPR2 silenced and VEGF treated HPAECs with the MLCK specific inhibitor (ML-7) resulted in decreased proliferation and migration further promoting the novel identification of nmMLCK playing an important role in PAH pathogenesis. The upregulated nmMLCK expression was associated with increased ERK production in endothelial cells as a likely mechanism of increased proliferation and migration under BMPR2 deficiency and VEGF stimulation. This work demonstrates that non-muscle MLCK is likely an important contributor to endothelial dysfunction recognized in PAH and may represent a unique therapeutic target (Figure 7).

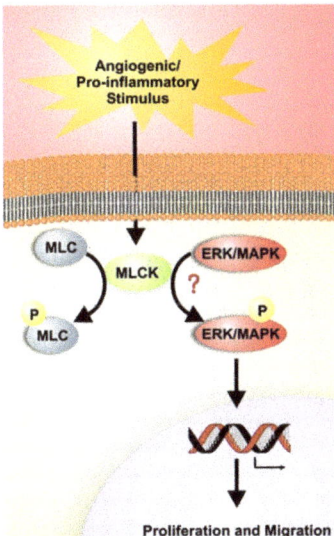

Figure 7. Schematic representation of our findings demonstrating that nmMLCK activation due to BMPR2 downregulation or VEGF stimulation is associated with increased ERK phosphorylation, potentially directly or indirectly, contributing to the pathogenesis of PAH by stimulating cellular proliferation and migration.

The human MYLK gene spanning 217.6 kb on chromosome 3q21.1 encodes three isoforms including non-muscle MLCK isoform (nmMLCK), smooth muscle isoform (smMLCK) and telokin (KRP), a small myosin filament-binding protein [50]. Vascular endothelial cells, only express the non-muscle MLCK isoform, which contains a novel NH2-terminus stretch (amino acid 1–922) not present in the open reading frame of smooth muscle MLCK [39,51]. Furthermore, the chromosome location of MYLK (3q21) is an active site for several inflammatory disorders including asthma, allergic rhinitis, COPD and atopic dermatitis. Both smMLCK and nmMLCK phosphorylate myosin light chains to regulate cellular contraction and relaxation along with barrier function in turn playing an important role in the pathogenesis of various disease processes including asthma and acute lung injury [38,52,53]. Since the identification of non-muscle MLCK, various studies have elucidated that nmMLCK is unique in structure [38].

We explored the significance of BMPR2 signaling involvement in cytoskeletal structure and function and its link to nmMLCK. Non muscle isoform of MLCK has itself been described as vital in the rapid dynamic coordination of the cytoskeleton involved in cancer cell proliferation and migration in ways similar to the tumor like growth of pulmonary endothelial cells in PAH [54]. Previous literature has demonstrated that both patients with PAH and endothelial cell models have increased levels of ERK [55]. Recently, Awad et al. show that Raf family members and ERK1/2 are activated after BMPR2 knockdown [35]. We were able to show that BMPR2 silencing in HPAECs is linked to increased expression of nmMLCK along with an increase ERK phosphorhylation indicative of a potential association between ERK, BMPR2 and nmMLCK. In previous studies, VEGF stimulation has led to increased angiogenesis in endothelial cells through various mechanisms [56–58]. It is well recognized that VEGF expression is elevated in arterial cells of the characteristic plexogenic lesions of patients with advanced pulmonary hypertension [59]. Similarly, cell proliferation and migration were both ascertained to be significantly increased in our in vitro model when VEGF was employed as an angiogenic stimulus. Our data provides insight into cellular migration enhanced as a function of cytoskeletal reorganization mediated by nmMLCK activation in the presence of VEGF stimulation in HPAECs.

ML-7, a selective inhibitor of MLCK, acts on the adenosine triphosphate (ATP)-binding site of the active center of MLCK [60–62]. We were able to validate our results with the decreased activity of nmMLCK by showing reduced expression of p-MLC with the use of ML-7 along with a concurrent decrease in ERK/MAPK pathway. Hence, establishing the link between nmMLCK and ERK in the various processes of cell proliferation and migration which are the cornerstone of PAH pathogenesis. In the future, exploring the transcriptional link between the two and further corroborating findings with additional nmMLCK inhibitors will be beneficial. Previous work from our group with use of MLCK specific inhibition has also implicated the cytoskeleton and prevention of MLC phosphorylation in hemin-induced endothelial dysfunction in the context of PH due to chronic hemolysis [60]. These two studies, in conjunction, support the need for further exploration of MLCK inhibition as a potential therapy for PAH.

This work is limited to the use of cell models that recapitulate the endothelial dysfunction that is seen during vascular remodeling in the pathogenesis of PAH. We utilized cell models that we think have strong relevance to human disease, particularly with reduction in BMPR2 levels as seen in numerous PAH patient subgroups as well as stimulation of the cells with a growth factor, VEGF, that is known to be elevated in patients with some forms of PAH [15,25–27]. Further study will require the use of experimental animal models of pulmonary hypertension as well as cells and tissues from human PAH in order to better understand the role of nmMLCK in endothelial dysfunction as well as development and progression of human PAH.

In summary, the data from this study is the first in our knowledge to recognize the importance of increased expression of nmMLCK contributing to endothelial cellular proliferation and migration with downstream activation of Ras/Raf/ERK pathway. The enhanced nmMLCK activity appears to play a crucial role in the pathobiology of PAH. The nmMLCK-Raf/ERK link presents a novel pathway for development of more efficient potential targets for treatment of pulmonary hypertension.

4. Materials and Methods

4.1. Cell Culture

Normal HPAECs were cultured at passages 5–8 in endothelial basal medium-2 (EBM-2) supplemented with growth factors [endothelial growth medium (EGM)-2 Single Quot kit from Lonza (Basel, Switzerland)] and containing 10% fetal bovine serum (FBS). Cells were maintained at 37 °C in a humidified incubator with 5% CO_2 and 95% air. Primary HPAECs were seeded at a density of 180,000–200,000 cells/well in 6-well plates for RNA and protein analysis, respectively. For all experiments, basal media was replaced with EGM-2 containing 2% FBS with added treatment or control conditions as described in each individual experiment.

4.2. BMPR2 Transfection

HPAECs were transfected with gene-specific siRNA pools targeting BMPR2 non-specific siRNA (Ambion, Austin, TX, USA) at a final concentration based on the culture vessel surface area per XFECT-1 (Clontech, Mountain View, CA, USA) protocol for 4 h followed by growth in EGM-2 containing 2% charcoal-stripped serum. Nontargeting siRNA pool-1 (siGENOME, Dharmacon, Lafayette, CO, USA) was used as a control. After incubating for an additional 46 h (48 h from the start of transfection), total RNA or protein lysates were collected for Western blot analysis. BMPR2 silencing was repeated with addition of nmMLCK inhibitor, ML-7 hydrochloride [1-(5-iodonaphthalene-1-sulfonyl)-[1H]-hexahydro-1,4-diazepine hydrochloride; (Tocris Bioscience, Bristol, UK) at 10 µM at 4 h time point followed by incubation for 42 h.

4.3. Western Blotting

Whole cell protein lysates were isolated from HPAEC with RIPA buffer (Millipore, Burlington, MA, USA) supplemented with protease and phosphatase inhibitor cocktail (Thermo Fisher Scientific, Waltham, MA, USA). Protein lysates were resolved by SDS-PAGE and transferred to a nitrocellulose membrane with Bio-Rad Laboratories (Hercules, CA, USA) Western blotting system. Membranes were incubated with 5% blocking solution (non-fat milk) in PBS-Tween 20 (0.1%) for 1 h at room temperature and then incubated overnight at 4 °C with primary antibody. The following day, membranes were incubated in secondary antibodies and then visualized with chemiluminescence (Thermo Fisher Scientific). Antibodies against p-MLC (1:1000), p-ERK (1:1000), total ERK (1:1000), and proliferating cell nuclear antigen (PCNA) (1:500) were purchased from Proteintech Group Inc (Rosemont, IL, USA). Anti-BMPR2 antibody (1:500) was obtained from Abcam (Cambridge, MA, USA). Quantitative data was obtained using Image J (Version 1.46r, National Institutes of Health, Bethesda, MA, USA) and data is presented as mean relative protein expression ± standard error (SE). n is defined as protein lysates extracted from a single well for a given condition.

4.4. Cell Proliferation and Viability Assay

HPAECs were plated at $0.1–5 \times 10^4$/well in a 96-well microplate with a final volume of 100 µL/well. Cell proliferation and viability was assessed after treatment with addition of 10 µL/well WST-1/ECS (Millipore Sigma) solution to each well for 4 h. Cell proliferation in this assay is based on cleavage of the tetrazolium salt WST-1 to formazan by cellular mitochondrial dehydrogenases and an increase in formazan dye. Cell proliferation was quantified by a multi well spectrophotometer (microplate reader) by measuring the absorbance of dye solution at 450 nm and reference wavelength 650 nm. Quantitative data is presented as mean ± SE. n is defined a single well for a given condition.

4.5. Enzyme Linked Immunosorbent Assay (ELISA)

HPAECs were grown on 6-well plates and treated with respective conditions once confluent. The cell culture media was removed and centrifuged and the supernatant was used for ELISA analysis. Commercially available sandwich ELISA kits were purchased from BioLegend (San Diego, CA, USA) and measurement of IL-6 and IL-8 were obtained according to manufacturers instructions. A microplate reader was used to measure corresponding absorbance. Data is presented as mean ± SE. n is defined a single well for a given condition.

4.6. Electric Cell-Substrate Impedance Sensing (ECIS) Wound Assay

ECIS system (Applied Biophysics, Inc., USA) and 8W1E well arrays were used to detect and track HPAECs migration during electrical wound-healing assay. HPAECs were seeded at a density of 200,000–250,000 cells/well and treated with respective conditions 24 h after attachment. ECIS plate was then transferred to the humidified 5% CO_2 incubator at 37 °C and wells were allowed to equilibrate in the incubator. Wound was then applied at 40,000 Hz and 3 mA for 10 s per well. Over the next 24 h, the resistance from each well was measured every 15 min and then analyzed. n is defined as each independent well for a given condition. Area under the curve (AUC) was calculated for each condition and presented as mean ± SE.

4.7. Scratch Assay

Cell motility was assessed with cell scratch assay [61]. The cell monolayer was scratched with a sterile P20 pipette tip and the debris was removed by washing twice with warm working media, and then replaced with fresh working medium after which images were captured at 0 and 24 h with inverted microscope with a digital camera (Nikon Eclipse TE2000-s, Nikon Instruments Inc., Melville, NY, USA) at 4× g or 10× g magnification [62]. Quantification for gap closure was completed using Image J software, as described previously [63] and data is presented as mean ± SE. n is defined a single well for a given condition.

4.8. Statistical Analysis

SigmaPlot software (v14, Systat Software Inc., Palo Alto, CA, USA) was used for Student's t-test to calculate significance between two groups and One-Way ANOVA for many groups. Composite data are shown as the mean ± standard error. A p value of ≤ 0.05 was considered significant.

Author Contributions: Conceptualization, M.A. and D.R.F.; methodology, M.A., J.G., R.H., N.B. and Y.E.; software, D.R.F.; validation, M.A., J.G. and D.R.F.; formal analysis, M.A., J.G., R.H., N.B. and Y.E.; investigation, M.A., J.G., R.H., N.B., C.H., R.D. and Y.E.; resources, J.R.J. and D.R.F.; data curation, J.G. and D.R.F.; writing—original draft preparation, M.A. and J.G.; writing—review and editing, M.A., J.G., R.H., N.B., C.H., R.D., Y.E., J.R.J. and D.R.F.; visualization, J.G. and D.R.F.; supervision, J.R.J. and D.R.F.; funding acquisition, D.R.F. All authors have read and agreed to the published version of the manuscript.

Funding: This research was supported in part by the grants from the National Heart, Lung, and Blood Institute of the National Institutes of Health (HL133474 and HL144909).

Institutional Review Board Statement: Not applicable.

Informed Consent Statement: Not applicable.

Data Availability Statement: Not applicable.

Conflicts of Interest: The authors declare no conflict of interest.

References

1. Hoeper, M.M.; Bogaard, H.J.; Condliffe, R.; Frantz, R.; Khanna, D.; Kurzyna, M.; Langleben, D.; Manes, A.; Satoh, T.; Torres, F.; et al. Definitions and diagnosis of pulmonary hypertension. *J. Am. Coll. Cardiol.* **2013**, *62*, D42–D50. [CrossRef] [PubMed]
2. Chang, K.Y.; Duval, S.; Badesch, D.B.; Bull, T.M.; Chakinala, M.M.; de Marco, T.; Frantz, R.P.; Hemnes, A.; Mathai, S.C.; Rosenzweig, E.B.; et al. Mortality in pulmonary arterial hypertension in the modern era: Early insights from the pulmonary hypertension association registry. *J. Am. Heart Assoc.* **2022**, *11*, e024969. [CrossRef] [PubMed]

3. Ferreira, A.J.; Shenoy, V.; Yamazato, Y.; Sriramula, S.; Francis, J.; Yuan, L.; Castellano, R.K.; Ostrov, D.A.; Oh, S.P.; Katovich, M.J.; et al. Evidence for angiotensin-converting enzyme 2 as a therapeutic target for the prevention of pulmonary hypertension. *Am. J. Respir. Crit. Care Med.* **2009**, *179*, 1048–1054. [CrossRef]
4. International PPH Consortium; Lane, K.B.; Machado, R.D.; Pauciulo, M.W.; Thomson, J.R.; Phillips, J.A., 3rd; Loyd, J.E.; Nichols, W.C.; Trembath, R.C. Heterozygous germline mutations in BMPR2, encoding a TGF-beta receptor, cause familial primary pulmonary hypertension. *Nat. Genet.* **2000**, *26*, 81–84.
5. Lane, K.B.; Blackwell, T.R.; Runo, J.; Wheeler, L.; Phillips, J.A., 3rd; Loyd, J.E. Aberrant signal transduction in pulmonary hypertension. *Chest* **2005**, *128*, 564S–565S. [CrossRef] [PubMed]
6. Machado, R.D.; Aldred, M.A.; James, V.; Harrison, R.E.; Patel, B.; Schwalbe, E.C.; Gruenig, E.; Janssen, B.; Koehler, R.; Seeger, W.; et al. Mutations of the TGF-beta type II receptor BMPR2 in pulmonary arterial hypertension. *Hum. Mutat.* **2006**, *27*, 121–132. [CrossRef]
7. Rajkumar, R.; Konishi, K.; Richards, T.J.; Ishizawar, D.C.; Wiechert, A.C.; Kaminski, N.; Ahmad, F. Genomewide RNA expression profiling in lung identifies distinct signatures in idiopathic pulmonary arterial hypertension and secondary pulmonary hypertension. *Am. J. Physiol. Heart Circ. Physiol.* **2010**, *298*, H1235–H1248. [CrossRef]
8. West, J.; Harral, J.; Lane, K.; Deng, Y.; Ickes, B.; Crona, D.; Albu, S.; Stewart, D.; Fagan, K. Mice expressing BMPR2R899X transgene in smooth muscle develop pulmonary vascular lesions. *Am. J. Physiol.-Lung Cell Mol. Physiol.* **2008**, *295*, L744–L755. [CrossRef]
9. Atkinson, C.; Stewart, S.; Upton, P.D.; Machado, R.; Thomson, J.R.; Trembath, R.C.; Morrell, N.W. Primary pulmonary hypertension is associated with reduced pulmonary vascular expression of type II bone morphogenetic protein receptor. *Circulation* **2002**, *105*, 1672–1678. [CrossRef]
10. Thomson, J.R.; Machado, R.D.; Pauciulo, M.W.; Morgan, N.V.; Humbert, M.; Elliott, G.C.; Ward, K.; Yacoub, M.; Mikhail, G.; Rogers, P.; et al. Sporadic primary pulmonary hypertension is associated with germline mutations of the gene encoding BMPR-II, a receptor member of the TGF-beta family. *J. Med. Genet.* **2000**, *37*, 741–745. [CrossRef]
11. Humbert, M.; Morrell, N.W.; Archer, S.L.; Stenmark, K.R.; MacLean, M.R.; Lang, I.M.; Christman, B.W.; Weir, E.K.; Eickelberg, O.; Voelkel, N.F.; et al. Cellular and molecular pathobiology of pulmonary arterial hypertension. *J. Am. Coll. Cardiol.* **2004**, *43*, 13S–24S. [CrossRef] [PubMed]
12. Foletta, V.C.; Lim, M.A.; Soosairajah, J.; Kelly, A.P.; Stanley, E.G.; Shannon, M.; He, W.; Das, S.; Massague, J.; Bernard, O. Direct signaling by the BMP type II receptor via the cytoskeletal regulator LIMK1. *J. Cell Biol.* **2003**, *162*, 1089–1098. [CrossRef] [PubMed]
13. Machado, R.D.; Rudarakanchana, N.; Atkinson, C.; Flanagan, J.A.; Harrison, R.; Morrell, N.W.; Trembath, R.C. Functional interaction between BMPR-II and Tctex-1, a light chain of Dynein, is isoform-specific and disrupted by mutations underlying primary pulmonary hypertension. *Hum. Mol. Genet.* **2003**, *12*, 3277–3286. [CrossRef] [PubMed]
14. Wong, W.K.; Knowles, J.A.; Morse, J.H. Bone morphogenetic protein receptor type II C-terminus interacts with c-Src: Implication for a role in pulmonary arterial hypertension. *Am. J. Respir. Cell Mol. Biol.* **2005**, *33*, 438–446. [CrossRef]
15. Burton, V.J.; Ciuclan, L.I.; Holmes, A.M.; Rodman, D.M.; Walker, C.; Budd, D.C. Bone morphogenetic protein receptor II regulates pulmonary artery endothelial cell barrier function. *Blood* **2011**, *117*, 333–341. [CrossRef]
16. Gamell, C.; Osses, N.; Bartrons, R.; Ruckle, T.; Camps, M.; Rosa, J.L.; Ventura, F. BMP2 induction of actin cytoskeleton reorganization and cell migration requires PI3-kinase and Cdc42 activity. *J. Cell Sci.* **2008**, *121 Pt 23*, 3960–3970. [CrossRef]
17. Xu, W.; Erzurum, S.C. Endothelial cell energy metabolism, proliferation, and apoptosis in pulmonary hypertension. *Compr. Physiol.* **2011**, *1*, 357–372.
18. Guignabert, C.; Tu, L.; Girerd, B.; Ricard, N.; Huertas, A.; Montani, D.; Humbert, M. New molecular targets of pulmonary vascular remodeling in pulmonary arterial hypertension: Importance of endothelial communication. *Chest* **2015**, *147*, 529–537. [CrossRef]
19. Voelkel, N.F.; Vandivier, R.W.; Tuder, R.M. Vascular endothelial growth factor in the lung. *Am. J. Physiol.-Lung Cell. Mol. Physiol.* **2006**, *290*, L209–L221. [CrossRef]
20. Voelkel, N.F.; Gomez-Arroyo, J. The role of vascular endothelial growth factor in pulmonary arterial hypertension. The angiogenesis paradox. *Am. J. Respir. Cell Mol. Biol.* **2014**, *51*, 474–484. [CrossRef]
21. Claesson-Welsh, L.; Welsh, M. VEGFA and tumour angiogenesis. *J. Intern. Med.* **2013**, *273*, 114–127. [CrossRef] [PubMed]
22. Tuder, R.M.; Zhen, L.; Cho, C.Y.; Taraseviciene-Stewart, L.; Kasahara, Y.; Salvemini, D.; Voelkel, N.F.; Flores, S.C. Oxidative stress and apoptosis interact and cause emphysema due to vascular endothelial growth factor receptor blockade. *Am. J. Respir. Cell Mol. Biol.* **2003**, *29*, 88–97. [CrossRef] [PubMed]
23. Asai, K.; Kanazawa, H.; Kamoi, H.; Shiraishi, S.; Hirata, K.; Yoshikawa, J. Increased levels of vascular endothelial growth factor in induced sputum in asthmatic patients. *Clin. Exp. Allergy* **2003**, *33*, 595–599. [CrossRef] [PubMed]
24. Thickett, D.R.; Armstrong, L.; Christie, S.J.; Millar, A.B. Vascular endothelial growth factor may contribute to increased vascular permeability in acute respiratory distress syndrome. *Am. J. Respir. Crit. Care Med.* **2001**, *164*, 1601–1605. [CrossRef]
25. Kumpers, P.; Nickel, N.; Lukasz, A.; Golpon, H.; Westerkamp, V.; Olsson, K.M.; Jonigk, D.; Maegel, L.; Bockmeyer, C.L.; David, S.; et al. Circulating angiopoietins in idiopathic pulmonary arterial hypertension. *Eur. Heart J.* **2010**, *31*, 2291–2300. [CrossRef]
26. Papaioannou, A.I.; Zakynthinos, E.; Kostikas, K.; Kiropoulos, T.; Koutsokera, A.; Ziogas, A.; Koutroumpas, A.; Sakkas, L.; Gourgoulianis, K.I.; Daniil, Z.D. Serum VEGF levels are related to the presence of pulmonary arterial hypertension in systemic sclerosis. *BMC Pulm. Med.* **2009**, *9*, 18. [CrossRef]

27. Selimovic, N.; Bergh, C.H.; Andersson, B.; Sakiniene, E.; Carlsten, H.; Rundqvist, B. Growth factors and interleukin-6 across the lung circulation in pulmonary hypertension. *Eur. Respir. J.* **2009**, *34*, 662–668. [CrossRef]
28. Tuder, R.M.; Chacon, M.; Alger, L.; Wang, J.; Taraseviciene-Stewart, L.; Kasahara, Y.; Cool, C.D.; Bishop, A.E.; Geraci, M.; Semenza, G.L.; et al. Expression of angiogenesis-related molecules in plexiform lesions in severe pulmonary hypertension: Evidence for a process of disordered angiogenesis. *J. Pathol.* **2001**, *195*, 367–374. [CrossRef]
29. Hirose, S.; Hosoda, Y.; Furuya, S.; Otsuki, T.; Ikeda, E. Expression of vascular endothelial growth factor and its receptors correlates closely with formation of the plexiform lesion in human pulmonary hypertension. *Pathol. Int.* **2000**, *50*, 472–479. [CrossRef]
30. Tuder, R.M.; Flook, B.E.; Voelkel, N.F. Increased gene-expression for Vegf and the Vegf Receptors Kdr/Flk and Flt in lungs exposed to acute or to chronic hypoxia. Modulation of gene-expression by nitric-oxide. *J. Clin. Investig.* **1995**, *95*, 1798–1807. [CrossRef]
31. Partovian, C.; Adnot, S.; Eddahibi, S.; Teiger, E.; Levame, M.; Dreyfus, P.; Raffestin, B.; Frelin, C. Heart and lung VEGF mRNA expression in rats with monocrotaline- or hypoxia-induced pulmonary hypertension. *Am. J. Physiol.-Heart Circ. Physiol.* **1998**, *275*, H1948–H1956. [CrossRef] [PubMed]
32. Christou, H.; Yoshida, A.; Arthur, V.; Morita, T.; Kourembanas, S. Increased vascular endothelial growth factor production in the lungs of rats with hypoxia-induced pulmonary hypertension. *Am. J. Respir. Cell Mol. Biol.* **1998**, *18*, 768–776. [CrossRef] [PubMed]
33. Tan, W.H.; Popel, A.S.; Gabhann, F.M. Computational model of VEGFR2 pathway to ERK activation and modulation through receptor trafficking. *Cell. Signal.* **2013**, *25*, 2496–2510. [CrossRef] [PubMed]
34. Song, M.; Finley, S.D. Mechanistic insight into activation of MAPK signaling by pro-angiogenic factors. *BMC Syst. Biol.* **2018**, *12*, 1–17. [CrossRef]
35. Awad, K.S.; Elinoff, J.M.; Wang, S.; Gairhe, S.; Ferreyra, G.A.; Cai, R.; Sun, J.; Solomon, M.A.; Danner, R.L. Raf/ERK drives the proliferative and invasive phenotype of BMPR2-silenced pulmonary artery endothelial cells. *Am. J. Physiol.-Lung Cell Mol. Physiol.* **2016**, *310*, L187–L201. [CrossRef]
36. Barnes, E.A.; Chen, C.H.; Sedan, O.; Cornfield, D.N. Loss of smooth muscle cell hypoxia inducible factor-1alpha underlies increased vascular contractility in pulmonary hypertension. *FASEB J.* **2017**, *31*, 650–662. [CrossRef]
37. Wang, T.; Zhou, T.; Saadat, L.; Garcia, J.G. A MYLK variant regulates asthmatic inflammation via alterations in mRNA secondary structure. *Eur. J. Hum. Genet.* **2015**, *23*, 874–876. [CrossRef]
38. Garcia, J.G.; Davis, H.W.; Patterson, C.E. Regulation of endothelial cell gap formation and barrier dysfunction: Role of myosin light chain phosphorylation. *J. Cell Physiol.* **1995**, *163*, 510–522. [CrossRef]
39. Lazar, V.; Garcia, J.G. A single human myosin light chain kinase gene (MLCK; MYLK) transcribes multiple nonmuscle isoforms. *Genomics* **1999**, *57*, 256–267. [CrossRef]
40. Sawada, H.; Saito, T.; Nickel, N.P.; Alastalo, T.P.; Glotzbach, J.P.; Chan, R.; Haghighat, L.; Fuchs, G.; Januszyk, M.; Cao, A.; et al. Reduced BMPR2 expression induces GM-CSF translation and macrophage recruitment in humans and mice to exacerbate pulmonary hypertension. *J. Exp. Med.* **2014**, *211*, 263–280. [CrossRef]
41. Diebold, I.; Hennigs, J.K.; Miyagawa, K.; Li, C.G.; Nickel, N.P.; Kaschwich, M.; Cao, A.; Wang, L.; Reddy, S.; Chen, P.I.; et al. BMPR2 preserves mitochondrial function and DNA during reoxygenation to promote endothelial cell survival and reverse pulmonary hypertension. *Cell Metab.* **2015**, *21*, 596–608. [CrossRef] [PubMed]
42. de Jesus Perez, V.A.; Alastalo, T.P.; Wu, J.C.; Axelrod, J.D.; Cooke, J.P.; Amieva, M.; Rabinovitch, M. Bone morphogenetic protein 2 induces pulmonary angiogenesis via Wnt-beta-catenin and Wnt-RhoA-Rac1 pathways. *J. Cell Biol.* **2009**, *184*, 83–99. [CrossRef] [PubMed]
43. Alastalo, T.P.; Li, M.; Vde, J.P.; Pham, D.; Sawada, H.; Wang, J.K.; Koskenvuo, M.; Wang, L.; Freeman, B.A.; Chang, H.Y.; et al. Disruption of PPARgamma/beta-catenin-mediated regulation of apelin impairs BMP-induced mouse and human pulmonary arterial EC survival. *J. Clin. Investig.* **2011**, *121*, 3735–3746. [CrossRef]
44. Massague, J. Integration of Smad and MAPK pathways: A link and a linker revisited. *Genes Dev.* **2003**, *17*, 2993–2997. [CrossRef] [PubMed]
45. Hansmann, G.; Perez, V.A.d.; Alastalo, T.P.; Alvira, C.M.; Guignabert, C.; Bekker, J.M.; Schellong, S.; Urashima, T.; Wang, L.; Morrell, N.W.; et al. An antiproliferative BMP-2/PPARgamma/apoE axis in human and murine SMCs and its role in pulmonary hypertension. *J. Clin. Investig.* **2008**, *118*, 1846–1857. [CrossRef]
46. West, J.D.; Austin, E.D.; Gaskill, C.; Marriott, S.; Baskir, R.; Bilousova, G.; Jean, J.C.; Hemnes, A.R.; Menon, S.; Bloodworth, N.C.; et al. Identification of a common Wnt-associated genetic signature across multiple cell types in pulmonary arterial hypertension. *Am. J. Physiol. Cell Physiol.* **2014**, *307*, C415–C430. [CrossRef]
47. Cui, W.J.; Liu, Y.; Zhou, X.L.; Wang, F.Z.; Zhang, X.D.; Ye, L.H. Myosin light chain kinase is responsible for high proliferative ability of breast cancer cells via anti-apoptosis involving p38 pathway. *Acta Pharmacol. Sin.* **2010**, *31*, 725–732. [CrossRef]
48. Hopper, R.K.; Feinstein, J.A.; Manning, M.A.; Benitz, W.; Hudgins, L. Neonatal pulmonary arterial hypertension and Noonan syndrome: Two fatal cases with a specific RAF1 mutation. *Am. J. Med. Genet. A* **2015**, *167A*, 882–885. [CrossRef]
49. Chamorro-Jorganes, A.; Lee, M.Y.; Araldi, E.; Landskroner-Eiger, S.; Fernandez-Fuertes, M.; Sahraei, M.; del Rey, M.Q.; van Solingen, C.; Yu, J.; Fernandez-Hernando, C.; et al. VEGF-Induced Expression of miR-17-92 Cluster in Endothelial Cells Is Mediated by ERK/ELK1 Activation and Regulates Angiogenesis. *Circ. Res.* **2016**, *118*, 38–47. [CrossRef]
50. Shen, Q.; Rigor, R.R.; Pivetti, C.D.; Wu, M.H.; Yuan, S.Y. Myosin light chain kinase in microvascular endothelial barrier function. *Cardiovasc. Res.* **2010**, *87*, 272–280. [CrossRef]

51. Masri, F.A.; Xu, W.; Comhair, S.A.; Asosingh, K.; Koo, M.; Vasanji, A.; Drazba, J.; Anand-Apte, B.; Erzurum, S.C. Hyperproliferative apoptosis-resistant endothelial cells in idiopathic pulmonary arterial hypertension. *Am. J. Physiol.-Lung Cell Mol. Physiol.* **2007**, *293*, L548–L554. [CrossRef] [PubMed]
52. Toshner, M.; Voswinckel, R.; Southwood, M.; Al-Lamki, R.; Howard, L.S.; Marchesan, D.; Yang, J.; Suntharalingam, J.; Soon, E.; Exley, A.; et al. Evidence of dysfunction of endothelial progenitors in pulmonary arterial hypertension. *Am. J. Respir. Crit. Care Med.* **2009**, *180*, 780–787. [CrossRef] [PubMed]
53. Teichert-Kuliszewska, K.; Kutryk, M.J.; Kuliszewski, M.A.; Karoubi, G.; Courtman, D.W.; Zucco, L.; Granton, J.; Stewart, D.J. Bone morphogenetic protein receptor-2 signaling promotes pulmonary arterial endothelial cell survival: Implications for loss-of-function mutations in the pathogenesis of pulmonary hypertension. *Circ. Res.* **2006**, *98*, 209–217. [CrossRef] [PubMed]
54. Chang, Y.S.; Munn, L.L.; Hillsley, M.V.; Dull, R.O.; Yuan, J.; Lakshminarayanan, S.; Gardner, T.W.; Jain, R.K.; Tarbell, J.M. Effect of vascular endothelial growth factor on cultured endothelial cell monolayer transport properties. *Microvasc. Res.* **2000**, *59*, 265–277. [CrossRef]
55. Ferrara, N. Vascular endothelial growth factor: Basic science and clinical progress. *Endocr. Rev.* **2004**, *25*, 581–611. [CrossRef]
56. Geiger, R.; Berger, R.M.; Hess, J.; Bogers, A.J.; Sharma, H.S.; Mooi, W.J. Enhanced expression of vascular endothelial growth factor in pulmonary plexogenic arteriopathy due to congenital heart disease. *J. Pathol.* **2000**, *191*, 202–207. [CrossRef]
57. Saitoh, M.; Ishikawa, T.; Matsushima, S.; Naka, M.; Hidaka, H. Selective inhibition of catalytic activity of smooth muscle myosin light chain kinase. *J. Biol. Chem.* **1987**, *262*, 7796–7801. [CrossRef]
58. Ma, T.Y.; Nguyen, D.; Bui, V.; Nguyen, H.; Hoa, N. Ethanol modulation of intestinal epithelial tight junction barrier. *Am. J. Physiol.* **1999**, *276*, G965–G974. [CrossRef]
59. Usatyuk, P.V.; Singleton, P.A.; Pendyala, S.; Kalari, S.K.; He, D.; Gorshkova, I.A.; Camp, S.M.; Moitra, J.; Dudek, S.M.; Garcia, J.G.; et al. Novel role for non-muscle myosin light chain kinase (MLCK) in hyperoxia-induced recruitment of cytoskeletal proteins, NADPH oxidase activation, and reactive oxygen species generation in lung endothelium. *J. Biol. Chem.* **2012**, *287*, 9360–9375. [CrossRef]
60. Gonzales, J.; Holbert, K.; Czysz, K.; George, J.; Fernandes, C.; Fraidenburg, D.R. Hemin-induced endothelial dysfunction and endothelial to mesenchymal transition in the pathogenesis of pulmonary hypertension due to chronic hemolysis. *Int. J. Mol. Sci.* **2022**, *23*, 4763. [CrossRef]
61. Trepat, X.; Chen, Z.; Jacobson, K. Cell migration. *Compr. Physiol.* **2012**, *2*, 2369–2392. [PubMed]
62. Guo, S.; Lok, J.; Liu, Y.; Hayakawa, K.; Leung, W.; Xing, C.; Ji, X.; Lo, E.H. Assays to examine endothelial cell migration, tube formation, and gene expression profiles. *Methods Mol. Biol.* **2014**, *1135*, 393–402. [PubMed]
63. Liang, C.C.; Park, A.Y.; Guan, J.L. In Vitro scratch assay: A convenient and inexpensive method for analysis of cell migration in Vitro. *Nat. Protoc.* **2007**, *2*, 329–333. [CrossRef] [PubMed]

Article

Cobra Venom Factor Boosts Arteriogenesis in Mice

Philipp Götz [1,2,†], Sharon O. Azubuike-Osu [1,2,3,†], Anna Braumandl [1,2], Christoph Arnholdt [1,2], Matthias Kübler [1,2], Lisa Richter [4], Manuel Lasch [1,2,5], Lisa Bobrowski [1,2], Klaus T. Preissner [6] and Elisabeth Deindl [1,2,*]

[1] Walter-Brendel-Centre of Experimental Medicine, University Hospital, Ludwig-Maximilians-Universität München, 81377 Munich, Germany; p.goetz@med.uni-muenchen.de (P.G.); sharon.azubuike-osu@med.uni-muenchen.de or sharon.eboagwu@funai.edu.ng (S.O.A.-O.); anna.braumandl@med.uni-muenchen.de (A.B.); christoph.arnholdt@med.uni-muenchen.de (C.A.); matthias.kuebler@med.uni-muenchen.de (M.K.); manuel_lasch@gmx.de (M.L.); lisa.bobrowski@med.uni-muenchen.de (L.B.)
[2] Biomedical Center, Institute of Cardiovascular Physiology and Pathophysiology, Ludwig-Maximilians-Universität München, 82152 Planegg-Martinsried, Germany
[3] Department of Physiology, Faculty of Basic Medical Sciences, College of Medicine, Alex Ekwueme Federal University Ndufu Alike, Abakaliki 482131, Ebonyi, Nigeria
[4] Flow Cytometry Core Facility, Biomedical Center, Ludwig-Maximilians-Universität München, 82152 Planegg-Martinsried, Germany; l.richter@med.uni-muenchen.de
[5] Department of Otorhinolaryngology, Head and Neck Surgery, University Hospital, Ludwig-Maximilians-Universität München, 81377 Munich, Germany
[6] Department of Cardiology, Kerckhoff-Heart Research Institute, Faculty of Medicine, Justus Liebig University, 35392 Giessen, Germany; klaus.t.preissner@biochemie.med.uni-giessen.de
* Correspondence: elisabeth.deindl@med.uni-muenchen.de; Tel.: +49-(0)-89-2180-76504
† These authors contributed equally to this work.

Citation: Götz, P.; Azubuike-Osu, S.O.; Braumandl, A.; Arnholdt, C.; Kübler, M.; Richter, L.; Lasch, M.; Bobrowski, L.; Preissner, K.T.; Deindl, E. Cobra Venom Factor Boosts Arteriogenesis in Mice. Int. J. Mol. Sci. 2022, 23, 8454. https://doi.org/10.3390/ijms23158454

Academic Editor: Steve Peigneur

Received: 23 June 2022
Accepted: 27 July 2022
Published: 30 July 2022

Publisher's Note: MDPI stays neutral with regard to jurisdictional claims in published maps and institutional affiliations.

Copyright: © 2022 by the authors. Licensee MDPI, Basel, Switzerland. This article is an open access article distributed under the terms and conditions of the Creative Commons Attribution (CC BY) license (https://creativecommons.org/licenses/by/4.0/).

Abstract: Arteriogenesis, the growth of natural bypass blood vessels, can compensate for the loss of arteries caused by vascular occlusive diseases. Accordingly, it is a major goal to identify the drugs promoting this innate immune system-driven process in patients aiming to save their tissues and life. Here, we studied the impact of the Cobra venom factor (CVF), which is a C3-like complement-activating protein that induces depletion of the complement in the circulation in a murine hind limb model of arteriogenesis. Arteriogenesis was induced in C57BL/6J mice by femoral artery ligation (FAL). The administration of a single dose of CVF (12.5 μg) 24 h prior to FAL significantly enhanced the perfusion recovery 7 days after FAL, as shown by Laser Doppler imaging. Immunofluorescence analyses demonstrated an elevated number of proliferating (BrdU$^+$) vascular cells, along with an increased luminal diameter of the grown collateral vessels. Flow cytometric analyses of the blood samples isolated 3 h after FAL revealed an elevated number of neutrophils and platelet-neutrophil aggregates. Giemsa stains displayed augmented mast cell recruitment and activation in the perivascular space of the growing collaterals 8 h after FAL. Seven days after FAL, we found more CD68$^+$/MRC-1$^+$ M2-like polarized pro-arteriogenic macrophages around growing collaterals. These data indicate that a single dose of CVF boosts arteriogenesis by catalyzing the innate immune reactions, relevant for collateral vessel growth.

Keywords: arteriogenesis; complement system; C3; cobra venom factor; mast cells; macrophages; complement activation; neutrophils; platelets; platelets-neutrophil aggregates

1. Introduction

The vascular system has the fundamental role of delivering blood with its nutrients and oxygen to peripheral tissues. The devastating consequences of vascular occlusive diseases such as myocardial infarction, peripheral artery disease, stroke, or even loss of an artery due to abdominal aortic aneurysm (AAA) surgery might be prevented by the timely induction of collateral artery growth to form natural bypasses [1–4]. This highly complex and multifactorial process, which is particularly driven by changes in fluid shear stress of affected vessels, is known as arteriogenesis [5,6]. This multistep process is described as the remodeling of pre-existing arterio-arteriolar anastomoses into completely developed functional arteries, characterized by the proliferation of endothelial and smooth muscle cells and promoted by the cellular and humoral components of the innate immune system [7,8].

In detail, it has been shown that, upon occlusion of a supplying artery, the blood flow is redirected into pre-existing arteriolar connections where the arising increased fluid shear stress elicits a local and timely well-coordinated sterile inflammatory process. Upon this mechanical stress, endothelial cells release the von Willebrand factor (vWF), which promotes the activation of platelets. Subsequently, these activated platelets interact with neutrophils to form platelet-neutrophil aggregates (PNA) with the concomitant activation of NADPH oxidase 2 (Nox-2) in neutrophils. Upon adhesion on intercellular adhesion molecule-1 (ICAM-1) and urokinase plasminogen activator (uPA)-mediated extravasation of neutrophils—both proteins are increased, expressed in activated endothelial cells of growing collaterals [9,10]—neutrophil-derived reactive oxygen species (ROS) activate perivascular mast cells. These cells, in turn, create an inflammatory microenvironment, resulting in the further recruitment of neutrophils, as well as macrophages, the latter supplying growth factors and cytokines to the growing collateral vessel (for a review, see [11]). Interestingly, it has been demonstrated that especially CD68+/MRC-1+ M2-like polarized macrophages, which play a major role in tissue remodeling and the resolution of inflammation, are of major relevance for an effective arteriogenesis [12,13].

The complement system is an important part of the innate immune system playing a role in the resistance against infection and clearance of altered host cells. Complement components also have regulating properties in inflammatory and immune responses and thereby are key players in the pathogenesis of several diseases, like shock, stroke, immune complex diseases, autoimmune hemolytic anemia, or asthma [14]. The complement system is composed of about 30 plasma and cell surface proteins, numerous receptors, and regulatory factors expressed in the membranes of various cell types, such as platelets, endothelial cells, and cells of the immune system [15,16]. These proteins allow to recognize pathogens and eradicate them from the host's system [17].

The complement system can be canonically activated via three specific, well-defined pathways: the classical, the lectin and the alternative pathway. These three pathways share a similar molecular architecture, including somewhat different primary recognition and activation events that are magnified by proteolytic reactions converging at the point of C3 cleavage/activation [18]. As an effective chemoattractant, the resulting small C3a peptide serves to concentrate neutrophils at the inflamed site, whereas the large C3b fragment is incorporated into the proteolytic convertases to amplify the complement activation. Finally, in the terminal phase, the membrane attack complex (MAC) is assembled on the surface of pathogens to obtain cytolytic poly-C9 pores as an effective killing mechanism [18]. However, it should be mentioned that the complement system can be activated noncanonically [19].

Cobra venom factor (CVF) is a complement-activating protein, which has been isolated from venom of the cobra [15]. It is functionally homologous to C3b, the activated form of C3, and accordingly binds factor B, to be subsequently cleaved by factor D to form the bimolecular complex CVF–Bb, which is a C3/C5 convertase. CVF–Bb cleaves both C3 and C5 but is resistant to inactivation by the regulating factors of the complement system such as Factor H and due to its special structure, more stable than its physiological pendant [20]. As CVF can cleave C3 almost completely, it has been used for many years to deplete the serum complement (C3) in animals in a bid to delineate the biological functions of the complement system and its role in the pathogenesis of many diseases [18]. CVF is used as an important tool to target complement activation, and no adverse effects were observed in animals, except neutrophil sequestration to the lungs [21]. CVF is not a toxin, but it can activate the alternative pathway. However, it only consumes complement (C3) and effectively depletes the complement activity in the serum of treated animals [15]. A great deal of effort has been invested in the development of drugs for complement inhibition [22]. These drugs inhibit complement activation, whereas CVF depletes complement C3.

Despite the fact that humoral and cellular components of the innate immune system play a major role in arteriogenesis, the contribution of the complement remains obscure. Since the generation of collateral vessels is required in situations of artery occlusion, one is left to promote collateral artery growth in clinical practice. This study will provide information about the application and usefulness of CVF for the process of arteriogenesis as an option for therapeutical intervention in patients with vascular occlusive diseases. We investigated the possible mechanisms by which CVF through complement activation and consumption may have an impact on arteriogenesis. The obtained results indicate that treatment with a single dose of CVF 24 h prior to artery occlusion results in an enhanced growth and formation of collateral vessels in mice. While these data provide further insight into the contribution of innate immunity for the process of arteriogenesis, it remains to be investigated whether CVF administration could be available as a medical treatment.

2. Results

2.1. Treatment with CVF in the Experimental Mouse Model of Arteriogenesis

To study the influence of the complement system and the impact of CVF on arteriogenesis, we used a well-established murine hind limb model to induce collateral vessel formation [8]. In this model, the unilateral ligation of the right femoral artery results in the growth of collateral blood vessels (arteriogenesis) in the adductor muscle of the upper leg. As an internal control, the left femoral artery was sham-operated. To ascertain the efficacy of CVF in our animal model, 24 h prior to femoral artery ligation (FAL), C57Bl/6J mice underwent i.p. injection of either a single dose of 12.5 µg of CVF dissolved in 50 µL PBS or an equivalent volume of PBS (control group). The complement consumption impact of the treatment was proven by measuring the 50% hemolytic complement activity of the serum before starting the FAL (Figure 1).

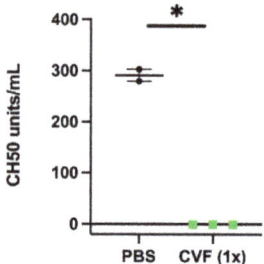

Figure 1. Single-dose application of 12.5 µg of the cobra venom factor (CVF (1x)) leads to the loss of hemolytic complement activity of the serum 24 h after injection. The scatter plot represents the results of a CH50 assay. Data are the means ± S.E.M., $n = 3$ for the CVF-treated group, $n = 2$ for the PBS-treated control group, * $p < 0.05$ (PBS vs. CVF (1x)) by an unpaired Student's t-test.

2.1.1. Influence of CVF on Perfusion Recovery

To measure the influence of CVF on the perfusion recovery after FAL, mice underwent laser Doppler perfusion measurements of their hindlimbs before, directly after, 3 days, and 7 days after ligation. In mice receiving a single dose of CVF 24 h before the induction of arteriogenesis via FAL, we observed a significantly improved perfusion recovery compared to the PBS-treated control group 7 days after the surgical procedure (Figure 2a,b). However, when the mice were treated twice, i.e., 24 h before and again 3 days after FAL, with CVF, a negative impact on the perfusion recovery was seen (Figure 2c,d).

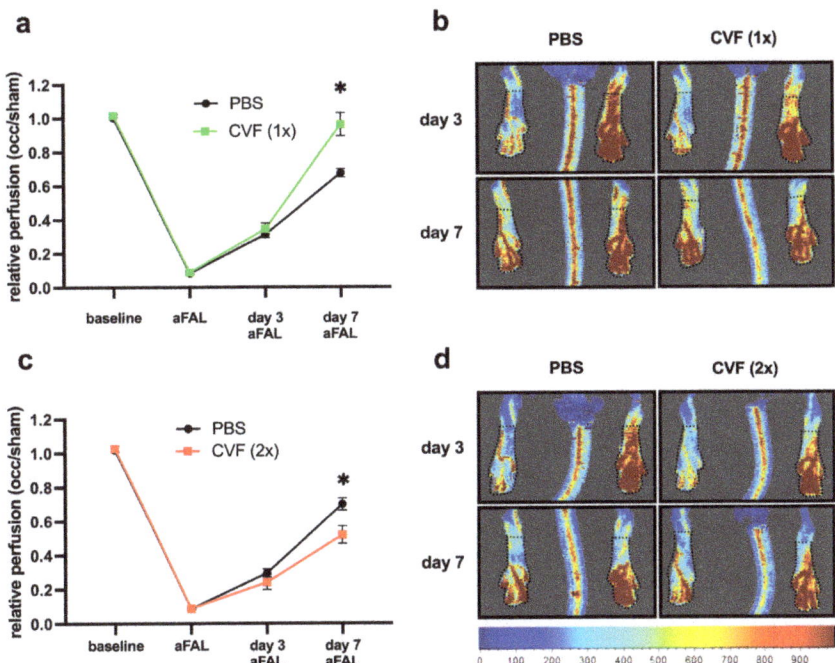

Figure 2. Impact of CVF administration on the perfusion after femoral artery ligation (FAL). (**a**,**b**) Single-dose application of the cobra venom factor (CVF (1x)) promotes perfusion recovery. (**a**) The line graph describes laser Doppler perfusion measurements of the hindlimbs from mice treated with a single dose of phosphate-buffered saline (PBS) or CVF (1x) 24 h prior to FAL. The relative perfusion was calculated using an occluded/sham (right to left hind limb) ratio before FAL (baseline), directly after FAL (aFAL), at day 3, and at day 7 aFAL. Data are the means ± S.E.M., $n = 5$ mice per group, * $p < 0.05$ (PBS vs. CVF (1x)) by two-way analysis of variance (ANOVA) with the Bonferroni multiple comparison test. (**b**) Representative flux images of laser Doppler measurements from mice treated with a single dose of PBS (left images) or CVF (1x) (right images) showing color-coded the relative perfusion of hind limbs at day 3 and 7 aFAL (blue color indicates low perfusion, and red color indicates high perfusion). Black dotted lines indicate the regions of interest (ROI), which were used for perfusion analyses. (**c**,**d**) Repeated treatment of mice with CVF (2x) counteracts the perfusion recovery. (**c**) The line graph depicts the results of the laser Doppler perfusion measurements of the hind limbs of repeated PBS- or CVF (2x)-treated mice. Some (12.5 μg CVF) were administered by i.p. injection 24 h prior FAL and again 3 days aFAL. The relative perfusion was calculated using the occluded/sham (right to left hind limb) ratio before FAL (baseline), directly aFAL, at day 3, and at day 7 aFAL. Data are the means ± S.E.M., $n = 5$ per group, * $p < 0.05$ (PBS vs. CVF (2x)) by two-way analysis of variance (ANOVA) with the Bonferroni multiple comparison test. (**d**) Representative flux images of laser Doppler measurements from repeated PBS- (left images) or CVF (2x)-treated mice (right images) showing in color code the relative hindlimb perfusion at day 3 and 7 aFAL. Black dotted lines indicate the ROI, which were used for perfusion analyses.

2.1.2. Influence of Treatment with a Single Dose of CVF on Luminal Diameter and Vascular Cell Proliferation

To confirm that the improved perfusion recovery observed after administration of a single dose of CVF was due to enhanced collateral artery growth (and not due to simple vasodilation), we performed immunofluorescence analyses of the adductor muscle by studying the inner luminal diameter and the number of proliferating vascular cells 7 days after FAL. Lectin was implemented to mark the inner vascular cell layer for the analysis of the luminal diameter. Bromodeoxyuridine (BrdU) served as a proliferation marker, while ACTA2 as smooth muscle cell marker was used to mark the outer vascular boundary for the proliferation analysis. Thus, BrdU$^+$ cells within the vascular structure were counted as proliferating vascular cells, whereby BrdU signals in ACTA2$^+$ cells were counted as proliferating vascular smooth muscle cells, and BrdU signals at luminal ACTA2$^−$ structures as proliferating endothelial cells. However, for quantification, we did not distinguish between these subpopulations of vascular cells. CVF (1x)-treated mice showed a significantly increased inner luminal diameter of the growing collaterals compared to the PBS-treated control group (Figure 3a,b). Additionally, there was a significant increase in the number of proliferating collateral vascular cells in the CVF (1x)-treated group compared to the PBS-treated controls (Figure 3b,d). Investigations on sham-operated legs showed no differences either in the luminal diameter or proliferating vascular cells between CVF- and PBS-treated mice (data not shown).

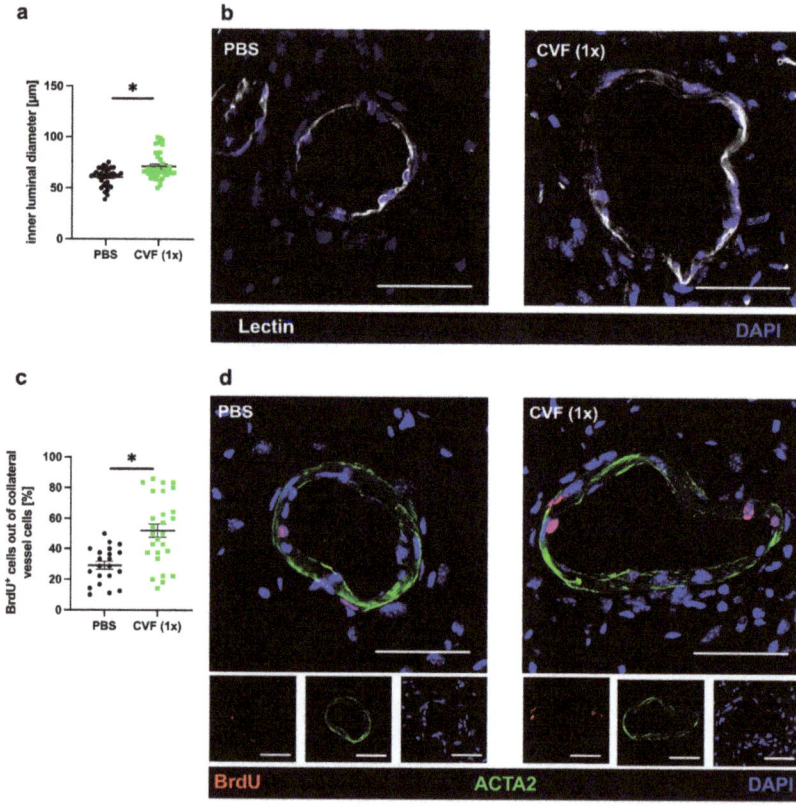

Figure 3. Single-dose application of the cobra venom factor (CVF (1x)) 24 h prior to femoral artery ligation (FAL) enhances collateral growth. The scatter plots describe (**a**) the inner luminal diameter of proliferating collaterals and (**c**) the number of proliferating vascular cells (BrdU+, bromodeoxyuridine+ cells) per growing collateral in percent of phosphate-buffered saline (PBS)- and CVF-treated mice 7 days after FAL. Data are the means ± S.E.M., n = 5 mice per group, n > 10 values per group, * p < 0.05 (PBS vs. CVF (1x)) by the unpaired Student's t-test. (**b,d**) Representative immunofluorescence pictures of growing collaterals collected from PBS-treated (left panels) and CVF (1x)-treated (right panels) mice 7 days after ligation stained with lectin (white), indicating the luminal vessel boundary (**b**, **upper** images), as well as antibodies against BrdU (red), marking proliferating cells and ACTA2 (green) staining smooth muscle cells (**d**, **lower** images). DAPI (blue) was used to label the nuclei; scale bar: 50 µm.

2.1.3. Single-Dose Application of CVF Enhances Neutrophil Mobilization and PNA Formation

As PNA formation is an important prerequisite for effective arteriogenesis, we investigated the number of platelets and neutrophils, as well as their complexes (expressed as PNAs), as early as 3 h after FAL and 27 h after treatment with a single dose of CVF or PBS, respectively. The influence of CVF on neutrophil and platelet counts, as well as PNA formation, was analyzed in blood samples of PBS- and CVF (1x)-treated mice that either underwent ligation of the right and left femoral arteries or received a sham operation of both legs. Interestingly, we found a significantly increased number of neutrophils and PNAs in mice treated with a single dose of CVF compared to the PBS-treated control groups, irrespective of whether the mice experienced femoral artery ligation or a sham operation (Figure 4a,c). The platelet count was slightly increased in CVF (1x)-treated mice; however, it did not show significantly different values (Figure 4b).

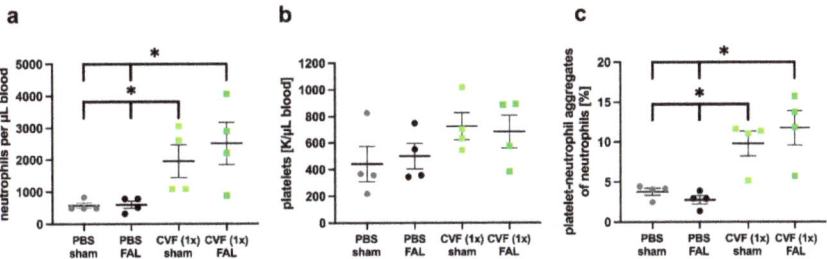

Figure 4. Single-dose application of the cobra venom factor (CVF (1x)) enhances the formation of platelet-neutrophil aggregates (PNAs) and the mobilization of neutrophils to the peripheral blood. Scatter plots show flow cytometry analyses of (**a**) the number of neutrophils per microliter of blood, (**b**) the number of platelets per microliter of blood measured by the differential blood count, and (**c**) the percentage of PNA formation relative to the total number of neutrophils in phosphate-buffered saline (PBS)- and CVF (1x)-treated mice 27 h after injection, respectively, 3 h after double femoral artery ligation (FAL) or a double sham operation (sham) of both femoral arteries. For the PNA analysis, platelets were detected by an anti-CD41 antibody, and neutrophils were identified by anti-CD11b and anti-Gr-1 antibodies. All values are the means ± S.E.M., $n = 4$ mice per group, * $p < 0.05$ by one-way analysis of variance (ANOVA) with Tukey's multiple comparison test.

2.1.4. Single-Dose CVF Treatment Results in Increased Perivascular Mast Cell Recruitment and Activation

As already shown in a previous study, mast cell activation is essential for arteriogenesis [8]. Here, we assessed the total number, as well as the number of degranulating mast cells, in the perivascular space of the proliferating and resting collaterals by Giemsa staining of the adductor muscle tissue collected 8 h after the FAL or sham operation. Comparing the adductor muscles from the femoral artery ligated site of single-dose CVF- with PBS-treated mice, we found significantly more mast cells in the perivascular space of the collaterals collected from CVF (1x)-treated mice (Figure 5a,c). Furthermore, we observed an increase of degranulated mast cells per collateral in mice that underwent the single-dose CVF treatment (Figure 5b,c). However, no significant difference concerning the number of mast cells was seen in the perivascular space of single-dose CVF vs. PBS-treated sham-operated mice, although, here, the number of degranulated mast cells was increased as well (Supplementary Materials Figure S1).

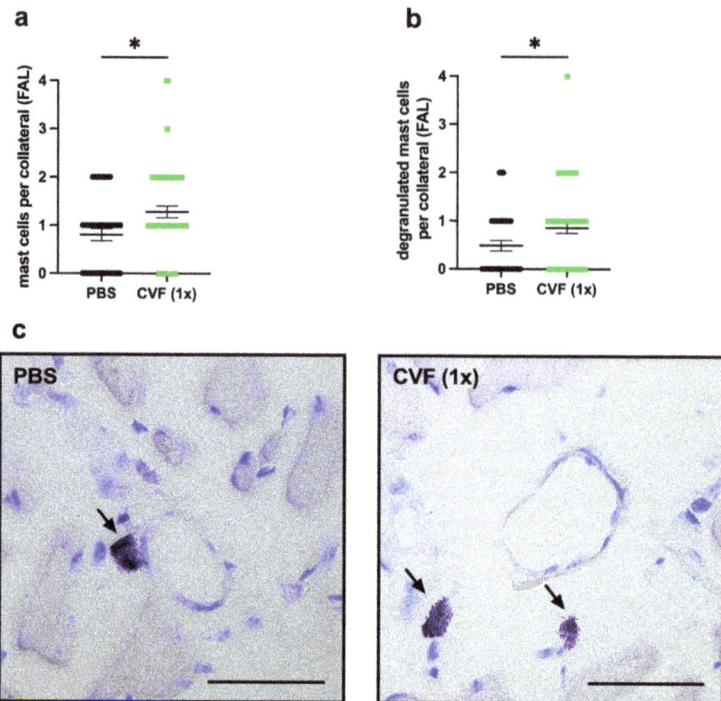

Figure 5. Single-dose application of the cobra venom factor (CVF 1x)) increases the number of perivascular mast cells and promotes the degranulation of mast cells around growing collateral arteries. The scatter plots represent (**a**) the number of perivascular mast cells per collateral in the adductor muscles 8 h after femoral artery ligation (FAL) of phosphate-buffered saline (PBS)- and CVF-treated mice, as well as (**b**) the number of degranulated mast cells per collateral 8 h after FAL. Data are the means ± S.E.M., n = 3 mice per group, n > 10 values per group, * $p < 0.05$ (PBS vs. CVF (1x)) by the unpaired Student's t-test. (**c**) Representative Giemsa stains depicting mast cells (arrows) in the perivascular space of collateral vessels of PBS- (left image) and CVF (1x)-treated (right image) mice 8 h after FAL. Scale bar: 50 µm.

2.1.5. Single-Dose CVF Treatment Promotes Regenerative M2-like Macrophage Polarisation

Macrophages and their polarization state play a pivotal role in arteriogenesis. To analyze macrophages in the perivascular space of the growing and resting collaterals, we used an anti-CD68 antibody to stain macrophages, together with an anti-mannose receptor c-type 1 (MRC1) antibody as a marker for the anti-inflammatory and regenerative M2-like polarization phenotype. Hence, CD68$^+$/MRC1$^-$ cells were quantified as proinflammatory M1-like polarized macrophages, whereas CD68$^+$/MRC1$^+$ cells were counted as anti-inflammatory M2-like polarized macrophages. We found a significantly increased number of perivascular macrophages per growing collateral in the adductor muscles of mice treated with a single dose of CVF compared to control mice 7 days after femoral artery ligation (Figure 6a). Hereby, the number of M1-like polarized macrophages (CD68$^+$/MRC1$^-$) per collateral did not differ between both experimental groups, whereas a significant increase in the number of M2-like polarized perivascular macrophages (CD68$^+$/MRC1$^+$) per collateral in CVF (1x)-treated mice compared to PBS-treated mice was observed (Figure 6b,c). No differences regarding the number of perivascular macrophages per resting collateral or their polarization state was noted in the sham-operated adductor muscles from both experimental groups (Supplementary Materials Figure S2).

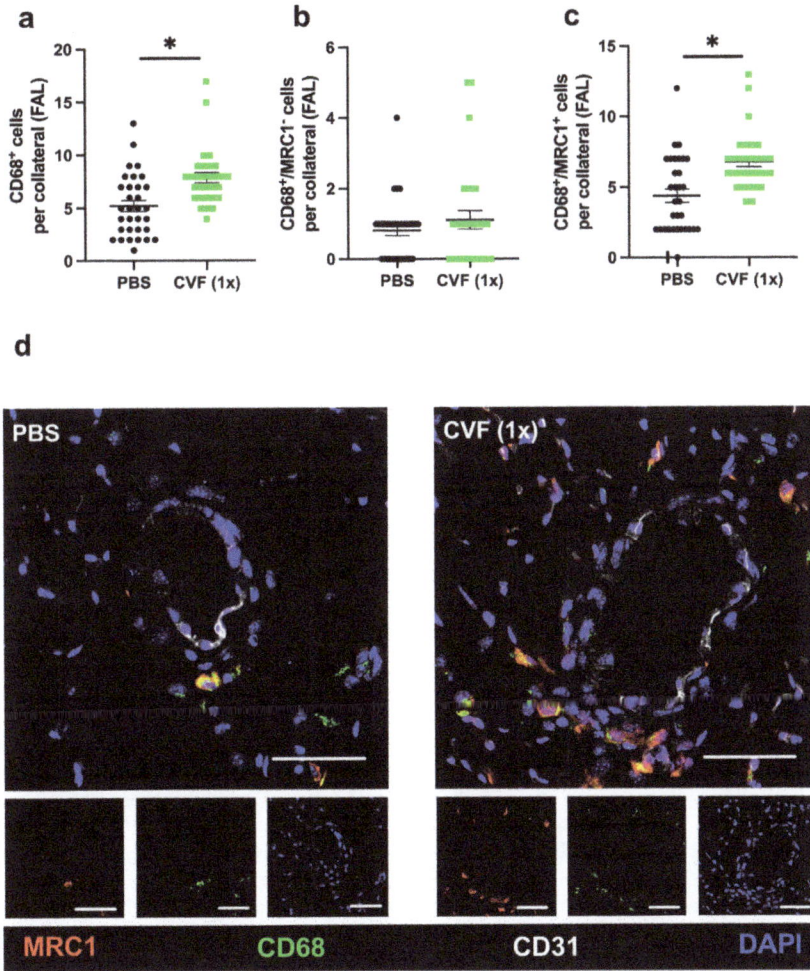

Figure 6. Single-dose cobra venom factor (CVF (1x)) application leads to an increased number of M2-like polarized macrophages per growing collateral. The scatter plots display (**a**) the number of perivascular CD68$^+$ cells (macrophages), (**b**) the number of CD68$^+$/MRC1$^-$ (mannose receptor C-type 1) cells per growing collateral, and (**c**) the number of CD68$^+$/MRC1$^+$ cells per growing collateral in the adductor muscle of single-dose phosphate-buffered saline (PBS)- or CVF-treated mice collected 7 days after femoral artery ligation (FAL). Data are the means ± S.E.M., n = 5 mice per group, n > 10 values per group, * p < 0.05 (PBS vs. CVF (1x)) by the unpaired Student's t-test. (**d**) Representative immunofluorescence images of adductor muscles of single-dose PBS- (left images) or CVF-treated (right images) mice collected 7 days after FAL are presented. Images of single and merged channels show macrophages labeled with antibodies against CD68 (green) and MRC1 (red) in the perivascular space of the collaterals. The endothelial cell marker CD31 (white) was used to depict collaterals and DAPI (blue) to label nucleic DNA; scale bar: 50 μm.

3. Discussion

Arteriogenesis is a shear stress driven process that is supported by the immune system. Using a murine hind limb model, we show that a single dose of CVF boosts arteriogenesis by activating innate immune reactions. We demonstrate that a single-dose treatment with CVF results in the mobilization of neutrophils to the peripheral blood and an increased PNA formation, as well as an enhanced recruitment and activation of the mast cells, all of which are prerequisites for effective arteriogenesis. Moreover, we report a raised number of pro-arteriogenic M2-like polarized macrophages in the perivascular space of growing collateral arteries. Consequently, mice treated with a single dose of CVF displayed increased vascular cell proliferation, enlarged collateral diameters, and increased perfusion recovery upon femoral artery occlusion, making CVF a potent drug to foster collateral artery growth.

The influence of the complement system and different complement factors has been extensively studied in the context of angiogenesis [23–25]. In addition, CVF as a complement consuming drug has been investigated in various settings of inflammation in animal experiments [18,26]. However, the role of the complement system in arteriogenesis has only started to be looked at [27], and CVF treatment has never been a topic in settings of experimental collateral growth.

There is plethora of data available in the literature suggesting that the complement system is involved in the process of arteriogenesis. For example, it has been described that endothelial cells are an extrahepatic source of components of the complement system, such as properdin, and that shear stress can cause their increased expression and release [28–30]. Properdin positively regulates the alternative pathway of the complement system by stabilizing C3 and C5 convertases. C3b can bind to ultra-large vWF—a protein that is released due to shear stress in growing collaterals [7]—and thereby locally activates the alternative pathway to generate a large amount of effector proteins [31,32]. Moreover, high concentrations of C3a and C5a can stimulate endothelial cells locally to express adhesion molecules for leukocytes such as intercellular adhesion molecule-1 (ICAM-1) [33] and leukocyte-recruiting chemokines such as monocyte chemoattractant protein-1 (MCP-1) [34]. Indeed, we have previously shown that an increased expression of ICAM-1 occurs in growing collaterals, contributing to arteriogenesis [9,35]. Moreover, we have shown that arteriogenesis is associated with increased levels of MCP-1 and that the administration of MCP-1 strongly promotes arteriogenesis [35].

To investigate the impact of CVF on arteriogenesis, we used a murine hind limb model of collateral artery growth to administer CVF either once or twice prior to FAL. When mice were treated twice with CVF, 24 h before induction of arteriogenesis by FAL and again 3 days after FAL, reduced arteriogenesis was noted, as evidenced by the LDI measurements. However, when mice were treated only once with CVF, namely 24 h before the induction of collateral artery growth, arteriogenesis was significantly improved, as confirmed by an increased luminal diameter and cell proliferation in growing collateral arteries 7 days after FAL.

The administration of CVF results within hours in decomplementation and has been demonstrated to reach its maximum after 24 h in mice [36]. A feasible test to investigate the effectiveness of CVF is a hemolytic assay [37,38]. Upon activation of the complement system, the membrane attack complex (MAC) is formed, which lyses erythrocytes. Accordingly, decomplementation results in a severe reduction in hemolytic activity. The decomplementating effect of CVF compared to PBS was proven in our study in the serum of mice 24 h after administration of a single dose of CVF.

In terms of the mobilization of leukocytes in New Zealand white rabbits, CVF administration showed a biphasic and opposing effect, since the application of CVF resulted in leukopenia within seconds, followed by leukocytosis with an almost fivefold increase in cell number (mainly of neutrophils) within 24 h [39]. Additional studies evidenced that the increased neutrophil count is due to the chemotactic recruitment of cells from the bone marrow as a consequence of increased C5a upon CVF treatment [40–42]. We here report a significant increase in the number of neutrophils 27 h after administration of a single dose CVF in mice, which was associated with an increased platelet–neutrophil aggregate formation. Interestingly, at that time point, which corresponds to 3 h after the onset of the surgical procedure of FAL, we found no changes in the number of PNA formation compared to the sham-operated PBS-treated mice. However, our results evidenced significant rises in PNA formation in both FAL and sham-operated mice after treatment with a single dose of CVF compared to the PBS treatment.

Increased PNA formation is a prerequisite for effective arteriogenesis, as it results in neutrophil, as well as subsequent mast cell, activation [7,8]. Indeed, CVF treatment resulted in an increased recruitment of mast cells in the perivascular space of growing collaterals already 8 h after FAL, which was not seen in sham-operated animals. In a previous study, we demonstrated that an increasing mast cell recruitment by treating mice with diprotin A (a dipeptidyl-peptidase IV (DPPIV) inhibitor that retards stroma cell-derived factor-1α, SDF-1α, degradation) resulted in improved arteriogenesis by recruiting SDF-1 receptor CXCR-4-expressing mast cells to the growing collaterals [8]. Additionally, PNA formation is a marker for platelet activation [43], and platelets are a rich source of SDF-1α [44,45], indicating that, upon CVF treatment, increased mast cell recruitment may be due to an enhanced PNA formation. Yet, as C3a and C5a are also chemo-attractants for mast cells [46], the CVF treatment could have contributed to the emergence of these anaphylatoxins to contribute to mast cell recruitment as well.

Under (patho-)physiological conditions, mast cell activation in the perivascular space of growing collaterals is largely mediated by neutrophil-derived ROS [8]. This is also very likely the case in CVF-treated mice, as these mice showed up with more PNA formation, which is associated with increased Nox-2 activation and, hence, ROS formation in neutrophils [8]. Moreover, anaphylatoxins produced by CVF application are able to trigger an oxidative burst of neutrophils [47]. Additionally, by binding to their respective receptors, C3a and C5a can directly promote mast cell activation [48,49]. We indeed found an increased number of activated mast cells in the perivascular space of sham-operated mice; however, mast cell activation did not result in vascular cell proliferation and subsequent collateral artery growth. The initial trigger for arteriogenesis is increased fluid shear stress, which is elicited in preexisting collateral arteries by redirected blood flow and results in endothelial cell activation [1]. Inter alia, this results in the increased expression of uPA in the wall of preexisting collaterals, which is essential not only for the extravasation of neutrophils to the perivascular space [50] but also important for monocyte transmigration, which, once matured to macrophages, are crucial for effective arteriogenesis [1,6]. Accordingly, CVF treatment did not result in an increased accumulation of CD68[+] macrophages around collateral arteries of sham-operated legs, and the sham operation was not associated with increased vascular cell proliferation or the luminal diameter in preexisting collaterals, indicating that the CVF treatment itself is not a direct growth stimulus for collaterals under non-pathophysiological conditions.

The pivotal role of macrophages in arteriogenesis is well-known, as has been shown in several previous studies [51,52]. By providing growth factors and remodeling the perivascular space, they have a huge impact on vascular cell proliferation and artery growth [53,54]. In the current study, mice treated with a single dose of CVF showed a higher macrophage accumulation in the perivascular space of growing collaterals, which was based on a higher number of M2-like polarized macrophages.

4. Materials and Methods

4.1. Animals and Experimental Procedures

Permissions for in vivo experiments and animal use were obtained from the Bavarian Animal Care and Use Committee (ethical approval code: ROB-55.2Vet-2532.Vet_02-17-99), and the performances of all experiments were in strict accordance with German and NIH animal legislation guidelines. The experiments included the use of C57BL/6J mice provided by Charles River (Sulzfeld, Germany). Eight-to-twelve-week-old mice were anesthetized with a combination of fentanyl (0.05 mg/kg, CuraMED Pharma, Karlsruhe, Germany), midazolam (5.0 mg/kg, Ratiopharm GmbH, Ulm, Germany), and medetomidine (0.5 mg/kg, Pfister Pharma, Berlin, Germany). The right femoral artery was ligated to set the stimulus, leading to arteriogenesis in the adductor muscle, whereas the left leg was sham-operated, as previously described [55]. As the proliferation marker, bromodeoxyuridine

(BrdU) was administered daily i.p. (1.25 mg per day, Sigma-Aldrich, St. Louis, MO, USA) dissolved in 100 µL phosphate-buffered saline (PBS, 148 mM Na$^+$, 1.8 mM K$^+$, pH 7.2) starting immediately after femoral artery ligation (FAL). One day prior to surgery, cobra venom factor (REF A600, Quidel Co., Athens, OH, USA) was administered as a single-dose intraperitoneal (i.p.) injection with 12.5 µg diluted in 50 µL PBS. A PBS i.p. dose of the same volume served as the control. The perfusion of the hind limb was measured by Laser Doppler Imaging (LDI) using a Moor LDI 5061 and Moor Software Version 3.01 (Moor Instruments, Remagen, Germany). Prior to measurements, the body temperature was controlled for 10 min and kept between 36 °C and 38 °C during the experiments. LDI was performed before ligation (baseline) and directly after FAL, as well as 3 and 7 days after FAL. A flux mean value of a defined region (0.45 cm^2) starting from the ankle to the toes of each animal was calculated. The perfusion was calculated by flux means of the ligated (right)-to-sham (left)-operated ratios.

Tissues were collected for histological analyses at intervals of 8 h or 7 days after FAL. During surgery and the experimental period until the collection of the tissue, none of the mice died or showed signs of infection or necrosis. To study the formation of platelet neutrophil aggregates, blood was collected 3 h after ligation of both femoral arteries (right) and the sham operation (left). Mice were sacrificed using cardiac puncture under deep narcosis. The bodies were perfused with 1% adenosine buffer (Sigma-Aldrich, Taufkirchen, Germany) containing 5% bovine serum albumin (BSA, Sigma-Aldrich, Taufkirchen, Germany) and dissolved in PBS, followed by 3% paraformaldehyde (PFA, Merck, Darmstadt, Germany). Finally, the adductor muscles were collected, kept in a 30% sucrose solution overnight, and then stored in vinyl molds (REF 4566, Sakura Finetek, Torrance, CA, USA) on Tissue-Tek® (REF 4583, Sakura Finetek, Torrance, CA, USA) at −80 °C until further use.

4.2. Histological and Immunofluorescence Analyses

For histology, staining was performed with 8-µm-thick cryosections of the adductor muscle. Tissues of day 7 after FAL were used to stain for vascular cell proliferation, inner luminal diameter analysis, and the presence of macrophages. For cell proliferation, the sections were incubated with 1N HCl for 30 min at 37 °C to bare BrdU in the nuclei, blocked with 10% goat serum, dissolved in 4% BSA PBS/0.1% Tween-20 (Tween 20, AppliChem GmbH, Darmstadt, Germany) for 1 h at room temperature (RT), and then incubated with an anti-BrdU antibody (1:50, Abcam, Cambridge, UK, ab6326) at 4 °C overnight. Goat anti-rat Alexa Fluor® 546 antibody (1:100, Thermo Fischer, A-11081, Rockford, IL, USA) was used as a secondary antibody, and anti-ACTA2-Alexa Flour® 488 (anti-actin alpha 2, 1:400, Sigma-Aldrich, F3777, Saint-Louis, MO, USA) was used to mark the outer vessel layer.

Macrophages in tissues were labeled with anti-CD68-Alexa Fluor®488 (1:200, Abcam, ab201844) and anti-MRC1 antibody (1:200, Abcam, ab64693) at 4 °C overnight, followed by the secondary antibody donkey anti-rabbit-Alexa Fluor®546 (1:200, Thermo Fisher, A10040). An Alexa Fluor®647 anti-mouse CD31 antibody (1:100, BioLegend, San Diego, CA, USA, 102516) was used to label the vascular cells. Additionally, all tissues were incubated with DAPI (1:1000, Thermo Fisher, ord. no. 62248, Rockford, IL, USA) for 10 min at RT to label the nucleic DNA. Dako mounting medium (Dako, Agilent, Santa Clara, CA, USA) was used to mount the stained slides. Giemsa stain was performed following the standard protocols. The mounted tissues of CVF-treated mice and control mice were analyzed using an epifluorescence microscope DM6 B (Leica microsystems, Wetzlar, Germany) for dark and bright field imaging. The open-source program ImageJ was used for counting analyses and inner luminal diameter analyses. Several sections were analyzed per muscle.

To measure the luminal diameter of the collaterals, the sections were stained with lectin (5 µg/mL, Sigma-Aldrich, L4895, Saint-Louis, MO, USA) for 2 h at RT.

4.3. Blood Analyses

Blood was collected via cardiac puncture with a standardized amount of anticoagulation (10 UE heparin per mL blood). For the flow cytometry analysis, 100 µL of full blood was lysed in 2 mL lysing solution (1:10 in aqua, BD Biosciences, 349202), centrifuged, and resuspended in a staining solution containing FITC anti-mouse CD41 (1:400, BioLegend, 133903), PE anti-mouse CD11b (1:300, BioLegend, 101208), Brilliant Violet 421™ anti-mouse CD115 (1:300, BioLegend, 135513), APC anti-mouse Ly-6G/Ly-6C (1:800, BioLegend, 108412), and eBioscience™ Fixable Viability Dye eFluor™ 780 (1:1000, Invitrogen, 65-0865-14). The samples were incubated for 20 min at 4 °C, washed, and subsequently analyzed with a BD LSRFortessa™ cell analyzer. Gating and analysis were performed using FlowJo V10. A full blood count was done with a ProCyte Dx using mouse-specific settings.

For the CH50 assay, measuring the complement lysing capacity, blood was collected without anticoagulant 24 h after treatment with a single dose of CVF or PBS to achieve clotting and serum generation for 1 h at room temperature. The probes were centrifuged at $706 \times g$ for 10 min at 4 °C. The supernatant was collected and stored at -80 °C. The CH50 assay was performed using 100 million antibody-sensitized sheep erythrocytes (Complement Technology, Inc., Tyler, TX, USA, B200). Five microliters of a Mouse Complement Assay Reagent (MCAR, Complement Technology, Inc., Tyler, TX, USA, B250) was used to improve the mouse CH50 titers. The serum samples were diluted to a final range from 1/163 to 1/650 in a veronal buffer (GVB++, Complement Technology, Inc., Tyler, TX, USA, B102). Controls included two samples with no serum as the background control and two samples containing sheep erythrocytes diluted in deionized water in place of the buffer for the 100% lysis control. Samples were handled on wet ice, and lysis was performed at 37 °C for 30 min. The remaining cells were spun down at $500 \times g$ for 3 min, and the absorbance of the supernatant was determined at 541 nm in a 1-cm cuvette. CH50 values (CH50 units/mL) represent the reciprocal value of the amount of serum able to lyse 50% of the sheep erythrocytes. Samples without lysing activities were set at 0 CH50 units/mL.

4.4. Statistical Analyses

The results were analyzed with GraphPad Prism 8 (GraphPad Software, LA Jolla, CA, USA) using the unpaired Student's *t*-test or two-way analysis of variance (ANOVA) with the Bonferroni multiple comparison test. All data were presented as the mean values ± standard error of the mean (SEM). The findings were considered statistically significant at $p < 0.05$.

5. Conclusions

In summary, our data indicate that the application of a single dose of CVF 24 h prior to induction of arteriogenesis by FAL significantly promotes collateral vessel growth, as shown by increased perfusion recovery based on enhanced vascular cell proliferation and, hence, an increased collateral diameter. The administration of CVF already prior to the induction of arteriogenesis resulted in a massive mobilization of neutrophils to the peripheral blood and an increased PNA formation already at the time point when arteriogenesis was induced by FAL. Accordingly, we conclude from study that the prerequisites for efficient arteriogenesis, which are especially important for mast cell recruitment and activation—which, in turn, are essential for monocyte recruitment during the process of arteriogenesis—were already present and available at the moment when collateral artery growth was elicited by increased fluid shear stress. Thus, in situations where arteriogenesis is artificially promoted, e.g., prior to abdominal aortic aneurism surgery [3], the timely administration of CVF may be of particular interest. This could become a particular pro-arteriogenic treatment in patients since a humanized form of CVF already exists and would be ready to use [15,56–58].

Supplementary Materials: The following supporting information can be downloaded at: https://www.mdpi.com/article/10.3390/ijms23158454/s1.

Author Contributions: Surgeries, A.B., P.G., C.A., M.K. and L.B.; histology, P.G. and S.O.A.-O.; conceptualization, P.G., S.O.A.-O. and E.D.; methodology, P.G., A.B., S.O.A.-O. and E.D.; software P.G.; validation, P.G., S.O.A.-O., M.L. and E.D.; formal analysis, A.B., P.G., S.O.A.-O. and C.A.; investigation, P.G., S.O.A.-O. and E.D.; resources, S.O.A.-O., P.G., A.B. and E.D.; data curation, A.B., P.G. and S.O.A.-O.; writing—original draft preparation, S.O.A.-O. and P.G.; writing—review and editing, P.G., S.O.A.-O., C.A., M.K., L.R., L.B., K.T.P., M.L. and E.D.; visualization, P.G. and S.O.A.-O.; supervision, M.L. and E.D.; and project administration and funding acquisition, E.D. All authors have read and agreed to the published version of the manuscript.

Funding: This research was funded by The World Academy of Science and DFG (TWAS-DFG), grant number DFG DE 685/6-1 (S.O.A.-O.), and by Münchner Universitätsgesellschaft (E.D.), Förderprogramm für Forschung und Lehre (FöFoLe) (P.G.), the Lehre@LMU program, all from the Ludwig-Maximilians-Universität, Munich, Germany.

Institutional Review Board Statement: The study was approved by the Institutional Review Board of Walter-Brendel-Centre of Experimental Medicine and the Bavarian Animal Care and Use Committee (ethical approval code: ROB-55.2Vet-2532.Vet_02-17-99).

Informed Consent Statement: Not applicable.

Data Availability Statement: The data presented in this study are available on request from P.G. and A.B.

Acknowledgments: The authors wish to thank C. Eder and D. van den Heuvel for the technical support. We thank L. Richter as Manager of the Core Facility Flow Cytometry, Biomedical Center (BMC) of the Ludwig-Maximilians-Universität München for her assistance and access to the analytical instruments of the institute.

Conflicts of Interest: The authors declare no conflict of interest.

References

1. Deindl, E.; Schaper, W. The art of arteriogenesis. *Cell Biochem. Biophys.* **2005**, *43*, 1–15. [CrossRef]
2. Petroff, D.; Czerny, M.; Kölbel, T.; Melissano, G.; Lonn, L.; Haunschild, J.; Von Aspern, K.; Neuhaus, P.; Pelz, J.; Epstein, D.M.; et al. Paraplegia prevention in aortic aneurysm repair by thoracoabdominal staging with 'minimally invasive staged segmental artery coil embolisation' (MIS²ACE): Trial protocol for a randomised controlled multicentre trial. *BMJ Open* **2019**, *9*, e025488. [CrossRef] [PubMed]
3. Etz, C.D.; Kari, F.A.; Mueller, C.S.; Brenner, R.M.; Lin, H.-M.; Griepp, R.B. The collateral network concept: Remodeling of the arterial collateral network after experimental segmental artery sacrifice. *J. Thorac. Cardiovasc. Surg.* **2011**, *141*, 1029–1036. [CrossRef]
4. Stoller, M.; Seiler, C. Effect of Permanent Right Internal Mammary Artery Closure on Coronary Collateral Function and Myocardial Ischemia. *Circ. Cardiovasc. Interv.* **2017**, *10*, e004990. [CrossRef] [PubMed]
5. Faber, J.E.; Chilian, W.M.; Deindl, E.; van Royen, N.; Simons, M. A brief etymology of the collateral circulation. *Arter. Thromb. Vasc. Biol.* **2014**, *34*, 1854–1859. [CrossRef] [PubMed]
6. Lasch, M.; Nekolla, K.; Klemm, A.H.; Buchheim, J.-I.; Pohl, U.; Dietzel, S.; Deindl, E. Estimating hemodynamic shear stress in murine peripheral collateral arteries by two-photon line scanning. *Mol. Cell. Biochem.* **2019**, *453*, 41–51. [CrossRef]
7. Lasch, M.; Kleinert, E.C.; Meister, S.; Kumaraswami, K.; Buchheim, J.-I.; Grantzow, T.; Lautz, T.; Salpisti, S.; Fischer, S.; Troidl, K.; et al. Extracellular RNA released due to shear stress controls natural bypass growth by mediating mechanotransduction in mice. *Blood* **2019**, *134*, 1469–1479. [CrossRef]
8. Chillo, O.; Kleinert, E.C.; Lautz, T.; Lasch, M.; Pagel, J.-I.; Heun, Y.; Troidl, K.; Fischer, S.; Caballero-Martinez, A.; Mauer, A.; et al. Perivascular Mast Cells Govern Shear Stress-Induced Arteriogenesis by Orchestrating Leukocyte Function. *Cell Rep.* **2016**, *16*, 2197–2207. [CrossRef]
9. Hoefer, I.E.; van Royen, N.; Rectenwald, J.E.; Deindl, E.; Hua, J.; Jost, M.; Grundmann, S.; Voskuil, M.; Ozaki, C.K.; Piek, J.J.; et al. Arteriogenesis proceeds via ICAM-1/Mac-1- mediated mechanisms. *Circ. Res.* **2004**, *94*, 1179–1185. [CrossRef] [PubMed]
10. Deindl, E.; Ziegelhöffer, T.; Kanse, S.M.; Fernandez, B.; Neubauer, E.; Carmeliet, P.; Preissner, K.T.; Schaper, W. Receptor-independent role of the urokinase-type plasminogen activator during arteriogenesis. *FASEB J.* **2003**, *17*, 1174–1176. [CrossRef]
11. Kluever, A.K.; Braumandl, A.; Fischer, S.; Preissner, K.T.; Deindl, E. The Extraordinary Role of Extracellular RNA in Arteriogenesis, the Growth of Collateral Arteries. *Int. J. Mol. Sci.* **2019**, *20*, 6177. [CrossRef] [PubMed]
12. Du Cheyne, C.; Tay, H.; De Spiegelaere, W. The complex TIE between macrophages and angiogenesis. *Anat. Histol. Embryol.* **2020**, *49*, 585–596. [CrossRef] [PubMed]
13. Troidl, C.; Jung, G.; Troidl, K.; Hoffmann, J.; Mollmann, H.; Nef, H.; Schaper, W.; Hamm, C.W.; Schmitz-Rixen, T. The temporal and spatial distribution of macrophage subpopulations during arteriogenesis. *Curr. Vasc. Pharm.* **2013**, *11*, 5–12. [CrossRef]

14. Holers, V.M.; Thurman, J.M. The alternative pathway of complement in disease: Opportunities for therapeutic targeting. *Mol. Immunol.* **2004**, *41*, 147–152. [CrossRef] [PubMed]
15. Vogel, C.W.; Fritzinger, D.C. Humanized cobra venom factor: Experimental therapeutics for targeted complement activation and complement depletion. *Curr. Pharm. Des.* **2007**, *13*, 2916–2926. [CrossRef] [PubMed]
16. Merle, N.S.; Church, S.E.; Fremeaux-Bacchi, V.; Roumenina, L.T. Complement System Part I-Molecular Mechanisms of Activation and Regulation. *Front. Immunol.* **2015**, *6*, 262. [CrossRef] [PubMed]
17. Walport, M.J. Complement. First of two parts. *N. Engl. J. Med.* **2001**, *344*, 1058–1066. [CrossRef] [PubMed]
18. Vogel, C.W. The Role of Complement in Myocardial Infarction Reperfusion Injury: An Underappreciated Therapeutic Target. *Front. Cell Dev. Biol.* **2020**, *8*, 606407. [CrossRef]
19. Huber-Lang, M.; Sarma, J.V.; Zetoune, F.S.; Rittirsch, D.; A Neff, T.; McGuire, S.R.; Lambris, J.; Warner, R.L.; A Flierl, M.; Hoesel, L.M.; et al. Generation of C5a in the absence of C3: A new complement activation pathway. *Nat. Med.* **2006**, *12*, 682–687. [CrossRef] [PubMed]
20. Vogel, C.W.; Bredehorst, R.; Fritzinger, D.C.; Grunwald, T.; Ziegelmüller, P.; Kock, M.A. Structure and function of cobra venom factor, the complement-activating protein in cobra venom. *Adv. Exp. Med. Biol.* **1996**, *391*, 97–114.
21. Till, G.O.; Morganroth, M.L.; Kunkel, R.; Ward, P.A. Activation of C5 by cobra venom factor is required in neutrophil-mediated lung injury in the rat. *Am. J. Pathol.* **1987**, *129*, 44–53. [PubMed]
22. Morgan, B.P.; Harris, C.L. Complement therapeutics; history and current progress. *Mol. Immunol.* **2003**, *40*, 159–170. [CrossRef]
23. Bossi, F.; Tripodo, C.; Rizzi, L.; Bulla, R.; Agostinis, C.; Guarnotta, C.; Munaut, C.; Baldassarre, G.; Papa, G.; Zorzet, S.; et al. C1q as a unique player in angiogenesis with therapeutic implication in wound healing. *Proc. Natl. Acad. Sci. USA* **2014**, *111*, 4209–4214. [CrossRef] [PubMed]
24. Langer, H.F.; Chung, K.-J.; Orlova, V.V.; Choi, E.Y.; Kaul, S.; Kruhlak, M.J.; Alatsatianos, M.; DeAngelis, R.A.; Roche, P.A.; Magotti, P.; et al. Complement-mediated inhibition of neovascularization reveals a point of convergence between innate immunity and angiogenesis. *Blood* **2010**, *116*, 4395–4403. [CrossRef]
25. Nozaki, M.; Raisler, B.J.; Sakurai, E.; Sarma, J.V.; Barnum, S.R.; Lambris, J.D.; Chen, Y.; Zhang, K.; Ambati, B.K.; Baffi, J.Z.; et al. Drusen complement components C3a and C5a promote choroidal neovascularization. *Proc. Natl. Acad. Sci. USA* **2006**, *103*, 2328–2333. [CrossRef]
26. Candinas, D.; Lesnikoski, B.-A.; Robson, S.C.; Miyatake, T.; Scesney, S.M.; Marsh, H.C.; Ryan, U.S.; Dalmasso, A.P.; Hancock, W.W.; Bach, F.H. Effect of repetitive high-dose treatment with soluble complement receptor type 1 and cobra venom factor on discordant xenograft survival. *Transplantation* **1996**, *62*, 336–342. [CrossRef]
27. Nording, H.; Baron, L.; Haberthür, D.; Emschermann, F.; Mezger, M.; Sauter, M.; Sauter, R.; Patzelt, J.; Knoepp, K.; Nording, A.; et al. The C5a/C5a receptor 1 axis controls tissue neovascularization through CXCL4 release from platelets. *Nat. Commun.* **2021**, *12*, 3352. [CrossRef]
28. Bongrazio, M.; Pries, A.R.; Zakrzewicz, A. The endothelium as physiological source of properdin: Role of wall shear stress. *Mol. Immunol.* **2003**, *39*, 669–675. [CrossRef]
29. Chen, J.Y.; Cortes, C.; Ferreira, V.P. Properdin: A multifaceted molecule involved in inflammation and diseases. *Mol. Immunol.* **2018**, *102*, 58–72. [CrossRef]
30. Fischetti, F.; Tedesco, F. Cross-talk between the complement system and endothelial cells in physiologic conditions and in vascular diseases. *Autoimmunity* **2006**, *39*, 417–428. [CrossRef]
31. Turner, N.A.; Moake, J. Assembly and activation of alternative complement components on endothelial cell-anchored ultra-large von Willebrand factor links complement and hemostasis-thrombosis. *PLoS ONE* **2013**, *8*, e59372. [CrossRef] [PubMed]
32. Turner, N.; Nolasco, L.; Nolasco, J.; Sartain, S.; Moake, J. Thrombotic microangiopathies and the linkage between von Willebrand factor and the alternative complement pathway. *Semin. Thromb. Hemost.* **2014**, *40*, 544–550. [CrossRef] [PubMed]
33. Wu, F.; Zou, Q.; Ding, X.; Shi, D.; Zhu, X.; Hu, W.; Liu, L.; Zhou, H. Complement component C3a plays a critical role in endothelial activation and leukocyte recruitment into the brain. *J. Neuroinflamm.* **2016**, *13*, 23. [CrossRef]
34. Laudes, I.J.; Chu, J.C.; Huber-Lang, M.; Guo, R.-F.; Riedemann, N.C.; Sarma, J.V.; Mahdi, F.; Murphy, H.S.; Speyer, C.; Lu, K.T.; et al. Expression and function of C5a receptor in mouse microvascular endothelial cells. *J. Immunol.* **2002**, *169*, 5962–5970. [CrossRef]
35. Scholz, D.; Ito, W.; Fleming, I.; Deindl, E.; Sauer, A.; Wiesnet, M.; Busse, R.; Schaper, J. Ultrastructure and molecular histology of rabbit hind-limb collateral artery growth (arteriogenesis). *Virchows Arch.* **2000**, *436*, 257–270. [CrossRef] [PubMed]
36. Vogel, C.W.; Fritzinger, D.C. Cobra venom factor: Structure, function, and humanization for therapeutic complement depletion. *Toxicon* **2010**, *56*, 1198–1222. [CrossRef] [PubMed]
37. Vogel, C.W.; Müller-Eberhard, H.J. Cobra venom factor: Improved method for purification and biochemical characterization. *J. Immunol. Methods* **1984**, *73*, 203–220. [CrossRef]
38. Pickering, R.J.; Wolfson, M.R.; Good, R.A.; Gewurz, H. Passive hemolysis by serum and cobra venom factor: A new mechanism inducing membrane damage by complement. *Proc. Natl. Acad. Sci. USA* **1969**, *62*, 521–527. [CrossRef] [PubMed]
39. McCall, C.E.; De Chatelet, L.R.; Brown, D.; Lachmann, P. New biological activity following intravascular activation of the complement cascade. *Nature* **1974**, *249*, 841–843. [CrossRef]
40. Schmid, E.; Warner, R.L.; Crouch, L.D.; Friedl, H.P.; Till, G.O.; Hugli, T.E.; Ward, P.A. Neutrophil chemotactic activity and C5a following systemic activation of complement in rats. *Inflammation* **1997**, *21*, 325–333. [CrossRef]

41. Mitchell, R.H.; McClelland, R.M.; Kampschmidt, R.F. Comparison of neutrophilia induced by leukocytic endogenous mediator and by cobra venom factor. *Proc. Soc. Exp. Biol. Med.* **1982**, *169*, 309–315. [CrossRef]
42. Xu, G.; Feng, Y.; Li, D.; Zhou, Q.; Chao, W.; Zou, L. Importance of the Complement Alternative Pathway in Serum Chemotactic Activity During Sepsis. *Shock* **2018**, *50*, 435–441. [CrossRef]
43. Yun, S.H.; Sim, E.-H.; Goh, R.-Y.; Park, J.-I.; Han, J.-Y. Platelet Activation: The Mechanisms and Potential Biomarkers. *Biomed. Res. Int.* **2016**, *2016*, 9060143. [CrossRef] [PubMed]
44. Bakogiannis, C.; Sachse, M.; Stamatelopoulos, K.; Stellos, K. Platelet-derived chemokines in inflammation and atherosclerosis. *Cytokine* **2019**, *122*, 154157. [CrossRef]
45. Chatterjee, M.; Gawaz, M. Platelet-derived CXCL12 (SDF-1α): Basic mechanisms and clinical implications. *J. Thromb. Haemost.* **2013**, *11*, 1954–1967. [CrossRef] [PubMed]
46. Yang, D. Chapter 85-Anaphylatoxins. In *Handbook of Biologically Active Peptides*, 2nd ed.; Kastin, A.J., Ed.; Academic Press: Boston, MA, USA, 2013; pp. 625–630.
47. Ehrengruber, M.U.; Geiser, T.; Deranleau, D.A. Activation of human neutrophils by C3a and C5A. Comparison of the effects on shape changes, chemotaxis, secretion, and respiratory burst. *FEBS Lett.* **1994**, *346*, 181–184. [PubMed]
48. Gaudenzio, N.; Sibilano, R.; Marichal, T.; Starkl, P.; Reber, L.L.; Cenac, N.; McNeil, B.D.; Dong, X.; Hernandez, J.D.; Sagi-Eisenberg, R.; et al. Different activation signals induce distinct mast cell degranulation strategies. *J. Clin. Investig.* **2016**, *126*, 3981–3998. [CrossRef]
49. Hartmann, K.; Henz, B.M.; Krüger-Krasagakes, S.; Köhl, J.; Burger, R.; Guhl, S.; Haase, I.; Lippert, U.; Zuberbier, T. C3a and C5a stimulate chemotaxis of human mast cells. *Blood* **1997**, *89*, 2863–2870. [CrossRef]
50. Reichel, C.A.; Uhl, B.; Lerchenberger, M.; Puhr-Westerheide, D.; Rehberg, M.; Liebl, J.; Khandoga, A.; Schmalix, W.; Zahler, S.; Deindl, E.; et al. Urokinase-type plasminogen activator promotes paracellular transmigration of neutrophils via Mac-1, but independently of urokinase-type plasminogen activator receptor. *Circulation* **2011**, *124*, 1848–1859. [CrossRef]
51. Pipp, F.; Heil, M.; Issbrücker, K.; Ziegelhoeffer, T.; Martin, S.; Van Den Heuvel, J.; Weich, H.; Fernandez, B.; Golomb, G.; Carmeliet, P.; et al. VEGFR-1-selective VEGF homologue PlGF is arteriogenic: Evidence for a monocyte-mediated mechanism. *Circ. Res.* **2003**, *92*, 378–385. [CrossRef]
52. Ito, W.D.; Arras, M.; Winkler, B.; Scholz, D.; Schaper, J.; Schaper, W. Monocyte chemotactic protein-1 increases collateral and peripheral conductance after femoral artery occlusion. *Circ. Res.* **1997**, *80*, 829–837. [CrossRef] [PubMed]
53. Wynn, T.A.; Vannella, K.M. Macrophages in Tissue Repair, Regeneration, and Fibrosis. *Immunity* **2016**, *44*, 450–462. [CrossRef] [PubMed]
54. Arras, M.; Ito, W.D.; Scholz, D.; Winkler, B.; Schaper, J.; Schaper, W. Monocyte activation in angiogenesis and collateral growth in the rabbit hindlimb. *J. Clin. Investig.* **1998**, *101*, 40–50. [CrossRef] [PubMed]
55. Limbourg, A.; Korff, T.; Napp, L.C.; Schaper, W.; Drexler, H.; Limbourg, F. Evaluation of postnatal arteriogenesis and angiogenesis in a mouse model of hind-limb ischemia. *Nat. Protoc.* **2009**, *4*, 1737–1746. [CrossRef]
56. Ing, M.; Hew, B.E.; Fritzinger, D.C.; Delignat, S.; Lacroix-Desmazes, S.; Vogel, C.-W.; Rayes, J. Absence of a neutralizing antibody response to humanized cobra venom factor in mice. *Mol. Immunol.* **2018**, *97*, 1–7. [CrossRef]
57. Vogel, C.W.; Gorsuch, B.; Stahl, G.; Vogel, C.-W. Complement depletion with humanised cobra venom factor: Efficacy in preclinical models of vascular diseases. *Thromb. Haemost.* **2015**, *113*, 548–552.
58. Vogel, C.W.; Finnegan, P.W.; Fritzinger, D.C. Humanized cobra venom factor: Structure, activity, and therapeutic efficacy in preclinical disease models. *Mol. Immunol.* **2014**, *61*, 191–203. [CrossRef] [PubMed]

Review

Small GTPases and Their Regulators: A Leading Road toward Blood Vessel Development in Zebrafish

Ritesh Urade [1], Yan-Hui Chiu [1], Chien-Chih Chiu [1,2,*] and Chang-Yi Wu [1,2,3,4,*]

1. Department of Biological Sciences, National Sun Yat-sen University, Kaohsiung 804, Taiwan; uraderit@gmail.com (R.U.); paty.121351@gmail.com (Y.-H.C.)
2. Department of Biotechnology, Kaohsiung Medical University, Kaohsiung 807, Taiwan
3. Doctoral Degree Program in Marine Biotechnology, National Sun Yat-sen University, Kaohsiung 804, Taiwan
4. Institute of Medical Science and Technology, National Sun Yat-sen University, Kaohsiung 804, Taiwan
* Correspondence: cchiu@kmu.edu.tw (C.-C.C.); cywu@mail.nsysu.edu.tw (C.-Y.W.); Tel.: +886-7-3121101 (ext. 2368) (C.-C.C.); +886-7-5252-000 (ext. 3627) (C.-Y.W.)

Abstract: Members of the Ras superfamily have been found to perform several functions leading to the development of eukaryotes. These small GTPases are divided into five major subfamilies, and their regulators can "turn on" and "turn off" signals. Recent studies have shown that this superfamily of proteins has various roles in the process of vascular development, such as vasculogenesis and angiogenesis. Here, we discuss the role of these subfamilies in the development of the vascular system in zebrafish.

Keywords: small GTPases; GTP-binding proteins; vascular development and zebrafish

1. Introduction

Small GTPases are GTP-binding proteins frequently found in eukaryotes. These are profoundly reported to have roles in processes such as differentiation, proliferation, morphology, adhesion, survival, migration, apoptosis, cytoskeletal reorganization, cellular polarity, cell cycle progression and many noteworthy biological functions in cells. These proteins cycle between their active form, which is GTP bound, and their inactive form, which is GDP bound, which can affect almost all cellular processes [1]. Approximately 160 members of the small GTPase family have been reported to date [2]. The Ras (rat sarcoma) subfamily of small GTPases contains the largest number of members; hence, sometimes it is called a Ras GTPase [3]. Depending on their structures and functions, these proteins are divided into five main categories: Ras, Ras homology (Rho), Ras proteins in the brain (Rab), Ras nuclear protein (Ran) and adenosine diphosphate ribosylation factor (Arf)/ secretion-associated and Ras-related (Sar) GTPase [4,5]. The regulation of small GTPases (Figure 1) is controlled by three groups of proteins, namely, GTPase-activating/accelerating proteins (GAPs) assisting in hydrolyzing GTP, guanine nucleotide exchange factors (GEFs) stimulating the exchange of GDP to GTP, and guanine nucleotide dissociation inhibitors (GDIs), which accumulate GDP- or GTP-bound small GTPase inside the cytoplasm (by masking their C-terminal isoprenyl group) and terminate its activation. When they are bound to GTP, Ras GTPase forms an association with effectors that lead the way for downstream signaling. These regulators work immediately upstream of the small GTPases to provide a link between small GTPase activation and their receptors [6]. Due to back-and-forth rotation of GTPases, their regulators coordinate and take part in many biological functions. Many regulators of small GTPases coexist in most cells to control the smooth coordination of small GTPases. Many recent studies have revealed that these regulators have their own regulatory mechanism by which they process cellular signals and accumulate specific cell responses. Small GTPases were found to assist the process of blood vessel development. Three pathways are critical for the process of blood vessel development in zebrafish: vascular endothelial growth factor (VEGF) signaling [7], Notch signaling [8] and bone morphogenetic protein (BMP) signaling [9]. Impaired small GTPases can contribute to serious threats such as cancer and developmental malfunctioning [4]. Despite this importance, their regulatory role in vascular development is unclear. Hence, in this review, we discuss them along with their regulators and in blood vessel development in zebrafish.

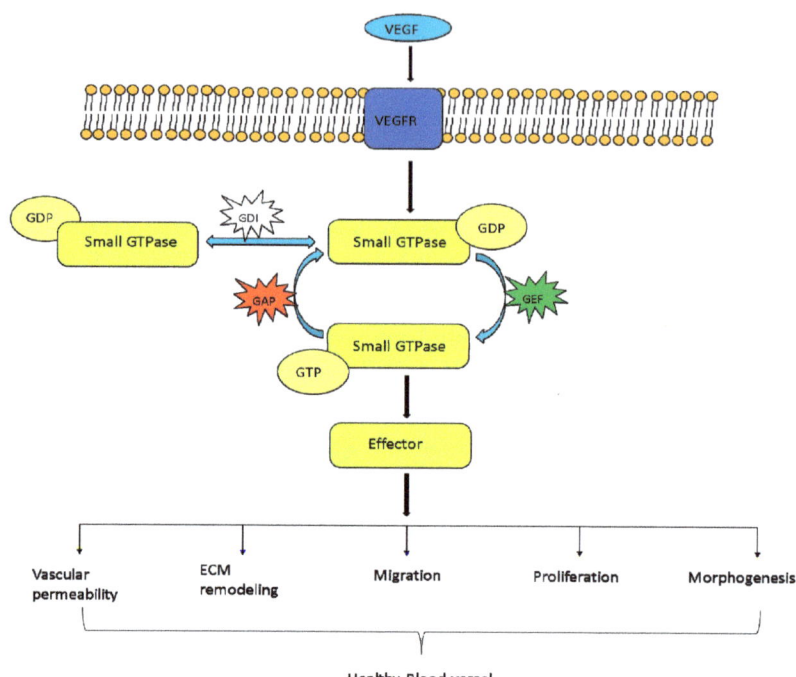

Figure 1. Small GTPase regulation leading to healthy vessels: Cycling of GTPases in the active state and inactive state. Their activation is governed by GEFs, which remove GDP and allow excess cytoplasmic GTP to attach. The binding of active GTPase to effector proteins aggravates the cell response to give rise to blood vessels. GAP, by increasing GTPase activity, turns off the switch for GTPases. Inactive GTPase aggregates in the cytosol via GDIs. By activating effector proteins, downstream processes led to the development of healthy blood vessels (adapted and modified from Cherfils and Zeghouf 2013 [10]).

2. Blood Vessel Development

Vasculature development is an important process for the survival of any organism. Despite this much importance, we still are unaware of how the process of its formation takes place. How blood vessels maintain their structure, diameter, permeability, shear stress, etc., these are some of the aspects we are trying to figure out. Some researchers have linked the answers to these questions to endothelial cells, since these cells have the potential to form new vessels with various mechanisms. Several studies on the in vitro culturing of embryonic stem cells (ESCs) showed an endothelial progenitor named hemangioblast [11,12]. Endothelial progenitor cells (EPCs) arise from hemangioblasts, which repair and revascularize the ischemic retina [13]. These EPCs, by two processes, form a blood vessel. The first is vasculogenesis, in which blood vessels form by de novo synthesis [14,15], and the other is angiogenesis, which uses preexisting vessels to extend and form new blood vessels [16]. Angiogenic cues or ischemia increase endothelial permeability, which gives a chance to matrix metalloproteins to debase the extracellular matrix, which relieves endothelial cell (EC)-pericyte contact and ultimately releases growth factors. EC permeability is controlled by various factors, such as thrombin, VEGF and sphingosine 1 phosphate. These factors guide the loss of junctional integrity, which is a reversible process [17]. This gives the bordering cells a space to influx fluids and small molecules due to the absence of cellular contacts. Due to coordinated activation of each GTPase, ECs tend to migrate to promigratory cues and then proliferate to reach their final destination, where they undergo morphogenesis to form a functional lumen and further branches if required [6].

Two forms of angiogenesis have been proposed explicitly: sprouting angiogenesis and intussusceptive angiogenesis. In sprouting angiogenesis, branching of primary blood vessels to form a new vessel takes place [18]. Sequential events are as follows: The desired site directs the ECs, which create a bipolar mode and align endothelial cells and form lumen and tip cells that sprout from distant sites and connect and initiate blood circulation [19]. In intussusceptive angiogenesis, longitudinal splitting of a primary vessel takes place into two different new branches, hence increasing the vascular surface area [20]. Both processes not only provide oxygen but also supply required nutrients to the desired sites and help eliminate waste products. Each angiogenesis is controlled by proangiogenic factors such as VEGF and its receptors VEGFR1 and VEGFR2. Activation of these tyrosine kinase receptors leads to activation of different pathways, such as MAPK, PI3K, and PLCγ, favoring angiogenesis [21]. Cancer cells take over some of these molecules to fulfil their own requirements, such as oxygen and nutrients, for metastatic spread. Where insufficient vessels or short growth leads to tissue ischemia, unnecessary vessel growth or abnormal repair can lead to cancer, inflammation disorders, and retinopathy [22]. The process of angiogenesis is governed by activators as well as inhibitors. The mechanism and location of angiogenic activators and inhibitors could lead us to design a specific drug.

3. Why Zebrafish and Our Recent Study in the Field of GTPase Related Protein

Zebrafish (*Danio rerio*) is a freshwater fish belonging to the *Cyprinidae* family native to Southeast Asia [23]. Increasing restriction on using animal model organisms in research has paved the way for zebrafish to become a popular vertebrate model in many fields, such as developmental biology, toxicology and oncology. Zebrafish provide a series of advantages over other vertebrate animal models, such as external fertilization, fecundity, rapid developmental ability, favorable forward and reverse genetic manipulation, availability of cell lines, availability of transgenic lines and tractability to genetic manipulation. Sequencing of the zebrafish genome revealed 70% similarity in its protein coding regions to humans and 84% genes linked with human diseases [24]. Several genetic studies on zebrafish revealed ferocious conservation of molecular pathways in vertebrates for the development and physiology of blood vessels [25]. Using zebrafish to study vascular development has intensively identified many molecules that control artery-vein identity, caudal vein plexus (CVP) formation, and pattern intersegmental vessel (ISV) due to their optical transparency and the availability of labeling techniques for endothelial cells with specific antibodies or tagging with specific fluorescence, allowing us to observe cellular localization, migration, division and rearrangement during vasculogenesis and angiogenesis [20]. Another advantage of the zebrafish is the ability to rapidly and inexpensively downregulate gene expression using morpholino (MO). Morpholinos are oligonucleotides with a modified nondegradable backbone designed to block translation or splicing of a specific mRNA, leading to dramatic reduction of gene expression (Figure 2).

We previously reported the role of the transcription factors Islet2 and Nr2f1b in specification of the vein and tip cell identity mediated by the Notch pathway in zebrafish (Figure 2) [26,27]. To further explore this possibility, we used an unbiased microarray approach and identified many novel genes related to vasculature development regulated by the Islet2 and Nr2f1b transcription factors. We noticed an interesting group of GTPase-related genes, including *G-coupled receptor-like (gpcrl)*, *septin 8b (sept8b)*, *Rho-related protein (ect2)*, *rhoub (ras homolog gene family)*, *RAS-like family 11 (rasl11b)* and *Wiskott-Aldrich syndrome protein (WASF1)*. The putative function related to GTPase signals is shown in Figure 3.

GTPases are key proteins in many critical biological processes, including hormonal and sensory signals, ribosomal protein synthesis, cytoskeletal organization, signal transduction cascades and motility. Small GTPases are hydrolase enzymes present in the cytosol that can bind and hydrolyze GTP and GDP. These enzymes have been shown to have diverse roles in the development of healthy vasculature. The small GTPase Rap1 has been shown to promote VEGFR2 activation and angiogenesis [28]. The Ras GTPase family has been shown to function in vascular patterning via semaphorin-Plexin signaling [29]. However, the GTPase genes we list above do not have any yet known functions in vessels, and we have currently addressed these questions. Since humans and zebrafish share a common mechanism for the process of vessel development [30], we will review available small GTPases and their regulators involved in the process of vascular development.

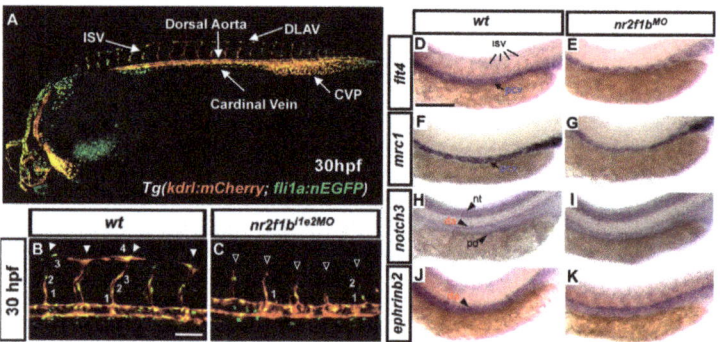

Figure 2. Study the function of genes in vascular development. (**A**) Confocal image of transgenic fish *Tg(kdrl:mCherry; fli1a:nEGFP)* where mCherry expression is in the endothelial cells and GFP expression is in the nucleus at 30 hpf. The image shows clear vessel structures of the dorsal aorta (da), cardinal vein (pcv), ISV, DLAV and CVP. (**B,C**) Knockdown of *nr2f1b* in transgenic fish results in vascular defects, i.e., Fewer ISVs migrated to the top of the embryo, and fewer ISV cells migrated per ISV in morphants (hollow arrowheads and fewer numbers) than in the wild-type control (arrowheads). Scale bars in panels (**B**) and (**C**) represent 50 μm. (**D–K**) In situ hybridization data showed that knockdown of *nr2f1b* reduced the expression of vascular markers. Scale bars in figures (**D–K**) represent 200 μm. Images (**B–K**) courtesy of R.-F. Li, reproduced/adapted from Li et al. (2015) [27] with permission from *J. Biomed. Sci.*

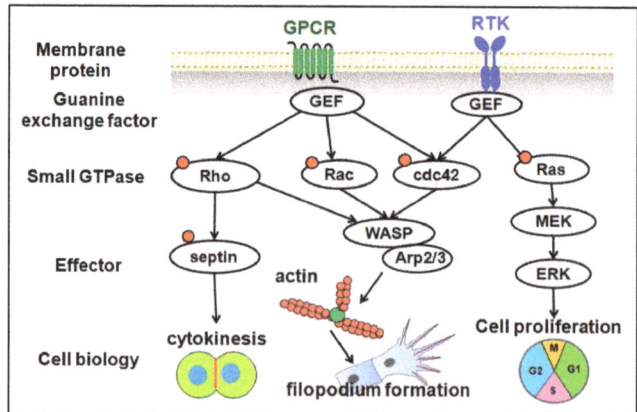

Figure 3. Schematic drawings of the proteins involved in GTPase signals and cellular function related to the cell biological process of vascular development.

4. Small GTPases and Their Regulators

Due to mutations in the GTPase domain of small GTPase (in various cancers) this family have approximately 160 members, which makes them a superfamily [31]. The process of angiogenesis is controlled by various angiogenic factors, including VEGF. VEGF binding to its tyrosine kinase receptors VEGFR1 and VEGFR2 stimulates downstream signaling cascades such as MAPK, PI3K and PLCγ, which can ultimately contribute to the process of angiogenesis [21]. There is significant evidence supporting the contribution of these proteins as downstream effectors of the VEGF signaling pathway in angiogenic processes. According to their sequence, structure, and functions, this wide-ranging superfamily has been further classified into five subfamilies of Ras, Rho, Ran, Rab, and Arf/Sar GTPase [4,5].

4.1. Ras Family

The Ras GTPase family is the first family among others and the most diversified family. Due to prenylation, most members of this family are present in the plasma membrane [32]. The activated Ras members interact with the effector moiety and play a different cellular role in the development, proliferation, differentiation, and survival of eukaryotes [4]. A total of 38 members have been reported in this family [5]. The conserved and ubiquitously distributed forms of the Ras family include H-ras, N-ras, and K-ras, which have different biological functions [31].

Pezeron et al. reported the first cytosolic small GTPase rasl11b in the development of zebrafish and showed that it has a zygotic and maternal origin. Rasl11b's dorso-marginal expression shows its role in the formation of the endodermal and/or mesodermal layer. Downregulation of rasl11b acts as a suppressor of the EGF-CEF factor *one-eyed pinhead* (oep) phenotype (such as an altered A-P axis, failing to develop endoderm, prechordal plate, and posterior mesoderm [33]) and showed that it can partially rescue prechordal plate and endoderm formation in oep-deficient embryos. However, loss of rasl11b function halted the formation mesendoderm without activation of Nodal signaling when attempted in other than oep mutants. This correlation between oep and rasl11b reveals that oep can influence mesendoderm formation without taking part in the Nodal-Smad2 signaling pathway [34].

Another frequently activated oncogene from the Ras family is K-Ras. Mouse knockout studies have already established their role in normal developmental processes [35,36]. In vivo studies by Liu et al. in zebrafish showed that K-Ras expression starts from the single-cell to throughout the embryo. Morpholino injection showed reduced blood circulation with a lower heart-beat rate, and the accumulation of blood cells was often found away from circulation sites when compared to the negative control morpholino. Apart from these, defects that increased in later stages showed a disorganized subintestinal vein (SIV) with a reduced number of vessel branches along with a reduction in size and/or ectopic blood vessels in K-Ras morpholino-injected embryos. All defects caused by morpholinos could effectively be rescued after coinjection with K-Ras mRNA, supporting its role in hematopoiesis and angiogenesis. Treatment with PI3K/Akt and Mek-Erk1/2 inhibitors provides direct evidence in vivo of the involvement of PI3K-Akt signaling in orchestrating K-Ras signaling for these two salient processes [37].

Semaphorin and its receptors Plexins have been associated with regulating angioblast behaviors [29]. The members of the plexin family co-interact with small GTPases, such as the Rnd, R-Ras, M-Ras and Rap families, and function as Ras-GAPs [38–41]. In zebrafish, only a single semaphorin3e is expressed in DA, ECs and primary motoneurons and is associated with delayed angioblast migration from DA to structural ISV [29]. Apart from semaphorin, its receptors PlexinD1 and PlexinB2 were found to be expressed in angioblasts. PlxnD1 was found to be expressed in angioblasts and within DA, PCV, and ISV, which shows its involvement in both processes of vasculogenesis and angiogenesis [29]. Loss of one of the receptors, PlxnB2, delayed ISV, which resembles the loss of sema3e morphants, while the loss of another receptor, PlxnD1, in an *out-of-bound* (obd) mutant results in precocious sprouting [42]. This riveting result shows that Sema3e and PlxnD1 do not act as ligand–receptor pairs here for vascular morphogenesis, but PlxnB2 and Sema3e do.

A genetic interaction study between PlxnB2 and Sema3e controls the time of sprouting of angioblasts [43]. The transplantation experiment showed that PlxnB2 and Sema3e act autonomously to control the timing of angioblast migration. ECs fail to sense repelling signals produced by semaphorin in the absence of Plexins. Torres et al. morpholino studies in the obd mutant show that loss of one of two Sema3e or PlxnB2 produces an intermediate phenotype, concluding the role of PlxnD1 and Sema3e/PlxnB2 in antagonizing each other's role in tuning the timing of ISV sprouting but following different signaling and independent pathways downstream of each receptor [43].

Integrins are extracellular matrix receptors present on endothelial cells that play crucial roles in the process of blood vessel development in zebrafish, especially $\alpha_5\beta_1$, $\alpha_v\beta_3$ and $\alpha_v\beta_8$, by binding to ECM components [41,44]. Lakshmikanthan et al. showed the role of Rap GTPase in the process of angiogenesis for the activation of VEGF signaling and paved the way for angiogenesis. Both isoforms Rap1a and Rap1b are required for the activation of VGFR2 kinase through integrin $\alpha_v\beta_3$. In zebrafish, Rab1b acts upstream of the VEGF signaling pathway and is expressed in ISV and has a role in initial events in ISV sprouting but does not contribute to vasculogenesis. Combinatorial effects of VEGFR2 inhibitors showed the role of Rap1bs in anterior as well as mid- trunk formation and ISV sensitivity for VEGF signaling [28].

One of the important family members is N-Ras. N-Ras signaling in zebrafish has a high degree of similarity to that in humans and is functionally conserved. N-Ras regulates venous fate of arterial-venous cell specification, hematopoiesis and EC proliferation. Overexpression of N-Ras does not have any impact on hematopoietic markers such as gata1, αe1, pu.1, l-plastin, and mpo, suggesting normal primitive hematopoiesis, although the absence of HSC markers such as cmyb and runx1 proved the complete absence of definitive hematopoiesis. Expression of N-Ras under the lmo2 promoter showed accumulation of blood cells at the axial vessel and heart chamber due to a lack of blood circulation in the head as well as in trunk vessels or could be due to defective cardiovascular development, although embryos did not survive after 5–8 dpf. Injection of fluorescein-coupled latex beads into the atrium proved the involvement of Ras signaling in this disruption of circulation. Apart from all these defects, there was defective assembly of vessels, especially DA or PCV, reduction in ISV length, defective head vasculature and slow heart beating rate in N-Ras embryos compared to control embryos.

As well as small GTPases, GAPs can also have a high impact on blood vessel development. A single allele of Ras GAP called Ras p21 protein activator 1 (RASA 1/p120-RasGAP) was sufficient to cause capillary malformation-arteriovenous malformation (CM-AVM) [45]. GAPs are negative regulators of small GTPase activity. Vascular defects have been noted, although there was no vascular-specific expression of the RASA1 gene. Lack of blood flow to the posterior part, incomplete formation of CVP and large caudal vascular deformities were noted in morphants. Due to this, arterial blood flow had to return to posterior cardinal vain abruptly. RASA1 works as a critical effector downstream of one of the endothelial receptors called the EPHB4 receptor, which promotes the segregation of endothelial cells to form the aorta as well as cardinal vein [46]. A knockdown study found very similar defects in vasculature; in fact, a reduction in RASA1 leads to compromised full function of the EPHB4 receptor. Compared to normal embryos, both morphants (RASA1 and EPBH4) sprouted more venous endothelial cells, and more venous connections were made at the expense of arterial connections. Inactivation of RAS was achieved by RASA1, proving that EPHB4-RASA1-TORC1 signaling could participate in the process of normal blood vessel development. The same phenotypes were noted when another small GTPase called RhebS16H was knocked down [46].

Lamellipodia formation and sprouting of endothelial cells from the ventral part of the dorsal aorta extend toward guiding cues in their environment to orchestrate growing blood vessels. Polo-like kinase 2 (PLK2) is a family protein conserved in ECs of vertebrates that regulates Rap1 activity to control the formation of tip cell lamellipodia but not filopodia and sprouting of endothelial cells. This lamellipodia formation and protrusion during angiogenesis was found to be dependent on focal adhesion kinase and integrin αVβ3 [47]. Knockdown of PLK2 by morpholino reduced sprouting of ECs as well as its migration and overexpression found to overcome these defects. ISVs did not reach DLAV due to failure of migration from the horizontal myoseptum. While doing so, PLK2 makes a contact with PDZ-GEF, a Rap1-GEF, to control the downstream activity of Rap1 to regulate the formation of EC focal adhesion and the growth of lamellipodia to maintain endothelial tip cell behavior.

The Ras family has often been linked to the regulation of neuronal functions. The study conducted by Yeh and Hsu, 2016 showed that members of the Ras family, such as diras1 (diras1a and diras1b), are expressed in the CNS and dorsal neuron ganglion and function in neuronal outgrowth and neuronal proliferation. Wild-type diras1 can elevate or downregulate the members of the Rho family of GTPases, Rac1 and RhoA. A knockdown study by Morpholino proved its involvement in axon guidance and maintaining the numbers of trigeminal ganglions [48].

Rap1b was found to be associated with hematopoietic stem cell development (HSC) development by promoting Notch signaling. Rap1b promotes specification of posterior lateral plate mesoderm (PLPMs) by encouraging notch signaling. However, while migrating to midline, fibronectin directs the PLPMs along the somite boundary via integrin β1. Rap1b induces the spreading, migration and adhesion of PLPMs to somites to stimulate HE specification. Rap1b was not found to be involved in the process of vascular development but was critical for HSC development. Rap1b is ubiquitously expressed and promotes HSC development by inducing hemogenic endothelium (HE) development in a cell autonomous manner [49].

4.2. Rho Family

This family, along with its regulators, controls various cellular processes, including cell polarity, cell proliferation, membrane transport, apoptosis, gene expression, and membrane transport [50,51]. Recently, the role of these small GTPases in the process of angiogenesis was reviewed by Bryan and D'Amore, 2007 [6]. The Rho family is an essential downstream effector of VEGF signaling that induces angiogenic development. Most studies of this family are associated with RhoA, Rac1 and Cdc42. Regulators of this family control various biological activities via activation or deactivation of small GTPases. The Rho family downstream of the VEGF receptor transmits various signals to activate MAPK, PI3K and PLCγ, which are the main signaling pathways that take place during the process of blood vessel development [21].

Vascular permeability is coordinated by loosening and creating a space between the cells to facilitate the influx of macromolecules. Rho GTPase was found to increase vascular ECs permeably, destabilizing adherens and tight junctions. The cell–cell contact junctions of ECs contain Rac1 and Cdc42, and these junctions dissociate during an increase in permeability [52]. RhoC negatively regulates vascular permeability in a VEGF-dependent manner by compensating for EC loss. It maintains homeostasis by creating a balance between vascular injury and repair. Apart from this, it prevents acute endothelial hyperpermeability in zebrafish. RhoC was found to be expressed in DA, PCV, ISVs and NT. However, when injected with morpholinos, no vascular defects were observed [53].

Remodeling and degradation of ECM pave the way for EC to proliferate by following angiogenic cues such as VEGF in the surroundings in the absence of cell–cell contacts to build a functional lumen. The interaction between Arhgap29, a RhoA-GAP, and its binding partner Ras interacting protein 1 (Rasip1) is necessary to modulate EC polarity and cell adhesion to the ECM to activate RhoA signaling to orchestrate the lumen [54]. RhoA's role in a study conducted by Zhu et al. showed its importance in embryonic survival [55]. The ubiquitous expression of RhoA during early embryogenesis and reduction in the level of RhoA can lead to shrinkage in overall body size along with reduced head size and body length [56]. These defects could be due to increases in the level of apoptosis during embryonic development. As a consequence, there is a reduction in two crucial factors: one is the reduction in the activation of Erk, a growth-promoting factor, and the reduction in bcl-2, an anti-apoptotic factor that could be due to an increase in apoptosis. Regulation of cell survival by RhoA is achieved via the Mek/Erk pathway during embryonic development [55]. Depletion of the RhoA-GAP called Arhgap29 increased RhoA GTPase signaling but repressed Cdc42 and Rac1 GTPase signaling.

Filopodia are thin finger-like protrusions present on the leading edge of ECs that sense their microenvironment and direct the tip EC toward promigratory signals. On the other end of the EC, adhesion should be released for the forward movement of EC. Several studies have shown that Cdc42 is associated with the formation of filopodia [57–59]. A recent study showed that these filopodia drive angiogenesis in response to activation of Cdc42 [58,59]. Ventral migration of these filopodia from caudal vein primordia leads to CVP formation. Filopodia are filled with linear F-actin filaments for CVP formation. Bmp signaling has been shown to be responsible for the migration of ECs toward the ventral side independent of EC fate determination. Given that Cdc42 regulates EC morphology, motility, proliferation and survival, this could regulate BMP signaling to bring about normal CVP formation. GAPs are negative regulators of tip cell angiogenesis, and they limit proangiogenic factors to stabilize the vasculature. One of the Rho-GAPs called ARHGAP18 was found to have a role as a fine tuner for vascular morphogenesis. It is an endogenous molecule that is expressed in ECs and curbs the formation of tip cells to promote junctional integrity [60,61]. It acts on Rho-C to destabilize EC junctions in a ROCK-dependent manner. When it is knocked down by morpholino, increased ISV lengths may be due to an increase in filopodia, supporting its role in hypersprouting [60]. It would be interesting to determine how these factors contribute to VEGF-mediated angiogenesis. Arhgef9b and fgd5 are the Cdc42 GEFs expressed in zebrafish. Apparently, Arhgef9b reduced the number of sprouts from caudal vein primordia and filopodia were noted, which shows that Arhgef9b but not fgd5 could act as a Cdc42-GEF to regulate Bmp signaling to form CVP [59]. The role of Cdc42 along with transporter proteins has been associated with the normal eye development and survival of cells in the eye [57]. However, Cdc42 inhibition severely reduces the speed at which ISVs are formed, and this reduction could be correlated with the reduction in the formation of filopodia and defects in EC proliferation; hence, inappropriate formation of tip cells occurred. Given that Cdc42 regulates the sprouting of EC to form ISVs, it would not be wrong to call it a positive regulator of vessel sprouting. Similar effects have been observed during retinal angiogenesis, showing that similar pathways are followed for vessel development in these two organs [58]. While orchestrating the patterns of vessels, RhoA GEF Syx interacts with angiomotin in the presence of VEGF-A to regulate EC migration [62,63]. A recent study showed that these two interact with a scaffold protein and form a ternary complex to promote the migration of endothelial cells [64]. Coordination between these scaffold protein is require to activate and regulate RhoA activity to lead the tip cell toward guiding cues. Wu et al., 2011 reported that Syx and RhoA regulate not only cell junctions but also EC directional migration by forming lamellipodia [65]. RhoA and Syx show localization in the gradient-dependent manner of VEGF-A toward the leading edge. Cotrafficking of RhoA and Syx is required for cell migration, which depends on another family of small GTPases Rab GTPase, showing that they work interdependently to maintain structure of embryos [65].

βPix is a scaffold protein, and a GEF for Rac and Cdc42 binds to p21-activated kinase (Pak) to regulate vascular stability. It is expressed in embryonic development in the brain as well as large blood vessels. It mainly contributes to embryonic vascular stability and hydrocephalus. Pak2a signaling works downstream of βPix to regulate cerebrovascular development. Loss of βPix led to hemorrhage in the head, signifying its part in cerebral vessel stability instead of vessel-specific breakage. Mutants were found to develop hydrocephalus [2]. Another study conducted by Liu et al. showed that βPix binds to an ARF-GAP called G-protein coupled receptor kinase interacting target (Git1); hence, Git1 functions as a molecular link between integrins and βPix, bringing about a stable vascular system [41].

The complex formed by βPix, integrin $α_vβ_8$ and Git1 regulates not only vascular stability but also endothelial cell proliferation and cerebral angiogenesis [41].

Engulfment and cell motility 1 (ELMO1) and dedicator of cytokinesis 180 (DOCK1) form an ELMO1/DOCK1 complex and work as a bipartite GEF to regulate monomeric GTPase Rac1 activity [66]. ELMO1 expressed in different developmental stages of embryogenesis is required for the formation of functional DA, PCV and ISVs, while DOCK180 is expressed predominantly in DA and PCV. Rac has previously been associated with embryonic vascular development [67] and is expressed ubiquitously throughout embryogenesis [68]. The ELMO1/DOCK1 complex works downstream of Netrin-1 (an axonal guiding molecule) and interacts with one of the endothelial receptors called the Unc5B receptor to specifically activate Rac1 to achieve vessel formation. Activation of Rac1 GTPase solely depends on ELMO1; without ELMO1, Rac1 is not activated [69]. ELMO1 found in DA and PCV activate vascular Rac1 to lead the migrating cell toward DLAV. ELMO1/DOCK1 in vitro data apparently did not support its role in VEGF-induced activation of Rac1 for the sprouting of ECs. Overexpression of ELM1 and DOCK1 reduced the total number of apoptotic endothelial cells, which encouraged blood vessel development and EC survival during embryonic development. This protection of ECs from apoptosis was achieved by the reduction in the number of caspase 3/7 molecules via activation of PI3K/AKT signaling to facilitate proper development of functional blood vessels [66]. These findings support the spatiotemporal activation of Rac GTPase by its GEF to bring about functional and healthy blood vessels.

Vascular pruning is a process of removing redundant vessels that form during early vascular growth by the process of apoptosis to form a normal vascular system. It is a crucial process to bring normal functional vasculature. FYVE, Rho-GEF, and PH domain–containing 5 (FGD5) is a Rho-GEF that is expressed in endothelial progenitors as well as mature ECs and regulates the function of Cdc42 small GTPase in both mice and zebrafish [70]. The expression of FGD5 is predominantly achieved in the endothelial lining of large blood vessels, such as DA, ISVs and PCV. To activate Cdc42, FGD5 binds to Cdc42 and activates Hey1-p53-mediated apoptosis in ECs. Overexpression of FGD5 leads to a reduction in the levels of RhoA, and Rac1 shows an indirect downstream relationship between them. Overexpression of FGD5 leads to activation of the Notch signaling pathway by Cdc42 via the MAPK kinase pathway. Hence, FGD5 could be the factor responsible for the aging and survival of vasculature [70].

In sprouting, one cell from the quiescent stage migrates and extends its filopodia toward guiding cues to the dorsal side from the ventral part of the DA to become a leading tip cell. This tip cell expresses genes such as dll4, flk-1 and flk-4 and suppresses bottom cells to become tip cells. GAPs are negative regulators of tip cell angiogenesis by limting proangiogenic factors to stabilize the vasculature.

Rho signaling has been elucidated in the regulation of atrioventricular canal (AV) and cardiac looping. RhoU is expressed in the atrioventricular canal (when it forms) and regulates cell adhesion molecules (such as N-cadherin and alcama) between cardiomyocytes through the Arhgef7/kinase pathway. Highly conserved RhoU in vertebrates was found to have gene duplication in zebrafish, resulting in Rhoua and Rhoub. Wnt signaling may regulate the expression of this atypical Rho-GTPase. RhoU/Arhgef7/Pak signaling drives the formation of cell junctions between cardiomyocytes and promotes cell–cell adhesion [71] and shapes the cells to bring a functional heart. To maintain cell-adhesion molecules, RhoU effectors such as Arhgef7 and Pak must be maintained for the functioning of AV cardiomyocyte cell junctions. RhoU primarily functions to change the shape of AV cardiomyocytes but does not necessarily affect their fate specification and patterning [71].

4.3. Rab Family

Rab-GTPases have a role in cell directional migration via endocytosis and trafficking [72]. The binding of VEGF-A to its receptor VEGFR2 triggers the endocytosis of transmembrane receptors by Rab13. Rab13 mRNA was found to be expressed in the vessels of the trunk [73]. Rab13 GTPase associates with Syx (a RhoA-GEF) at the leading edge of the tip cell. Depletion of Rab13 hampered the sprouting of ISVs and weakened the directionality of tip cells. Rab13 mediates tight junction recycling between the trans-Golgi network and recycling endosomes. VEGF guides Rab13 to direct cell migration. Knockdown of Rab13 not only reduced ISV length but also distorted the shape of tip cells, confirming its role in directional migration. There were no defects reported in DA, which shows its specificity for angiogenesis but not for vasculogenesis [65].

Rab4a and Rab4b have been found to regulate the endocytosis of VEGFR2 trafficking and signaling during the migration and proliferation of endothelial cells. In vitro data show that in early endosomes, VEGFR2 is coexpressed with Rab4a but not Rab11a. GDP-Rab4a increased the level of VEGFR2 in endosomes. Reduction in Rab4a increased intracellular VEGF-A, and its intracellular signaling resulted in increased endothelial cell proliferation. VEGF-A-induced endothelial cell migration is inhibited when Rab4a or Rab11a is reduced. Rab4a and Rab11a are both essential for the development of endothelial tubules and are required for the formation of blood vessels. Depletion of Rab4a in zebrafish caused defects in the formation of both ISVs and DLAV, apart from the fact that ISVs are often missing and terminate before they mature completely. Reduction in either Rab4a or Rab11a has morphological, developmental and detrimental effects [74].

Rab11 signaling has been shown to be involved in lumen formation in the gut of zebrafish. The formation of the lumen takes place through different processes, and membrane trafficking is one of them. In zebrafish, during gut development, multiple small lumens are formed that merge and form a single continuous lumen [75]. This single, continuous lumen formation takes place via Rab11-mediated signaling in the gut of zebrafish. Rab11a regulates the recycling of basolateral and apical membrane proteins, which is a critical step during lumen resolution to form a single continuous lumen [75]. Rab5 was found to be associated with nodal signaling in early embryonic development. Out of four orthologs, Rab5a is teleost-specific and is expressed in medaka. All four orthologs were mostly found to be expressed in the head region (brain). A Morpholino study showed that Rab5ab is involved in regulating nodal signaling [76].

VEGFR2 endocytosis requires the activation of Rab5A/Rab4A by being in the GTP-bound state to develop into zebrafish embryos. Physical interaction between the transporter protein (Sec14l3/SEC14L2), VEGFR2 and Rab5A/Rab4A leads to activation of VEGFR2 signaling by regulating angioblasts and venous progenitors to develop arteries and veins [77].

A balance in the endocytic trafficking of Rab5c is vital for the specification and production of hematopoietic stem and progenitor cells (HSPCs) [78]. Rab5c regulates endocytic trafficking of Notch ligand and its receptor for the cell fate transition from ECs to hemogenic endothelium (HE). Downregulation or overexpression of Rab5c led to HE specification, production, survival defects and HSPC development (via Notch signaling followed by Akt signaling for HE specification). Rab5c is highly expressed in the ventral wall of DA (VDA), a part where HE specification takes place and is restricted to definitive hematopoietic tissues; hence, it is speculated that it participates in the development of HSPCs [78]. A recent study has shown that Rab5c prevents the degradation of VEGFR2 in order to restore tip cell identity and control gene expression of VEGF target genes [79].

4.4. Arf Family

ADP-ribosylation factor-like 6 (Arl6) is a small GTPase that functions in cellular signaling and protein and membrane transport [80]. Arl6 interacts with another maternally expressed protein called Arl6 interacting protein (Arl6ip). Arl6ip was found to be expressed in various organs (for other organs, refer to [81]) and in the trunk of zebrafish. Knockdown of this particular protein (Arl6ip) showed defects in trunk formation suspected to have a role in heart development along with other organs [81].

Chen et al. showed a different role of Arf5. An organic contaminant called trimethyltin chloride (TMT) induces vascular toxicity, including a reduction in the distance between ISVs, leading to an overall reduction in body length. Arf5 is necessary and plays a significant role in inducing TMT-induced vascular deformities [82].

ArfGAP with a dual PH domain 2 (ADAP2) with GAP activity for Arf6 has a role in heart development. Knockdown of ADAP2 results in blood circulation defects and curved tails. The maternally and zygotically expressing ADAP2 is present in the heart and the region corresponding to the bulbus arteriosus in zebrafish [83]. Arf-GAP, called G protein-coupled receptor kinase interacting target (GIT1), interacts with Rho family GEF βpix (especially Rac and Cdc42) to stabilize blood vessels. Interaction between GIT1 and βpix with integrins regulates vascular stability, endothelial cell proliferation and cerebral angiogenesis. GIT1 is ubiquitously expressed in zebrafish, and its knockdown leads to an increase in hemorrhage, proving its role in vascular stabilization [41].

Apart from growth factors, integrins have been implicated in the process of angiogenesis. Brag2 is recognized as an Arf-GEF for Arf4, Arf5, and Arf6. An in vitro study showed its role in angiogenic sprouting, migration and adhesion of ECs. In vivo experimental silencing of Brag2 showed vascular and developmental defects in zebrafish. Silencing of Brag2 leads to defects mostly related to the formation of DLAV, ISVs and parachordal lymphangioblasts (PL-lymphatic system precursor), showing its role in vascular patterning and stability. Knockdown of both orthologs of Brag2 leads to severe defects in DLAV, ISVs and sometimes the absence of PL. Brag2-mediated activation of Arf5 and Arf6 leads to developmental and pathological angiogenic sprouting of ECs through regulation of adhesion mediated by β1- and β3-integrins [84].

Golgi brefeldin A-resistant factor 1 (Gbf1) is a maternally and zygotically expressed high molecular weight GEF for the Arf-GTPase family that regulates organelle structure and vesicle trafficking. Gbf1 is ubiquitously expressed in the early stage, but it was later found to be expressed in the head region. Isolation of cells showed its expression in ECs to develop vasculature in a cell autonomous manner. Mutated form of this specific GEF fail to activate Arf1 and are unable to recruit cargo complex COPI. A zebrafish mutant line was created by using the mutagen N-ethyl-N-nitrosourea (ENU), which carries the T→G transition on the 23rd exon of the Gbf1 locus. The mutant embryo displayed hemorrhage in the trunk and head regions. Mutants showed pigmentation reduction in the head region and short caudal fins in Mendelian inheritance. Blood cells leak into the head, eye and trunk, which leads to the death of an embryo within 96 hpf. Intracerebral vessels in the head and ISV in the trunk were broken or sometimes disappeared or disconnected, resulting in dissociation of ECs, which could be due to disruption of vascular integrity or homeostasis in mutants [85].

Brefeldin A inhibited guanine nucleotide exchange 1 and 2 (BIG1 & BIG2) protein 1 (arfgef1 and arfgef2 homolog in zebrafish) and is the GEF for two small GTPases, Arf1 and Arf2. Both GEEs are ubiquitously expressed in zebrafish. Knockdown of either BIG1 or BIG2 in zebrafish was associated with EC migration during blood vessel formation. An mboxin vitro study showed their involvement in the process of capillary tubule formation and EC migration by modulating actin cytoskeleton organization in HUVECs. Knockdown of BIG2 interferes with the completion of ISVs without reflecting on its numbers, and a reduction in PCV width was observed during embryonic development. BIG1 and BIG2 reduction suppressed the expression level of VEGF and EC migration in the process of blood vessel development [86].

5. Conclusions

These studies have shown that Ras superfamily of proteins has importance in many processes that are sufficient to develop completely functional and healthy vessels to carry different nutrients and macromolecules to the entire body. Future studies are still necessary to decode and stage the specific role of Ras-GTPases to fill the gap in vessel development. Their functional role in blood vessel development could guide us to form therapeutic strategies for diseases related to vascular development.

Author Contributions: Conceptualization, C.-Y.W. and C.-C.C.; methodology, R.U. and Y.-H.C.; data curation, Y.-H.C.; original draft preparation, R.U. and C.-Y.W.; review and editing, C.-C.C. and C.-Y.W.; visualization, R.U. and Y.-H.C. All authors have read and agreed to the published version of the manuscript.

Funding: This work was supported by grants from the Ministry of Science and Technology, Taiwan (MOST107-2311-B-110-002 and MOST108-2313-B-110-002-MY3) to CYW and from the NSYSU-KMU Joint Research Project (#NSYSUKMU106-P019 and #NSYSUKMU107-P002) to CYW and CCC.

Acknowledgments: We thank Ming-Hong Tai and Chun-Lin Chen for reading the manuscript and providing valuable comments.

Conflicts of Interest: The authors declare no conflict of interest.

Abbreviations

ISV, intersegmental vessel; CVP, caudal vein plexus; DA, dorsal aorta; PCV, posterior cardinal vein; DLAV, dorsal longitudinal anastomotic vessel; SIV, subintestinal vein; GAP, GTPase-activating protein; GEF, guanine nucleotide exchange factor; GDI, guanine nucleotide dissociation inhibitor; VEGF, vascular endothelial growth factor; BMP, bone morphogenetic protein; EC, endothelial cell.

References

1. Johnson, D.S.; Chen, Y.H. Ras family of small GTPases in immunity and inflammation. *Curr. Opin. Pharmacol.* **2012**, *12*, 458–463. [CrossRef] [PubMed]
2. Liu, W.N.; Yan, M.; Chan, A.M. A thirty-year quest for a role of R-Ras in cancer: From an oncogene to a multitasking GTPase. *Cancer Lett.* **2017**, *403*, 59–65. [CrossRef] [PubMed]
3. Pereira-Leal, J.B.; Seabra, M.C. The mammalian Rab family of small GTPases: Definition of family and subfamily sequence motifs suggests a mechanism for functional specificity in the Ras superfamily. *J. Mol. Biol.* **2000**, *301*, 1077–1087. [CrossRef] [PubMed]
4. Wennerberg, K.; Rossman, K.L.; Der, C.J. The Ras superfamily at a glance. *J. Cell Sci.* **2005**, *118*, 843–846. [CrossRef]
5. Goitre, L.; Trapani, E.; Trabalzini, L.; Retta, S.F. The Ras superfamily of small GTPases: The unlocked secrets. *Methods Mol. Biol.* **2014**, *1120*, 1–18. [CrossRef]
6. Bryan, B.A.; D'Amore, P.A. What tangled webs they weave: Rho-GTPase control of angiogenesis. *Cell Mol. Life Sci.* **2007**, *64*, 2053–2065. [CrossRef]
7. Liang, D.; Chang, J.R.; Chin, A.J.; Smith, A.; Kelly, C.; Weinberg, E.S.; Ge, R. The role of vascular endothelial growth factor (VEGF) in vasculogenesis, angiogenesis, and hematopoiesis in zebrafish development. *Mech. Dev.* **2001**, *108*, 29–43. [CrossRef]
8. Siekmann, A.F.; Lawson, N.D. Notch signalling and the regulation of angiogenesis. *Cell Adh. Migr.* **2007**, *1*, 104–106. [CrossRef]
9. Kim, J.D.; Lee, H.W.; Jin, S.W. Diversity is in my veins: Role of bone morphogenetic protein signaling during venous morphogenesis in zebrafish illustrates the heterogeneity within endothelial cells. *Arterioscler Thromb. Vasc. Biol.* **2014**, *34*, 1838–1845. [CrossRef]
10. Cherfils, J.; Zeghouf, M. Regulation of small GTPases by GEFs, GAPs, and GDIs. *Physiol. Rev.* **2013**, *93*, 269–309. [CrossRef]
11. Choi, K.; Kennedy, M.; Kazarov, A.; Papadimitriou, J.C.; Keller, G. A common precursor for hematopoietic and endothelial cells. *Development* **1998**, *125*, 725–732. [CrossRef] [PubMed]
12. Keller, G. Embryonic stem cell differentiation: Emergence of a new era in biology and medicine. *Genes Dev.* **2005**, *19*, 1129–1155. [CrossRef] [PubMed]
13. Medina, R.J.; O'Neill, C.L.; Humphreys, M.W.; Gardiner, T.A.; Stitt, A.W. Outgrowth endothelial cells: Characterization and their potential for reversing ischemic retinopathy. *Invest Ophthalmol. Vis. Sci.* **2010**, *51*, 5906–5913. [CrossRef]
14. Risau, W. Mechanisms of angiogenesis. *Nature* **1997**, *386*, 671–674. [CrossRef] [PubMed]
15. Barry, D.M.; Xu, K.; Meadows, S.M.; Zheng, Y.; Norden, P.R.; Davis, G.E.; Cleaver, O. Cdc42 is required for cytoskeletal support of endothelial cell adhesion during blood vessel formation in mice. *Development* **2015**, *142*, 3058–3070. [CrossRef] [PubMed]
16. Risau, W.; Flamme, I. Vasculogenesis. *Annu. Rev. Cell Dev. Biol.* **1995**, *11*, 73–91. [CrossRef]
17. Sukriti, S.; Tauseef, M.; Yazbeck, P.; Mehta, D. Mechanisms regulating endothelial permeability. *Pulm. Circ.* **2014**, *4*, 535–551. [CrossRef] [PubMed]
18. Patan, S. Vasculogenesis and angiogenesis as mechanisms of vascular network formation, growth and remodeling. *J. Neurooncol.* **2000**, *50*, 1–15. [CrossRef]
19. Ausprunk, D.H.; Folkman, J. Migration and proliferation of endothelial cells in preformed and newly formed blood vessels during tumor angiogenesis. *Microvasc. Res.* **1977**, *14*, 53–65. [CrossRef]
20. Ellertsdottir, E.; Lenard, A.; Blum, Y.; Krudewig, A.; Herwig, L.; Affolter, M.; Belting, H.G. Vascular morphogenesis in the zebrafish embryo. *Dev. Biol.* **2010**, *341*, 56–65. [CrossRef]
21. Ferrara, N.; Gerber, H.P.; LeCouter, J. The biology of VEGF and its receptors. *Nat. Med.* **2003**, *9*, 669–676. [CrossRef] [PubMed]

22. Pandya, N.M.; Dhalla, N.S.; Santani, D.D. Angiogenesis—A new target for future therapy. *Vasc. Pharmacol.* **2006**, *44*, 265–274. [CrossRef] [PubMed]
23. Spence, R.; Gerlach, G.; Lawrence, C.; Smith, C. The behaviour and ecology of the zebrafish, Danio rerio. *Biol. Rev. Camb. Philos. Soc.* **2008**, *83*, 13–34. [CrossRef] [PubMed]
24. Howe, K.; Clark, M.D.; Torroja, C.F.; Torrance, J.; Berthelot, C.; Muffato, M.; Collins, J.E.; Humphray, S.; McLaren, K.; Matthews, L.; et al. The zebrafish reference genome sequence and its relationship to the human genome. *Nature* **2013**, *496*, 498–503. [CrossRef]
25. Baldessari, D.; Mione, M. How to create the vascular tree? (Latest) help from the zebrafish. *Pharmacol. Ther.* **2008**, *118*, 206–230. [CrossRef]
26. Lamont, R.E.; Wu, C.Y.; Ryu, J.R.; Vu, W.; Davari, P.; Sobering, R.E.; Kennedy, R.M.; Munsie, N.M.; Childs, S.J. The LIM-homeodomain transcription factor Islet2a promotes angioblast migration. *Dev. Biol.* **2016**, *414*, 181–192. [CrossRef]
27. Li, R.F.; Wu, T.Y.; Mou, Y.Z.; Wang, Y.S.; Chen, C.L.; Wu, C.Y. Nr2f1b control venous specification and angiogenic patterning during zebrafish vascular development. *J. Biomed. Sci.* **2015**, *22*, 104. [CrossRef]
28. Lakshmikanthan, S.; Sobczak, M.; Chun, C.; Henschel, A.; Dargatz, J.; Ramchandran, R.; Chrzanowska-Wodnicka, M. Rap1 promotes VEGFR2 activation and angiogenesis by a mechanism involving integrin alphavbeta(3). *Blood* **2011**, *118*, 2015–2026. [CrossRef]
29. Torres-Vazquez, J.; Gitler, A.D.; Fraser, S.D.; Berk, J.D.; Van, N.P.; Fishman, M.C.; Childs, S.; Epstein, J.A.; Weinstein, B.M. Semaphorin-plexin signaling guides patterning of the developing vasculature. *Dev. Cell* **2004**, *7*, 117–123. [CrossRef]
30. Beis, D.; Stainier, D.Y. In vivo cell biology: Following the zebrafish trend. *Trends Cell Biol.* **2006**, *16*, 105–112. [CrossRef]
31. Song, S.; Cong, W.; Zhou, S.; Shi, Y.; Dai, W.; Zhang, H.; Wang, X.; He, B.; Zhang, Q. Small GTPases: Structure, biological function and its interaction with nanoparticles. *Asian J. Pharm. Sci.* **2019**, *14*, 30–39. [CrossRef] [PubMed]
32. Colicelli, J. Human RAS superfamily proteins and related GTPases. *Sci. STKE* **2004**, *2004*, RE13. [CrossRef] [PubMed]
33. Gritsman, K.; Zhang, J.; Cheng, S.; Heckscher, E.; Talbot, W.S.; Schier, A.F. The EGF-CFC protein one-eyed pinhead is essential for nodal signaling. *Cell* **1999**, *97*, 121–132. [CrossRef]
34. Pezeron, G.; Lambert, G.; Dickmeis, T.; Strahle, U.; Rosa, F.M.; Mourrain, P. Rasl11b knock down in zebrafish suppresses one-eyed-pinhead mutant phenotype. *PLoS ONE* **2008**, *3*, e1434. [CrossRef]
35. Umanoff, H.; Edelmann, W.; Pellicer, A.; Kucherlapati, R. The murine N-ras gene is not essential for growth and development. *Proc. Natl. Acad. Sci. USA* **1995**, *92*, 1709–1713. [CrossRef]
36. Koera, K.; Nakamura, K.; Nakao, K.; Miyoshi, J.; Toyoshima, K.; Hatta, T.; Otani, H.; Aiba, A.; Katsuki, M. K-ras is essential for the development of the mouse embryo. *Oncogene* **1997**, *15*, 1151–1159. [CrossRef]
37. Liu, L.; Zhu, S.; Gong, Z.; Low, B.C. K-ras/PI3K-Akt signaling is essential for zebrafish hematopoiesis and angiogenesis. *PLoS ONE* **2008**, *3*, e2850. [CrossRef]
38. Oinuma, I.; Katoh, H.; Negishi, M. Semaphorin 4D/Plexin-B1-mediated R-Ras GAP activity inhibits cell migration by regulating beta(1) integrin activity. *J. Cell Biol.* **2006**, *173*, 601–613. [CrossRef]
39. Saito, Y.; Oinuma, I.; Fujimoto, S.; Negishi, M. Plexin-B1 is a GTPase activating protein for M-Ras, remodelling dendrite morphology. *EMBO Rep.* **2009**, *10*, 614–621. [CrossRef]
40. Uesugi, K.; Oinuma, I.; Katoh, H.; Negishi, M. Different requirement for Rnd GTPases of R-Ras GAP activity of Plexin-C1 and Plexin-D1. *J. Biol. Chem.* **2009**, *284*, 6743–6751. [CrossRef]
41. Liu, J.; Zeng, L.; Kennedy, R.M.; Gruenig, N.M.; Childs, S.J. betaPix plays a dual role in cerebral vascular stability and angiogenesis, and interacts with integrin alphavbeta8. *Dev. Biol.* **2012**, *363*, 95–105. [CrossRef] [PubMed]
42. Childs, S.; Chen, J.N.; Garrity, D.M.; Fishman, M.C. Patterning of angiogenesis in the zebrafish embryo. *Development* **2002**, *129*, 973–982. [CrossRef]
43. Lamont, R.E.; Lamont, E.J.; Childs, S.J. Antagonistic interactions among Plexins regulate the timing of intersegmental vessel formation. *Dev. Biol.* **2009**, *331*, 199–209. [CrossRef] [PubMed]
44. Avraamides, C.J.; Garmy-Susini, B.; Varner, J.A. Integrins in angiogenesis and lymphangiogenesis. *Nat. Rev. Cancer* **2008**, *8*, 604–617. [CrossRef] [PubMed]
45. Eerola, I.; Boon, L.M.; Mulliken, J.B.; Burrows, P.E.; Dompmartin, A.; Watanabe, S.; Vanwijck, R.; Vikkula, M. Capillary malformation-arteriovenous malformation, a new clinical and genetic disorder caused by RASA1 mutations. *Am. J. Hum. Genet.* **2003**, *73*, 1240–1249. [CrossRef]
46. Kawasaki, J.; Aegerter, S.; Fevurly, R.D.; Mammoto, A.; Mammoto, T.; Sahin, M.; Mably, J.D.; Fishman, S.J.; Chan, J. RASA1 functions in EPHB4 signaling pathway to suppress endothelial mTORC1 activity. *J. Clin. Investig.* **2014**, *124*, 2774–2784. [CrossRef]
47. Zhao, X.; Guan, J.L. Focal adhesion kinase and its signaling pathways in cell migration and angiogenesis. *Adv. Drug Deliv. Rev.* **2011**, *63*, 610–615. [CrossRef]
48. Yeh, C.W.; Hsu, L.S. Zebrafish diras1 Promoted Neurite Outgrowth in Neuro-2a Cells and Maintained Trigeminal Ganglion Neurons In Vivo via Rac1-Dependent Pathway. *Mol. Neurobiol.* **2016**, *53*, 6594–6607. [CrossRef]
49. Rho, S.S.; Kobayashi, I.; Oguri-Nakamura, E.; Ando, K.; Fujiwara, M.; Kamimura, N.; Hirata, H.; Iida, A.; Iwai, Y.; Mochizuki, N.; et al. Rap1b Promotes Notch-Signal-Mediated Hematopoietic Stem Cell Development by Enhancing Integrin-Mediated Cell Adhesion. *Dev. Cell* **2019**, *49*, 681–696. [CrossRef]
50. Jaffe, A.B.; Hall, A. Rho GTPases: Biochemistry and biology. *Annu. Rev. Cell Dev. Biol.* **2005**, *21*, 247–269. [CrossRef]

51. Kather, J.N.; Kroll, J. Rho guanine exchange factors in blood vessels: Fine-tuners of angiogenesis and vascular function. *Exp. Cell Res.* **2013**, *319*, 1289–1297. [CrossRef] [PubMed]
52. Mehta, D.; Rahman, A.; Malik, A.B. Protein kinase C-alpha signals rho-guanine nucleotide dissociation inhibitor phosphorylation and rho activation and regulates the endothelial cell barrier function. *J. Biol. Chem.* **2001**, *276*, 22614–22620. [CrossRef] [PubMed]
53. Hoeppner, L.H.; Sinha, S.; Wang, Y.; Bhattacharya, R.; Dutta, S.; Gong, X.; Bedell, V.M.; Suresh, S.; Chun, C.; Ramchandran, R.; et al. RhoC maintains vascular homeostasis by regulating VEGF-induced signaling in endothelial cells. *J. Cell Sci.* **2018**, *131*. [CrossRef] [PubMed]
54. Xu, K.; Sacharidou, A.; Fu, S.; Chong, D.C.; Skaug, B.; Chen, Z.J.; Davis, G.E.; Cleaver, O. Blood vessel tubulogenesis requires Rasip1 regulation of GTPase signaling. *Dev. Cell* **2011**, *20*, 526–539. [CrossRef]
55. Zhu, S.; Korzh, V.; Gong, Z.; Low, B.C. RhoA prevents apoptosis during zebrafish embryogenesis through activation of Mek/Erk pathway. *Oncogene* **2008**, *27*, 1580–1589. [CrossRef]
56. Zhu, S.; Liu, L.; Korzh, V.; Gong, Z.; Low, B.C. RhoA acts downstream of Wnt5 and Wnt11 to regulate convergence and extension movements by involving effectors Rho kinase and Diaphanous: Use of zebrafish as an in vivo model for GTPase signaling. *Cell Signal.* **2006**, *18*, 359–372. [CrossRef] [PubMed]
57. Choi, S.Y.; Baek, J.I.; Zuo, X.; Kim, S.H.; Dunaief, J.L.; Lipschutz, J.H. Cdc42 and sec10 Are Required for Normal Retinal Development in Zebrafish. *Invest. Ophthalmol. Vis. Sci.* **2015**, *56*, 3361–3370. [CrossRef] [PubMed]
58. Fantin, A.; Lampropoulou, A.; Gestri, G.; Raimondi, C.; Senatore, V.; Zachary, I.; Ruhrberg, C. NRP1 Regulates CDC42 Activation to Promote Filopodia Formation in Endothelial Tip Cells. *Cell Rep.* **2015**, *11*, 1577–1590. [CrossRef]
59. Wakayama, Y.; Fukuhara, S.; Ando, K.; Matsuda, M.; Mochizuki, N. Cdc42 mediates Bmp-induced sprouting angiogenesis through Fmnl3-driven assembly of endothelial filopodia in zebrafish. *Dev. Cell* **2015**, *32*, 109–122. [CrossRef] [PubMed]
60. Chang, G.H.; Lay, A.J.; Ting, K.K.; Zhao, Y.; Coleman, P.R.; Powter, E.E.; Formaz-Preston, A.; Jolly, C.J.; Bower, N.I.; Hogan, B.M.; et al. ARHGAP18: An endogenous inhibitor of angiogenesis, limiting tip formation and stabilizing junctions. *Small GTPases* **2014**, *5*, 1–15. [CrossRef]
61. Garrett, T.A.; Van Buul, J.D.; Burridge, K. VEGF-induced Rac1 activation in endothelial cells is regulated by the guanine nucleotide exchange factor Vav2. *Exp. Cell Res.* **2007**, *313*, 3285–3297. [CrossRef] [PubMed]
62. Garnaas, M.K.; Moodie, K.L.; Liu, M.L.; Samant, G.V.; Li, K.; Marx, R.; Baraban, J.M.; Horowitz, A.; Ramchandran, R. Syx, a RhoA guanine exchange factor, is essential for angiogenesis in vivo. *Circ. Res.* **2008**, *103*, 710–716. [CrossRef]
63. Bratt, A.; Birot, O.; Sinha, I.; Veitonmaki, N.; Aase, K.; Ernkvist, M.; Holmgren, L. Angiomotin regulates endothelial cell-cell junctions and cell motility. *J. Biol. Chem.* **2005**, *280*, 34859–34869. [CrossRef] [PubMed]
64. Ernkvist, M.; Luna Persson, N.; Audebert, S.; Lecine, P.; Sinha, I.; Liu, M.; Schlueter, M.; Horowitz, A.; Aase, K.; Weide, T.; et al. The Amot/Patj/Syx signaling complex spatially controls RhoA GTPase activity in migrating endothelial cells. *Blood* **2009**, *113*, 244–253. [CrossRef] [PubMed]
65. Wu, C.; Agrawal, S.; Vasanji, A.; Drazba, J.; Sarkaria, S.; Xie, J.; Welch, C.M.; Liu, M.; Anand-Apte, B.; Horowitz, A. Rab13-dependent trafficking of RhoA is required for directional migration and angiogenesis. *J. Biol. Chem.* **2011**, *286*, 23511–23520. [CrossRef] [PubMed]
66. Schaker, K.; Bartsch, S.; Patry, C.; Stoll, S.J.; Hillebrands, J.L.; Wieland, T.; Kroll, J. The bipartite rac1 Guanine nucleotide exchange factor engulfment and cell motility 1/dedicator of cytokinesis 180 (elmo1/dock180) protects endothelial cells from apoptosis in blood vessel development. *J. Biol. Chem.* **2015**, *290*, 6408–6418. [CrossRef]
67. Tan, W.; Palmby, T.R.; Gavard, J.; Amornphimoltham, P.; Zheng, Y.; Gutkind, J.S. An essential role for Rac1 in endothelial cell function and vascular development. *FASEB J.* **2008**, *22*, 1829–1838. [CrossRef]
68. Srinivas, B.P.; Woo, J.; Leong, W.Y.; Roy, S. A conserved molecular pathway mediates myoblast fusion in insects and vertebrates. *Nat. Genet.* **2007**, *39*, 781–786. [CrossRef]
69. Epting, D.; Wendik, B.; Bennewitz, K.; Dietz, C.T.; Driever, W.; Kroll, J. The Rac1 regulator ELMO1 controls vascular morphogenesis in zebrafish. *Circ. Res.* **2010**, *107*, 45–55. [CrossRef]
70. Cheng, C.; Haasdijk, R.; Tempel, D.; van de Kamp, E.H.; Herpers, R.; Bos, F.; Den Dekker, W.K.; Blonden, L.A.; de Jong, R.; Burgisser, P.E.; et al. Endothelial cell-specific FGD5 involvement in vascular pruning defines neovessel fate in mice. *Circulation* **2012**, *125*, 3142–3158. [CrossRef]
71. Dickover, M.; Hegarty, J.M.; Ly, K.; Lopez, D.; Yang, H.; Zhang, R.; Tedeschi, N.; Hsiai, T.K.; Chi, N.C. The atypical Rho GTPase, RhoU, regulates cell-adhesion molecules during cardiac morphogenesis. *Dev. Biol.* **2014**, *389*, 182–191. [CrossRef] [PubMed]
72. Ulrich, F.; Heisenberg, C.P. Trafficking and cell migration. *Traffic* **2009**, *10*, 811–818. [CrossRef]
73. Kudoh, T.; Tsang, M.; Hukriede, N.A.; Chen, X.; Dedekian, M.; Clarke, C.J.; Kiang, A.; Schultz, S.; Epstein, J.A.; Toyama, R.; et al. A gene expression screen in zebrafish embryogenesis. *Genom. Res.* **2001**, *11*, 1979–1987. [CrossRef]
74. Jopling, H.M.; Odell, A.F.; Pellet-Many, C.; Latham, A.M.; Frankel, P.; Sivaprasadarao, A.; Walker, J.H.; Zachary, I.C.; Ponnambalam, S. Endosome-to-Plasma Membrane Recycling of VEGFR2 Receptor Tyrosine Kinase Regulates Endothelial Function and Blood Vessel Formation. *Cells* **2014**, *3*, 363–385. [CrossRef] [PubMed]
75. Alvers, A.L.; Ryan, S.; Scherz, P.J.; Huisken, J.; Bagnat, M. Single continuous lumen formation in the zebrafish gut is mediated by smoothened-dependent tissue remodeling. *Development* **2014**, *141*, 1110–1119. [CrossRef] [PubMed]
76. Kenyon, E.J.; Campos, I.; Bull, J.C.; Williams, P.H.; Stemple, D.L.; Clark, M.D. Zebrafish Rab5 proteins and a role for Rab5ab in nodal signalling. *Dev. Biol.* **2015**, *397*, 212–224. [CrossRef]

77. Gong, B.; Li, Z.; Xiao, W.; Li, G.; Ding, S.; Meng, A.; Jia, S. Sec14l3 potentiates VEGFR2 signaling to regulate zebrafish vasculogenesis. *Nat. Commun.* **2019**, *10*, 1606. [CrossRef]
78. Heng, J.; Lv, P.; Zhang, Y.; Cheng, X.; Wang, L.; Ma, D.; Liu, F. Rab5c-mediated endocytic trafficking regulates hematopoietic stem and progenitor cell development via Notch and AKT signaling. *PLoS Biol.* **2020**, *18*, e3000696. [CrossRef]
79. Kempers, L.; Wakayama, Y.; van der Bijl, I.; Furumaya, C.; De Cuyper, I.M.; Jongejan, A.; Kat, M.; van Stalborch, A.D.; van Boxtel, A.L.; Hubert, M.; et al. The endosomal RIN2/Rab5C machinery prevents VEGFR2 degradation to control gene expression and tip cell identity during angiogenesis. *Angiogenesis* **2021**, *24*, 695–714. [CrossRef]
80. Ikeda, S.; Ushio-Fukai, M.; Zuo, L.; Tojo, T.; Dikalov, S.; Patrushev, N.A.; Alexander, R.W. Novel role of ARF6 in vascular endothelial growth factor-induced signaling and angiogenesis. *Circ. Res.* **2005**, *96*, 467–475. [CrossRef]
81. Huang, H.Y.; Dai, E.S.; Liu, J.T.; Tu, C.T.; Yang, T.C.; Tsai, H.J. The embryonic expression patterns and the knockdown phenotypes of zebrafish ADP-ribosylation factor-like 6 interacting protein gene. *Dev. Dyn.* **2009**, *238*, 232–240. [CrossRef] [PubMed]
82. Chen, J.; Huang, C.; Truong, L.; La Du, J.; Tilton, S.C.; Waters, K.M.; Lin, K.; Tanguay, R.L.; Dong, Q. Early life stage trimethyltin exposure induces ADP-ribosylation factor expression and perturbs the vascular system in zebrafish. *Toxicology* **2012**, *302*, 129–139. [CrossRef] [PubMed]
83. Venturin, M.; Carra, S.; Gaudenzi, G.; Brunelli, S.; Gallo, G.R.; Moncini, S.; Cotelli, F.; Riva, P. ADAP2 in heart development: A candidate gene for the occurrence of cardiovascular malformations in NF1 microdeletion syndrome. *J. Med. Genet.* **2014**, *51*, 436–443. [CrossRef] [PubMed]
84. Manavski, Y.; Carmona, G.; Bennewitz, K.; Tang, Z.; Zhang, F.; Sakurai, A.; Zeiher, A.M.; Gutkind, J.S.; Li, X.; Kroll, J.; et al. Brag2 differentially regulates beta1- and beta3-integrin-dependent adhesion in endothelial cells and is involved in developmental and pathological angiogenesis. *Basic Res. Cardiol.* **2014**, *109*, 404. [CrossRef] [PubMed]
85. Chen, J.; Wu, X.; Yao, L.; Yan, L.; Zhang, L.; Qiu, J.; Liu, X.; Jia, S.; Meng, A. Impairment of Cargo Transportation Caused by gbf1 Mutation Disrupts Vascular Integrity and Causes Hemorrhage in Zebrafish Embryos. *J. Biol. Chem.* **2017**, *292*, 2315–2327. [CrossRef]
86. Lu, F.I.; Wang, Y.T.; Wang, Y.S.; Wu, C.Y.; Li, C.C. Involvement of BIG1 and BIG2 in regulating VEGF expression and angiogenesis. *FASEB J.* **2019**, *33*, 9959–9973. [CrossRef]

International Journal of
Molecular Sciences

Review

Interdependence of Angiogenesis and Arteriogenesis in Development and Disease

Ferdinand le Noble [1,2,3,*] and Christian Kupatt [4,5,*]

1. Department of Cell and Developmental Biology, Institute of Zoology (ZOO), Karlsruhe Institute of Technology (KIT), Fritz Haber Weg 4, 76131 Karlsruhe, Germany
2. Institute for Biological and Chemical Systems—Biological Information Processing, Karlsruhe Institute of Technology (KIT), P.O. Box 3640, 76021 Karlsruhe, Germany
3. Institute of Experimental Cardiology, Heidelberg Germany and German Center for Cardiovascular Research (DZHK), Partner Site Heidelberg/Mannheim, University of Heidelberg, 69117 Heidelberg, Germany
4. Klinik und Poliklinik für Innere Medizin I, Klinikum Rechts der Isar, Technical University Munich, 81675 Munich, Germany
5. DZHK (German Center for Cardiovascular Research), Munich Heart Alliance, 80802 Munich, Germany
* Correspondence: Ferdinand.noble@kit.edu (F.l.N.); christian.kupatt@tum.de (C.K.)

Citation: le Noble, F.; Kupatt, C. Interdependence of Angiogenesis and Arteriogenesis in Development and Disease. *Int. J. Mol. Sci.* **2022**, *23*, 3879. https://doi.org/10.3390/ijms23073879

Academic Editors: Paul Quax and Elisabeth Deindl

Received: 3 March 2022
Accepted: 27 March 2022
Published: 31 March 2022

Publisher's Note: MDPI stays neutral with regard to jurisdictional claims in published maps and institutional affiliations.

Copyright: © 2022 by the authors. Licensee MDPI, Basel, Switzerland. This article is an open access article distributed under the terms and conditions of the Creative Commons Attribution (CC BY) license (https://creativecommons.org/licenses/by/4.0/).

Abstract: The structure of arterial networks is optimized to allow efficient flow delivery to metabolically active tissues. Optimization of flow delivery is a continuous process involving synchronization of the structure and function of the microcirculation with the upstream arterial network. Risk factors for ischemic cardiovascular diseases, such as diabetes mellitus and hyperlipidemia, adversely affect endothelial function, induce capillary regression, and disrupt the micro- to macrocirculation cross-talk. We provide evidence showing that this loss of synchronization reduces arterial collateral network recruitment upon arterial stenosis, and the long-term clinical outcome of current revascularization strategies in these patient cohorts. We describe mechanisms and signals contributing to synchronized growth of micro- and macrocirculation in development and upon ischemic challenges in the adult organism and identify potential therapeutic targets. We conclude that a long-term successful revascularization strategy should aim at both removing obstructions in the proximal part of the arterial tree and restoring "bottom-up" vascular communication.

Keywords: angiogenesis; arteriogenesis; blood flow; shear stress; MRTF-A; AAV; endothelial cell shape; sFlt1; Trio

1. Introduction

Arterial networks are tree-like hierarchically branched structures that distribute blood flow via resistance-sized arteries and arterioles of gradually decreasing diameter to the distal capillary network, which is the exchange area for oxygen and nutrients. The arterial branching architecture and lumen dimensions are optimized to allow for efficient blood flow distribution while minimizing transport costs [1]. Continuous information transfer along the vascular tree is crucial for design optimization upon changes in hemodynamic conditions and organ metabolism. The formation of arterial collateral networks as occurs in ischemic cardiovascular diseases represents a specialized design optimization solution aimed at restoring flow delivery to compromised regions [2]. How collaterals are formed and how to selectively target their growth is an outstanding question in the field. Recent studies have shown that efficient induction and maintenance of stable arterial collateral networks requires precise retrograde information transfer between the microcirculation and the upstream arterial network in line with design optimization principles [3]. Unfortunately, patients with ischemic CVD typically have underlying risk factors that interfere with the retrograde information transfer process. Given the interdependence of arteries and capillaries for vessel network design optimization, we propose that successful collateralization strategies should consider microcirculation functionality. We highlight some new insights and collateralization approaches based on recent advances in understanding arterial network growth.

Conspicuously, analysis of arterial growth and adaptation to exercise, and arterial development during embryogenesis and neonatal stages, revealed in striking similarity, that arteries obtain their structure through inward growth. This is according to a "bottom-up" principle involving substantial retrograde communication against the direction of blood flow [3–5]. Taking the present clinical focus on removing obstructions in the upstream part of the arterial tree into account—through the insertion of a stent or bypass surgery—it becomes clear that the bottom-up principle is not heeded to a vast extent. In particular, most ischemic CVD patients harbor underlying risk factors, such as diabetes, hypertension, hyperlipidemia, or genetic diseases that adversely affect endothelial function and microvascular structure, prohibiting retrograde communication and structural optimization. Indeed, clinical evidence suggests that long-term end-organ function and survival upon ischemic insults not only requires removing the upstream arterial blockage with a stent, but also careful nurturing of the peripheral microcirculation [6]. We will advocate retrograde "bottom-up" communication to stimulate arterial growth, both of arterial bypass collaterals and of microvascular arterioles, as adjuvant therapy to the conventional revascularization strategies.

2. An Underestimated Problem—Capillary Rarefaction in Vascular Disease

The relevance of the microcirculatory vessel compartment may not seem obvious for ischemic heart disease, one of the most prevalent factors reducing health in Western societies and beyond. Becoming apparent as acute myocardial infarction due to atherosclerotic plaque rupture or hibernating myocardium due to prolonged plaque progression, coronary artery disease is treated as large artery disease by interventional or surgical means. In that perspective, the relevance of the interdependence of arteries and capillaries is easily underestimated. However, cardiovascular and genetic risk factors, which are responsible for the majority of cardiac events, are also affecting microvascular integrity. For example, capillary rarefaction in organs such as the eye and the kidney precede and predict macro vessel obstruction [7]. Capillary rarefaction in the coronary circulation is a hallmark of diabetes mellitus in small [8] and large animal models [9]. For this cardiovascular risk factor, revascularization therapy appears notoriously hard to apply via cardiac intervention studies [10], not least since microvascular hypodensity caused by capillary and arteriolar rarefaction, precludes a flow increase upon revascularization therapy. Hyperlipidemia, in particular hypercholesterinemia, is a major causative factor in atherogenesis and activation of inflammation [11]. Inflammatory processes induce rarefaction of small vessels, i.e., arterioles and capillaries [12], which profoundly alter the coronary flow pattern. Lipid-lowering pharmacological interventions, then resolve not only endothelial dysfunction, but also capillary rarefaction inflicted by chronic ischemia in a rabbit hindlimb model [13]. Hypertension, another major cardiovascular risk factor, has been associated with capillary and arteriolar rarefaction [14]. Sudden experimental onset of hypertension by transverse aortic banding (TAC) may induce a massive macrophage-driven hypertrophy, concomitant with induction of fibrosis and capillary rarefaction, which are, at least in part driven by miR21 [15,16]. Puzzling enough, hypertension is a common side effect of Vascular Endothelial Growth Factor (VEGF) neutralizing biologics, exemplifying the relevance of the microcirculation for hemodynamic homeostasis [17].

3. Angiogenic Vessel Building—Building back Better-Perfused Capillaries and Arterioles

If capillaries and arteries fall victim to cardiovascular risk factors, which options are biologically available to counteract this tendency? A look into vessel physiology is warranted. Oxygen delivery requires lumenized and perfused vessel segments, not blind-ending nonperfused angiogenic sprouts. It is therapeutically relevant to generate arterioles with a patent lumen that can carry blood flow. Titrating lumen diameter is a very important aspect. According to Poiseuille's law, blood flow through a vessel is proportional to diameter to the power four [18]. Hence a two-fold increase in diameter predicts a 16-fold increase in flow transporting capacity at a given pressure (Figure 1).

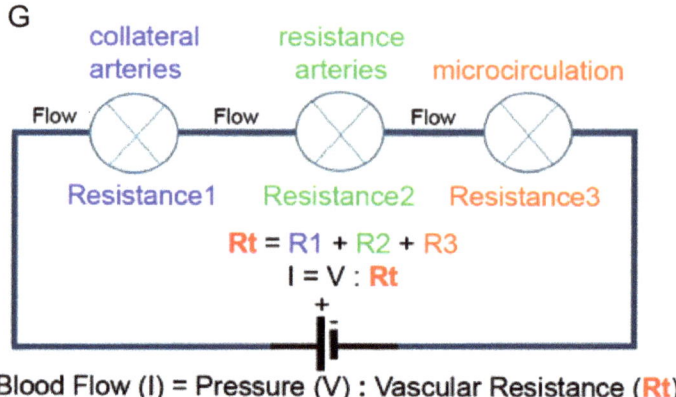

Figure 1. Schematic illustration of an arterial network indicated as electrical resistances coupled in series to a battery. In electrical terms: I (current) = V (voltage) divided by R (resistance), in biological terms Amount of Blood Flow = Pressure difference divided by Peripheral Resistance. The collateral arteries are upstream and coupled in series with the microcirculation. Therefore, changes in the resistance of the microcirculation affect the blood flow through the network. Lowering resistance R3 (microcirculation) will result in lower total resistance (Rt). As a consequence, at a given pressure difference (P), more flow will go through the arterial network. Subsequently, the upstream arterial collaterals (Resistance 1) will become exposed to more flow and shear stress levels, and increased shear stress stimulates outward lumen remodeling of collateral arteries. Diabetes and hypertension interfere with this feedback system.

Focusing on the bottom of the "bottom-up" principle, lumen formation ex nihilo is not a trivial process, and intense investigation has proposed several models, such as fusion of intracellular vacuoles [19], cord hollowing, and inverse blebbing [20]. Although these models nicely explain how a nonperfused angiogenic sprout generates its initial lumen, little is known how endothelial cells behave and rearrange to obtain a larger lumen during capillary maturation, later arteriologenesis, or how such diameter adaptation process can be stimulated in the context of a therapeutic strategy.

Conventional wisdom leaves a void here and suggests that the transition of a growing arteriole with a small diameter into a large caliber arteriole with a sizeable diameter occurs upon the blood flow driven change in a number of endothelial cells lining the arterial lumen [21,22]. An increase in flow promotes endothelial cell proliferation and migration collectively facilitating the transition of a small caliber vessel segment, with only a few endothelial cells, into a larger caliber arterial segment, with many endothelial cells. Flow activates the Akt-PI3-kinase signaling a pathway responsible for endothelial proliferation. In addition, flow activates the BMP-ALK-Smad signaling pathway, which restricts flow-induced Akt activation and promotes endothelial quiescence [21,23–27].

Attempting to fill the void between formation of a capillary sprout and its growth and integration into a perfused microvascular network, we recently described an alternative, flow-independent model, involving the enlargement of arteriolar endothelial cells, which resulted in the formation of large diameter arterioles [4]. Endothelial enlargement requires the GEF1 domain of the guanine nucleotide exchange factor Trio and activation of Rho-GTPases Rac1 and RhoG in the cell-cell junction region of endothelial cells. Cell domain specific activation of F-actin cytoskeleton remodeling events, and myosin-based tension at junction regions, provide physical forces for a structural enlargement of individual endothelial cells. Activation of Trio in developing arteries in vivo involves precise titration of the Vegf signaling strength in the arterial wall. Interestingly this signaling strength can be titrated by soluble Vegf receptor-1 (sFlt1). Moreover, this study suggests that sFlt1 may be used as a vehicle to deliver a physiologically relevant dosage of Vegf sufficient to stimulate endothelial cell enlargement while avoiding vascular leakage and unproductive angiogenesis, which are typically seen in Vegfa transgenic approaches [4].

The role of vascular sFlt1 acting as a rheostat to control Vegf signaling strength in the context of arterial endothelial cell enlargement may explain some of the pro-arteriogenic effects observed with the Flt1 specific ligands Vegfb and Plgf [28–31]. Vegfb acts as a cardiac-specific stimulator of arteriogenesis, yet the precise reason for this organ specificity is unclear. One explanation is based on the distribution of Flt1 and Kdr in and around the coronary vasculature. Assuming that Vegfb drives arteriogenesis by competing Vegfa away from Flt1, subsequently triggering Vegfa-Kdr mediated signaling, the tissue distribution of both Kdr and Flt1 may explain where Vegfb initiates the arteriogenesis response. In the coronaries, Flt1 and Kdr only colocalize in the most distal areas and capillaries, whereas coronary arteries in the proximal part express only Flt1. Based on the juxta positioning of Flt1 and Kdr, it is therefore conceivable that arterial enlargement and diameter increases commence in precapillary arterioles [29]. An increase in diameter lowers vascular resistance thereby attracting flow toward the distal regions. This simultaneously augments shear stress to promote outward lumen remodeling in the feed arteries, thereby augmenting blood flow conductance of the entire arterial network [18,29].

Deriving instructions from the observations in development and disease modeling in small and large animals, therapeutic angiogenesis has been applied before using protein and DNA wrapped in non-viral or viral vectors. Despite high pro-angiogenic potency and multiple clinical trials, VEGF-A has not been found to clearly improve ischemic heart function [32]. Owing to its powerful sprouting potential, VEGF-A temporarily increases capillarization up to hemangioma formation [33], without, however, providing mature microvascular networks. This signals back to conductance vessels sufficiently to induce (lumen diameter) growth of the conductance vessels [34]. For these reasons, VEGF-B may be used to greater therapeutic avail, thereby fine-tuning VEGF-A bioavailability by competition at Flt1, the weaker pro-angiogenic receptor, in effect releasing VEGF-A.

4. Capillary Maturation and De Novo Arteriologenesis—Linking Angiogenesis and Arteriogenesis

As outlined above, angiogenesis is essential to counteract capillary rarefaction, induced by three of five factors contributing most significantly to cardiovascular disease manifestations. However, evidence in preclinical and clinical studies revealed that stimulation of this process itself does not suffice to improve flow into an ischemic muscle area (reference), particularly in diabetic large animals [9]. Thus, an orchestrated sequence of sprouting of capillaries, potentially supported by the destruction of the surrounding extracellular matrix via Angiopoietin 2, is to be followed by a second phase, namely the recruitment of mural cells such as pericytes and smooth muscle cells for the function connection of capillaries to collaterals. Indeed, a couple of loss-of-function phenotypes indicate the existence of such a bottom-up vessel growth:

Lack of PDGF-B results in a hypercirculatory dilative heart and conductance vessel phenotype contrasted by capillary rarefaction [35], similar to overexpression of Angiopoietin-2, which counteracts pericyte-recruitment and vessel maturation driven by Angiopoietin-1 [36]. These findings indicate that in addition to initial capillary growth, subsequent stabilization by pericyte attachment is required for lasting microvessel structures which may be nurtured by growing collaterals. Thus, stabilization of growing capillaries and mural cell attachment in principle induce arteriogenesis. Of note, microvessel growth-dependent arteriogenesis is lost when the pericytes are intentionally detached, e.g., by Ang-2 overexpression [3]. Accordingly, Notch3 receptor mutations leading to hypomorphic activity and mural cell detachment (CADASIL disease) [37] also appear to provide shrinking of arteriolar and capillary beds leading to brain damage. In a zebrafish model, we have shown that Dll4 activating Notch stabilizes the branching pattern and prohibits aberrant angiogenic sprouting in the developing zebrafish [38]. It should be noted that prohibition of aberrant sprouting is also achieved independently from mural cells, e.g., fostering VE-cadherin-dependent endothelial cell contacts via sphingosin-1-phosphate interacting with its receptor S1pr1 [39].

Conversely, combining VEGF-A with maturation factors such as PDGF-B [40], a pericyte attractant factor, or angiopoietin-1, a microvessel stabilizing factor generated by pericytes themselves [41], suffices to improve flow into the ischemic muscle. Although formation of stabilized and mature capillary beds is likely to be followed by collateral growth, this interdependence is not trivial. Of note, arteriogenesis, forming a conductance vessel network out of preexisting collaterals, is distinct of de novo arteriologenesis, or formation of new arterioles in the microcirculation. Importantly, both processes may occur simultaneously. The molecular and cellular bases of the latter process are poorly understood but evidence is emerging showing similarities to the formation of arteries during early embryogenesis. Recruitment of native collaterals has been well documented and involves shear stress-driven activation of inflammatory processes and activation of monocyte and macrophage migration into the arterial wall collectively promoting outward remodeling, diameter growth, of the native collaterals. However, not all organs possess native collaterals. For example, the mouse heart is not equipped with preexisting collaterals at adulthood, and in such instances, de novo arteriologenesis is the predominant mode responsible for collateralization [42].

5. Arteriogenesis in the Developing Vasculature

Recent observations from developing arterial networks during embryogenesis and neonatal stages indicate that arterial trees form in reverse order. Initial arterial endothelial cell differentiation occurs outside of arterial vessels, and these pre-artery endothelial cells then build trees by following a migratory path from smaller into larger arteries, a process guided by the forces imparted by blood flow [5]. Endothelial cells polarize and subsequently migrate against the direction of blood flow, and thereby contribute to the growth and enlargement of the upstream arterial vessel segments. During artery formation, VEGF and Notch signaling act in a common pathway to induce arterial identity in endothelial cells. Notch and DLL ligands are furthermore important for differentiation, physiology, and function of vascular smooth cells in the arterial wall. Arterial endothelial cells use mechanoreceptors to sense the direction of blood flow followed by polarization and migration against this direction involving among others, the APJ, the Eng/Alk1/SMAD4, and DACH1/CXCL12/CXCR4 signaling pathways [5]. These observations from developing arterial networks clearly provide a cellular and molecular substrate for retrograde communication. Yet, it is unclear as to what extent these processes are active or can be re-activated in mature arterial networks in the adult with arterial walls consisting of several layers of smooth muscle. In mice application of CXCL12 can stimulate reassembly of arteries upon myocardial infarction. However, clinical trials thus far failed to demonstrate clinical efficacy [42].

Several studies have shown that in the brain and hindlimb, pre-existing arterial collateral networks develop prior to birth during critical phases of embryonic development [43]. In the brain pre-existing collateral number is highest at the time of birth after which it slowly decreases with age. Hypertension and diabetes accelerate the regression of pre-existing collaterals. In contrast, exercise, most likely via activation of shear stress-dependent mechanisms, prevents or reduces the regression of pre-existing collaterals [44]. Analysis of several inbred mouse strains has shown that arteriogenic capacity, in particular the extent of the pre-existing arterial collateral networks, differs greatly between the analyzed backgrounds suggesting a genetic component in the regulation of this process [38]. Mouse strains with small infarct size upon MCA occlusion showed significantly up to four-fold more and larger pre-existing collaterals when compared with mouse strains showing a relatively large infarct upon MCA occlusion [45]. Genetic linkage studies subsequently correlated the collateral network variation with Rabep2, a regulator of vesicular trafficking wherein cell surface receptors are internalized, and VEGFR2 signaling [46]. The VEGFA-VEGFR2/KDR signaling contributes importantly to the formation of pre-existing collaterals [43,47]. During embryogenesis, reduction of either the ligand or the receptor during the narrow time-window of collaterogenesis reduces collateral formation thus collateral number and diameter in the adult.

At present several known angiogenic proteins significantly impact the formation of collaterals and the extent of pre-existing collateral networks in pia and skeletal muscle including VEGFA, VEGFR2, Dll4, Notch, ADAM10, ADAM17, Gja4, Gja5, and molecules implied in macrophage behavior including Egln1 and NFkB1 [36,40–42]. Gja5, also known as connexin 40, is a gap junction protein that mediates electrical communication between endothelial cells thereby allowing vasodilatory signals that are initiated in the distal part of the microcirculation to travel to the proximal–feeding arterioles [48]. In this way, flow can be efficiently routed to the metabolically active regions. Dll4 and Notch are best known for their role in regulating sprouting angiogenesis downstream of VEGF signaling. Genetically or pharmacologically interfering with Dll4-Notch promoted formation of pre-existing collateral in part by enhancing arteriolar branching during late embryonic development [49]. However, in arterial occlusion models, the ischemic outcome did not improve. This was attributed to a defect in capillary functionality, vessel leakage, and impaired flow-induced outward remodeling of arterioles [49]. Perhaps a more physiological—non invasive—way to promote VEGF-Kdrl signaling is to reduce the ambient oxygen tension by moving from sea level to high altitude. As nicely demonstrated by the group of Jim Faber, high altitude rodents such as guinea pigs and deer mice featured a much higher number of collaterals than lowlander species, and were much better protected against cerebral ischemia [50]. The question then is how can lowlanders benefit from this vascular growth potential?

6. How Will Distal Microvessel Growth Induce Proximal Collateral Growth?

Since the seminal work of Wolfgang Schaper, direct growth of conductance vessels seemed to be an overarching goal: build the large roads, and the small networks will follow. Arteriovenous shunts providing an unprecedented level of shear stress locally to the vessel wall proved to be the optimal stimulus, the growth potential of which none of the classical or un-canonical growth factors could match [51]. Local inflammation, driven by MCP-1 responsive macrophages [52,53] induces arterial growth, which benefitted most from Il10-mediated M2-subtype polarization, but not from dexamethasone-inhibition of macrophage activation [54,55]. This is a relevant distinction since unselective macrophage-activation (e.g., by MCP-1) might accelerate atherosclerosis rather than solving its consequences, such as macrovascular obstruction [53]. However, even in this conductance-vessel centered model, microvessel networks may play a decisive role in providing pressure gradients required for collateral growth, the absence of which being prohibitive for this process (Figure 2).

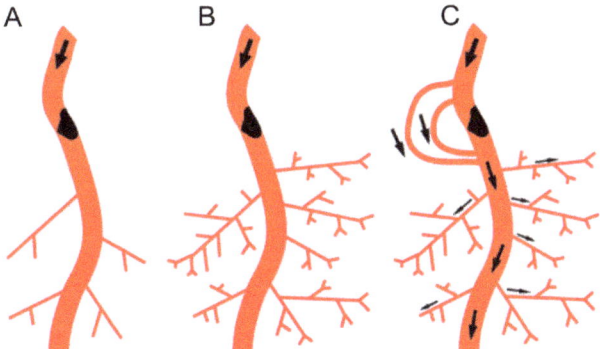

Figure 2. Cont.

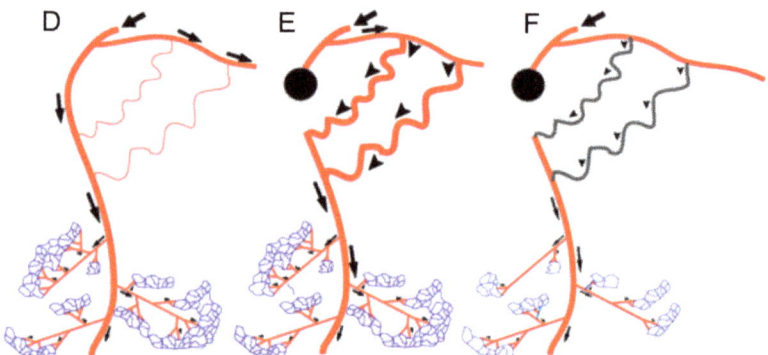

Figure 2. Interdependence of arteriogenesis and angiogenesis in healthy and arteriosclerotic vessels. (**A**): Schematic representation of the arterial network. Arterial occlusion disrupts blood flow to the microcirculation. (**B**): In the presence of the proximal occlusion, stimulation of angiogenesis in the microcirculation will not result in restoring peripheral blood flow. (**C**): Restoring blood flow upon occlusion requires the recruitment of native collaterals or de novo formation of collaterals that can bypass the occlusion in the arterial network. (**D**): Schematic representation of the arterial network in the leg with native arterial collaterals. Flow distribution indicated by arrows. (**E**): Occlusion in the main feed artery results in rerouting of blood flow and flow driven outward remodeling of pre-existing native arterial collaterals (arrowheads). (**F**): Diabetes and hypertension associated with rarefaction of small arterioles and capillary networks (indicated in dark blue) and impaired (flow driven) outward lumen remodeling (arteriogenesis) of arterial collateral networks (impaired collaterals in grey). In this scenario, only targeting upstream collaterals is not sufficient to achieve increased oxygen and nutrient delivery to the periphery. The perfusion of the microcirculation needs to be improved as well.

7. Angiogenesis, Maturation, and Arteriogenesis in Therapeutic Approaches

Are ischemia- and HIF1α-driven factors without alternative, when balanced angiogenesis/arteriogenesis is at stake? Recent observations by us and others have revealed that myogenic factors such as myocardin-related transcription factors (MRTFs), which in myocytes act as cofactors of SRF to keep up specific muscle protein stoichiometry, e.g., upon mechanosensing via the Rho-A pathway [56–58]. Of note, therapeutic agents derived from this pathway such as forced expression of MRTF-A or its depressor Thymosin ß4 (scavenging globular actin, which keeps MRTF-A in the cytosol), also provide strong vasoactive signals, synchronizing muscle with vessel growth or maintenance. Interestingly, the secreted factors found after MRTF-A activation are capable of providing capillary stabilization in addition to growth, indicating a program that may secure blood supply during increased demand [3]. Chronic ischemia reduces muscle regeneration and thereby RhoA-signaling dependent vessel maintenance. Indeed both MRTF-A and Thymosin ß4, when applied via a myotropic AAV9 vector, provide capillary growth and maturation in the peripheral and coronary microcirculation. Moreover, unless disrupted by Ang2, MRTF-A provides conductance vessel growth, highlighting the relevance of microvessel maturation for arteriogenesis [3].

Of note, the AAV-Tß4 approach in rabbits subjected to femoral artery excision improved arteriogenesis even when only locally applied to the ischemic lower limb. However, this was significantly less effective in increasing flow, when applied directly to the upper limb, i.e., the collateral growth area. Microvascular growth is signaling backward to collaterals to provide more flow, absent macrovessel obstruction. Since shear stress increase would be a likely candidate transmitting this growth signal, we applied L-NAME to inhibit its main effector, nitric oxide formation (Figure 3). Co-application of the NO-inhibitor abrogated solely arteriogenesis, but not angiogenesis. This finding indicates that flow-mediated vasodilation is a powerful signal for arteriogenesis, provided the endothelium is capable of forming the autacoid and is not impaired by atherosclerotic risk factors. To date, it is unclear whether other signals specifically enhancing arterial growth by targeting arterial endothelium, e.g., Notch [59], can therapeutically be utilized for arteriogenesis in addition to microvessel growth. Another potentially promising strategy consists of interference with microRNA regulation which may aid in overcoming negative feedback mechanisms in chronic ischemia of muscles and heart [60–62].

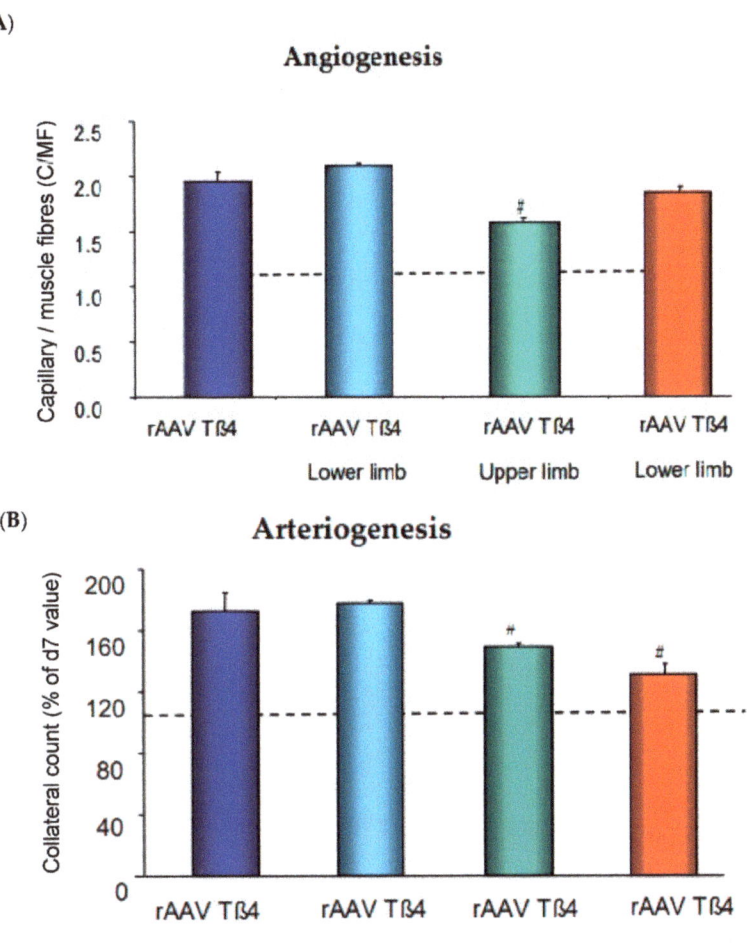

Figure 3. Angiogenesis and arteriogenesis upon recombinant AAV (rAAV) encoding for Thymosin ß4 (Tß4). (**A**) Capillary/muscle fiber ratio (C/MF) after application of rAAVTß4 either whole limb (retrograde venous infusion) or lower limb (intramuscular injection) with or without L-NAME coapplication. (**B**) Collateral count (% of d7 level) at d28. Dotted lines are control group levels. For methods cf. [22].

8. The Concept of Retrograde Communication and Arterial Growth

Retrograde communication has long been recognized in the context of exercise-induced vascular adaptation. Arterial networks dynamically adapt arterial branching patterning, vessel number, and lumen dimensions to exercise-induced changes in organ metabolism. These functional and structural alterations in the arterial network allow precise fine-tuning of flow delivery with the new level of metabolic activity of the organ it innervates. Metabolically active tissue regions are typically close to or in direct contact with the distal parts of the arteriolar microcirculation and the capillary network. In the exercise scenario, the stimulus for arterial network remodeling originates from the change in metabolic status of active cells in tissues surrounding the distal part of the microcirculation. A local mismatch between flow delivery and the new metabolic demand triggers the release of vascular growth factors and vasodilators which promote angiogenesis and expansion of the capillary network. They furthermore reduce vascular resistance in the periphery of the network to attract flow (cf. Figure 4). The local change in resistance and subsequent changes in hemodynamic factors ignite adaptive vascular remodeling responses that are propagated toward the more proximal parts of the arterial network. In these more proximal domains, increases in arterial lumen diameter and optimization of branching architecture augment blood flow conductance thereby increasing oxygen and nutrient delivery to the areas with the increased metabolic demand. The extent of the arterial lumen remodeling is a function of the amount of blood flow, the local shear stress setpoint, and the duration of the stimulus. Whereas an acute increase in flow causes functional endothelium-dependent vasodilation, long term upregulation in flow promotes structural outward arterial lumen remodeling, a process involving rearrangement of endothelial cells and vascular smooth muscle cells around an anatomically-structurally larger lumen. Conversely, restricting exercise and decreasing tissue metabolism and oxygen requirement results in pruning or rarefaction of microvessels, inward arterial lumen remodeling, and an overall reduction of blood flow conductance.

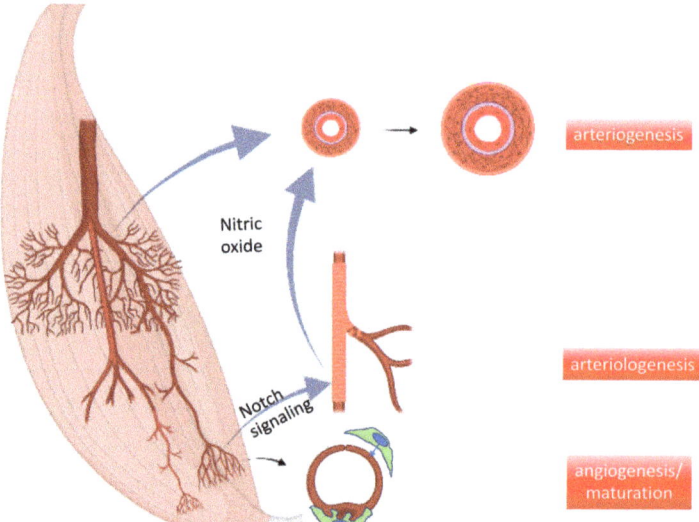

Figure 4. Interdependence of micro- and macrovessel growth. Key mechanisms of vessel growth in developing and adult organisms include capillary formation and stabilization by pericytes (=maturation) as well as resistance control by newly added smooth muscle cells to arteriolar endothelial cells (arteriologenesis), e.g., via Notch signaling. These functional units feedback demand to existing conductance vessels, e.g., via nitric oxide-mediated vessel dilation and subsequent growth (=arteriogenesis).

The lessons learned from exercise scenarios is that the initial trigger provoking the arterial network remodeling is caused by a change in resistance to flow at the pre-capillary to capillary level. Metabolic alterations induce the release of specific vascular growth factors (VEGF, angiopoietins), vasoactive metabolites (lactate, adenosine, Vegf), vasodilators (a.o. nitric oxide), as well as retrograde transfer—from the distal to proximal—of vasodilatory responses through electrical coupling in the vessel wall (connexins). Lower resistance in the distal areas will not only attract flow to this region but will also augment flow velocity and shear rates in the more proximal parts that feed the distal region. Increases in shear rates are by themselves a stimulus for dilation or—pending the duration of the stimulus—outward arterial remodeling. Hence the changes in the distal part are communicated to the proximal segments via changes in flow and translated into a gain of lumen dimension. Several risk factors for adverse vascular remodeling such as diabetes alter metabolism and expression of vascular growth factors. Thus they are most likely to affect retrograde communication efficiency.

In summary, microvessel growth requires maturation by the attraction of mural cells to provide long-lasting, functional capillary beds. In addition to providing blood to previously ischemic tissue, microvessel networks are capable of backward signaling to enlarge collaterals via nitric oxide. This interdependence can be exploited therapeutically at various entry points. Whether currently investigated vectors, such as AAVs or AdVs, and delivery systems, such as intramuscular or locoregional applications, hold up to newer developments such as modified RNA-carrying lipid nanoparticles, remains to be seen. However, it appears safe to state that neglecting microvessel network formation and maturation may lead to incomplete and unbalanced vessel growth, which is unlikely to improve the function of ischemic muscle tissue.

Funding: This research was funded by the Deutsche Forschungsgemeinschaft (DFG)—FOR2325 'Interactions at the Neurovascular Interface', by the DFG (Transregio 127 and 267) and by the BMBF (DZHK TRP GeneVaPad).

Institutional Review Board Statement: Not applicable.

Informed Consent Statement: Not applicable.

Data Availability Statement: Not applicable.

Conflicts of Interest: C.K. is holding IP on AAV-MRTF-A for therapeutic neovascularization. F.l.N. declares no conflict of interest.

References

1. Sherman, T.F. On connecting large vessels to small. The meaning of Murray's law. *J. Gen. Physiol.* **1981**, *78*, 431–453. [CrossRef]
2. Faber, J.E.; Chilian, W.M.; Deindl, E.; van Royen, N.; Simons, M. A brief etymology of the collateral circulation. *Arter. Thromb. Vasc. Biol.* **2014**, *34*, 1854–1859. [CrossRef] [PubMed]
3. Hinkel, R.; Trenkwalder, T.; Petersen, B.; Husada, W.; Gesenhues, F.; Lee, S.; Hannappel, E.; Bock-Marquette, I.; Theisen, D.; Leitner, L.; et al. MRTF-A controls vessel growth and maturation by increasing the expression of CCN1 and CCN2. *Nat. Commun.* **2014**, *5*, 3970. [CrossRef]
4. Klems, A.; van Rijssel, J.; Ramms, A.S.; Wild, R.; Hammer, J.; Merkel, M.; Derenbach, L.; Préau, L.; Hinkel, R.; Suarez-Martinez, I.; et al. The GEF Trio controls endothelial cell size and arterial remodeling downstream of Vegf signaling in both zebrafish and cell models. *Nat. Commun.* **2020**, *11*, 5319. [CrossRef] [PubMed]
5. Red-Horse, K.; Siekmann, A.F. Veins and Arteries Build Hierarchical Branching Patterns Differently: Bottom-Up versus Top-Down. *Bioessays* **2019**, *41*, e1800198. [CrossRef] [PubMed]
6. Yin, H.; Arpino, J.-M.; Lee, J.J.; Pickering, J.G. Regenerated Microvascular Networks in Ischemic Skeletal Muscle. *Front. Physiol.* **2021**, *12*, 662073. [CrossRef]
7. Rosenson, R.S.; Fioretto, P.; Dodson, P.M. Does microvascular disease predict macrovascular events in type 2 diabetes? *Atherosclerosis* **2011**, *218*, 13–18. [CrossRef] [PubMed]
8. Olivotto, I.; Oreziak, A.; Barriales-Villa, R.; Abraham, T.P.; Masri, A.; Garcia-Pavia, P.; Saberi, S.; Lakdawala, N.K.; Wheeler, M.T.; Owens, A.; et al. Mavacamten for treatment of symptomatic obstructive hypertrophic cardiomyopathy (EXPLORER-HCM): A randomised, double-blind, placebo-controlled, phase 3 trial. *Lancet* **2020**, *396*, 759–769. [CrossRef]
9. Hinkel, R.; Hoewe, A.; Renner, S.; Ng, J.; Lee, S.; Klett, K.; Kaczmarek, V.; Moretti, A.; Laugwitz, K.L.; Skroblin, P.; et al. Diabetes Mellitus-Induced Microvascular Destabilization in the Myocardium. *J. Am. Coll. Cardiol.* **2017**, *69*, 131–143. [CrossRef] [PubMed]
10. Schoos, M.M.; Dangas, G.D.; Mehran, R.; Kirtane, A.J.; Yu, J.; Litherland, C.; Clemmensen, P.; Stuckey, T.D.; Witzenbichler, B.; Weisz, G.; et al. Impact of Hemoglobin A1c Levels on Residual Platelet Reactivity and Outcomes After Insertion of Coronary Drug-Eluting Stents (from the ADAPT-DES Study). *Am. J. Cardiol.* **2016**, *117*, 192–200. [CrossRef]
11. Silvestre-Roig, C.; Braster, Q.; Wichapong, K.; Lee, E.Y.; Teulon, J.M.; Berrebeh, N.; Winter, J.; Adrover, J.M.; Santos, G.S.; Froese, A.; et al. Externalized histone H4 orchestrates chronic inflammation by inducing lytic cell death. *Nature* **2019**, *569*, 236–240. [CrossRef] [PubMed]

12. Ziegler, T.; Bähr, A.; Howe, A.; Klett, K.; Husada, W.; Weber, C.; Laugwitz, K.-L.; Kupatt, C.; Hinkel, R. Tβ4 Increases Neovascularization and Cardiac Function in Chronic Myocardial Ischemia of Normo- and Hypercholesterolemic Pigs. *Mol. Ther.* **2018**, *26*, 1706–1714. [CrossRef] [PubMed]
13. Kureishi, Y.; Luo, Z.; Shiojima, I.; Bialik, A.; Fulton, D.; Lefer, D.J.; Sessa, W.C.; Walsh, K. The HMG-CoA reductase inhibitor simvastatin activates the protein kinase Akt and promotes angiogenesis in normocholesterolemic animals. *Nat. Med.* **2000**, *6*, 1004–1010. [CrossRef] [PubMed]
14. Camici, P.G.; Tschöpe, C.; Di Carli, M.F.; Rimoldi, O.; Van Linthout, S. Coronary microvascular dysfunction in hypertrophy and heart failure. *Cardiovasc. Res.* **2020**, *116*, 806–816. [CrossRef] [PubMed]
15. Ramanujam, D.; Schon, A.P.; Beck, C.; Vaccarello, P.; Felician, G.; Dueck, A.; Esfandyari, D.; Meister, G.; Meitinger, T.; Schulz, C.; et al. MicroRNA-21-Dependent Macrophage-to-Fibroblast Signaling Determines the Cardiac Response to Pressure Overload. *Circulation* **2021**, *143*, 1513–1525. [CrossRef]
16. Hinkel, R.; Ramanujam, D.; Kaczmarek, V.; Howe, A.; Klett, K.; Beck, C.; Dueck, A.; Thum, T.; Laugwitz, K.L.; Maegdefessel, L.; et al. AntimiR-21 Prevents Myocardial Dysfunction in a Pig Model of Ischemia/Reperfusion Injury. *J. Am. Coll. Cardiol.* **2020**, *75*, 1788–1800. [CrossRef] [PubMed]
17. Yin, G.; Zhao, L. Risk of hypertension with anti-VEGF monoclonal antibodies in cancer patients: A systematic review and meta-analysis of 105 phase II/III randomized controlled trials. *J. Chemother.* **2021**, epub ahead of print. [CrossRef]
18. Zhao, E.; Barber, J.; Sen, C.K.; Arciero, J. Modeling acute and chronic vascular responses to a major arterial occlusion. *Microcirculation* **2022**, *29*, e12738. [CrossRef] [PubMed]
19. Kamei, M.; Saunders, W.B.; Bayless, K.J.; Dye, L.; Davis, G.E.; Weinstein, B.M. Endothelial tubes assemble from intracellular vacuoles in vivo. *Nature* **2006**, *442*, 453–456. [CrossRef] [PubMed]
20. Gebala, V.; Collins, R.; Geudens, I.; Phng, L.K.; Gerhardt, H. Blood flow drives lumen formation by inverse membrane blebbing during angiogenesis in vivo. *Nat. Cell Biol.* **2016**, *18*, 443–450. [CrossRef]
21. Jin, Y.; Muhl, L.; Burmakin, M.; Wang, Y.; Duchez, A.C.; Betsholtz, C.; Arthur, H.M.; Jakobsson, L. Endoglin prevents vascular malformation by regulating flow-induced cell migration and specification through VEGFR2 signalling. *Nat. Cell Biol.* **2017**, *19*, 639–652. [CrossRef] [PubMed]
22. Franco, C.A.; Jones, M.L.; Bernabeu, M.O.; Vion, A.C.; Barbacena, P.; Fan, J.; Mathivet, T.; Fonseca, C.G.; Ragab, A.; Yamaguchi, T.P.; et al. Non-canonical Wnt signalling modulates the endothelial shear stress flow sensor in vascular remodelling. *Elife* **2016**, *5*, e07727. [CrossRef] [PubMed]
23. Ola, R.; Künzel, S.H.; Zhang, F.; Genet, G.; Chakraborty, R.; Pibouin-Fragner, L.; Martin, K.; Sessa, W.; Dubrac, A.; Eichmann, A. SMAD4 Prevents Flow Induced Arteriovenous Malformations by Inhibiting Casein Kinase 2. *Circulation* **2018**, *138*, 2379–2394. [CrossRef] [PubMed]
24. Ola, R.; Dubrac, A.; Han, J.; Zhang, F.; Fang, J.S.; Larrivée, B.; Lee, M.; Urarte, A.A.; Kraehling, J.R.; Genet, G.; et al. PI3 kinase inhibition improves vascular malformations in mouse models of hereditary haemorrhagic telangiectasia. *Nat. Commun.* **2016**, *7*, 13650. [CrossRef] [PubMed]
25. Rochon, E.R.; Menon, P.G.; Roman, B.L. Alk1 controls arterial endothelial cell migration in lumenized vessels. *Development* **2016**, *143*, 2593–2602. [PubMed]
26. Laux, D.W.; Young, S.; Donovan, J.P.; Mansfield, C.J.; Upton, P.D.; Roman, B.L. Circulating Bmp10 acts through endothelial Alk1 to mediate flow-dependent arterial quiescence. *Development* **2013**, *140*, 3403–3412. [CrossRef] [PubMed]
27. Alsina-Sanchís, E.; García-Ibáñez, Y.; Figueiredo, A.M.; Riera-Domingo, C.; Figueras, A.; Matias-Guiu, X.; Casanovas, O.; Botella, L.M.; Pujana, M.A.; Riera-Mestre, A.; et al. ALK1 Loss Results in Vascular Hyperplasia in Mice and Humans Through PI3K Activation. *Arter. Thromb. Vasc. Biol.* **2018**, *38*, 1216–1229. [CrossRef] [PubMed]
28. Bry, M.; Kivelä, R.; Holopainen, T.; Anisimov, A.; Tammela, T.; Soronen, J.; Silvola, J.; Saraste, A.; Jeltsch, M.; Korpisalo, P.; et al. Vascular endothelial growth factor-B acts as a coronary growth factor in transgenic rats without inducing angiogenesis, vascular leak, or inflammation. *Circulation* **2010**, *122*, 1725–1733. [CrossRef] [PubMed]
29. Kivelä, R.; Bry, M.; Robciuc, M.R.; Räsänen, M.; Taavitsainen, M.; Silvola, J.M.; Saraste, A.; Hulmi, J.J.; Anisimov, A.; Mäyränpää, M.I.; et al. VEGF-B-induced vascular growth leads to metabolic reprogramming and ischemia resistance in the heart. *EMBO Mol. Med.* **2014**, *6*, 307–321. [CrossRef]
30. Lähteenvuo, J.E.; Lähteenvuo, M.T.; Kivelä, A.; Rosenlew, C.; Falkevall, A.; Klar, J.; Heikura, T.; Rissanen, T.T.; Vähäkangas, E.; Korpisalo, P.; et al. Vascular endothelial growth factor-B induces myocardium-specific angiogenesis and arteriogenesis via vascular endothelial growth factor receptor-1- and neuropilin receptor-1-dependent mechanisms. *Circulation* **2009**, *119*, 845–856. [CrossRef] [PubMed]
31. Dewerchin, M.; Carmeliet, P. PlGF: A multitasking cytokine with disease-restricted activity. *Cold Spring Harb. Perspect. Med.* **2012**, *2*, a011056. [CrossRef] [PubMed]
32. Ylä-Herttuala, S.; Baker, A.H. Cardiovascular Gene Therapy: Past, Present, and Future. *Mol. Ther.* **2017**, *25*, 1095–1106. [CrossRef] [PubMed]
33. Ozawa, C.R.; Banfi, A.; Glazer, N.L.; Thurston, G.; Springer, M.L.; Kraft, P.E.; McDonald, D.M.; Blau, H.M. Microenvironmental VEGF concentration, not total dose, determines a threshold between normal and aberrant angiogenesis. *J. Clin. Investig.* **2004**, *113*, 516–527. [CrossRef] [PubMed]

34. Carmeliet, P.; Dor, Y.; Herbert, J.M.; Fukumura, D.; Brusselmans, K.; Dewerchin, M.; Neeman, M.; Bono, F.; Abramovitch, R.; Maxwell, P.; et al. Role of HIF-1alpha in hypoxia-mediated apoptosis, cell proliferation and tumour angiogenesis. *Nature* **1998**, *394*, 485–490. [CrossRef] [PubMed]
35. Grunewald, M.; Kumar, S.; Sharife, H.; Volinsky, E.; Gileles-Hillel, A.; Licht, T.; Permyakova, A.; Hinden, L.; Azar, S.; Friedmann, Y.; et al. Counteracting age-related VEGF signaling insufficiency promotes healthy aging and extends life span. *Science* **2021**, *373*, eabc8479. [CrossRef] [PubMed]
36. Ziegler, T.; Horstkotte, J.; Schwab, C.; Pfetsch, V.; Weinmann, K.; Dietzel, S.; Rohwedder, I.; Hinkel, R.; Gross, L.; Lee, S.; et al. Angiopoietin 2 mediates microvascular and hemodynamic alterations in sepsis. *J. Clin. Investig.* **2013**, *123*, 3436–3445. [CrossRef] [PubMed]
37. Arboleda-Velasquez, J.F.; Manent, J.; Lee, J.H.; Tikka, S.; Ospina, C.; Vanderburg, C.R.; Frosch, M.P.; Rodríguez-Falcón, M.; Villen, J.; Gygi, S.; et al. Hypomorphic Notch 3 alleles link Notch signaling to ischemic cerebral small-vessel disease. *Proc. Natl. Acad. Sci. USA* **2011**, *108*, E128–E135. [CrossRef] [PubMed]
38. Jiang, Q.; Lagos-Quintana, M.; Liu, D.; Shi, Y.; Helker, C.; Herzog, W.; le Noble, F. miR-30a regulates endothelial tip cell formation and arteriolar branching. *Hypertension* **2013**, *62*, 592–598. [CrossRef]
39. Gaengel, K.; Niaudet, C.; Hagikura, K.; Siemsen, B.; Muhl, L.; Hofmann, J.-A.; Ebarasi, L.; Nystr+Âm, S.; Rymo, S.; Chen, L.-A.; et al. The Sphingosine-1-Phosphate Receptor S1PR1 Restricts Sprouting Angiogenesis by Regulating the Interplay between VE-Cadherin and VEGFR2. *Dev. Cell* **2012**, *23*, 587–599. [CrossRef] [PubMed]
40. Kupatt, C.; Hinkel, R.; Pfosser, A.; El-Aouni, C.; Wuchrer, A.; Fritz, A.; Globisch, F.; Thormann, M.; Horstkotte, J.; Lebherz, C.; et al. Cotransfection of Vascular Endothelial Growth Factor-A and Platelet-Derived Growth Factor-B Via Recombinant Adeno-Associated Virus Resolves Chronic Ischemic Malperfusion: Role of Vessel Maturation. *J. Am. Coll. Cardiol.* **2010**, *56*, 414–422. [CrossRef] [PubMed]
41. Arsic, N.; Zentilin, L.; Zacchigna, S.; Santoro, D.; Stanta, G.; Salvi, A.; Sinagra, G.; Giacca, M. Induction of functional neovascularization by combined VEGF and angiopoietin-1 gene transfer using AAV vectors. *Mol. Ther.* **2003**, *7*, 450–459. [CrossRef]
42. Red-Horse, K.; Das, S. New Research Is Shining Light on How Collateral Arteries Form in the Heart: A Future Therapeutic Direction? *Curr. Cardiol. Rep.* **2021**, *23*, 30. [CrossRef]
43. Lucitti, J.L.; Mackey, J.K.; Morrison, J.C.; Haigh, J.J.; Adams, R.H.; Faber, J.E. Formation of the collateral circulation is regulated by vascular endothelial growth factor-A and a disintegrin and metalloprotease family members 10 and 17. *Circ. Res.* **2012**, *111*, 1539–1550. [CrossRef] [PubMed]
44. Rzechorzek, W.; Zhang, H.; Buckley, B.K.; Hua, K.; Pomp, D.; Faber, J.E. Aerobic exercise prevents rarefaction of pial collaterals and increased stroke severity that occur with aging. *J. Cereb. Blood Flow Metab.* **2017**, *37*, 3544–3555. [CrossRef] [PubMed]
45. Zhang, H.; Prabhakar, P.; Sealock, R.; Faber, J.E. Wide genetic variation in the native pial collateral circulation is a major determinant of variation in severity of stroke. *J. Cereb. Blood Flow Metab.* **2010**, *30*, 923–934. [CrossRef]
46. Lucitti, J.L.; Sealock, R.; Buckley, B.K.; Zhang, H.; Xiao, L.; Dudley, A.C.; Faber, J.E. Variants of Rab GTPase-Effector Binding Protein-2 Cause Variation in the Collateral Circulation and Severity of Stroke. *Stroke* **2016**, *47*, 3022–3031. [CrossRef] [PubMed]
47. Clayton, J.A.; Chalothorn, D.; Faber, J.E. Vascular endothelial growth factor-A specifies formation of native collaterals and regulates collateral growth in ischemia. *Circ. Res.* **2008**, *103*, 1027–1036. [CrossRef] [PubMed]
48. Buschmann, I.; Pries, A.; Styp-Rekowska, B.; Hillmeister, P.; Loufrani, L.; Henrion, D.; Shi, Y.; Duelsner, A.; Hoefer, I.; Gatzke, N.; et al. Pulsatile shear and Gja5 modulate arterial identity and remodeling events during flow-driven arteriogenesis. *Development* **2010**, *137*, 2187–2196. [CrossRef]
49. Cristofaro, B.; Shi, Y.; Faria, M.; Suchting, S.; Leroyer, A.S.; Trindade, A.; Duarte, A.; Zovein, A.C.; Iruela-Arispe, M.L.; Nih, L.R.; et al. Dll4-Notch signaling determines the formation of native arterial collateral networks and arterial function in mouse ischemia models. *Development* **2013**, *140*, 1720–1729. [CrossRef]
50. Faber, J.E.; Storz, J.F.; Cheviron, Z.A.; Zhang, H. High-altitude rodents have abundant collaterals that protect against tissue injury after cerebral, coronary and peripheral artery occlusion. *J. Cereb. Blood Flow Metab.* **2021**, *41*, 731–744. [CrossRef] [PubMed]
51. Schierling, W.; Troidl, K.; Troidl, C.; Schmitz-Rixen, T.; Schaper, W.; Eitenmüller, I.K. The Role of Angiogenic Growth Factors in Arteriogenesis. *J. Vasc. Res.* **2009**, *46*, 365–374. [CrossRef] [PubMed]
52. Heil, M.; Ziegelhoeffer, T.; Wagner, S.; Fernandez, B.; Helisch, A.; Martin, S.; Tribulova, S.; Kuziel, W.A.; Bachmann, G.; Schaper, W. Collateral artery growth (arteriogenesis) after experimental arterial occlusion is impaired in mice lacking CC-chemokine receptor-2. *Circ. Res.* **2004**, *94*, 671–677. [CrossRef] [PubMed]
53. van Royen, N.; Hoefer, I.; Bottinger, M.; Hua, J.; Grundmann, S.; Voskuil, M.; Bode, C.; Schaper, W.; Buschmann, I.; Piek, J.J. Local Monocyte Chemoattractant Protein-1 Therapy Increases Collateral Artery Formation in Apolipoprotein E-Deficient Mice but Induces Systemic Monocytic CD11b Expression, Neointimal Formation, and Plaque Progression. *Circ. Res.* **2003**, *92*, 218–225. [CrossRef] [PubMed]
54. Troidl, C.; Jung, G.; Troidl, K.; Hoffmann, J.; Mollmann, H.; Nef, H.; Schaper, W.; Hamm, C.W.; Schmitz-Rixen, T. The temporal and spatial distribution of macrophage subpopulations during arteriogenesis. *Curr. Vasc. Pharmacol.* **2013**, *11*, 5–12. [CrossRef] [PubMed]
55. Götze, A.M.; Schubert, C.; Jung, G.; Dörr, O.; Liebetrau, C.; Hamm, C.W.; Schmitz-Rixen, T.; Troidl, C.; Troidl, K. IL10 Alters Peri-Collateral Macrophage Polarization and Hind-Limb Reperfusion in Mice after Femoral Artery Ligation. *Int. J. Mol. Sci.* **2020**, *21*, 2821. [CrossRef] [PubMed]

56. Lauriol, J.; Keith, K.; Jaffr, F.; Couvillon, A.; Saci, A.; Goonasekera, S.A.; McCarthy, J.R.; Kessinger, C.W.; Wang, J.; Ke, Q.; et al. RhoA signaling in cardiomyocytes protects against stress-induced heart failure but facilitates cardiac fibrosis. *Sci. Signal.* **2014**, *7*, ra100. [CrossRef] [PubMed]
57. Dorn, T.; Kornherr, J.; Parrotta, E.I.; Zawada, D.; Ayetey, H.; Santamaria, G.; Iop, L.; Mastantuono, E.; Sinnecker, D.; Goedel, A.; et al. Interplay of cell-cell contacts and RhoA/MRTF-A signaling regulates cardiomyocyte identity. *EMBO J.* **2018**, *37*, e98133. [CrossRef] [PubMed]
58. Eitenmuller, I.; Volger, O.; Kluge, A.; Troidl, K.; Barancik, M.; Cai, W.J.; Heil, M.; Pipp, F.; Fischer, S.; Horrevoets, A.J.G.; et al. The Range of Adaptation by Collateral Vessels After Femoral Artery Occlusion. *Circ. Res.* **2006**, *99*, 656–662. [CrossRef]
59. Limbourg, A.; Ploom, M.; Elligsen, D.; Sorensen, I.; Ziegelhoeffer, T.; Gossler, A.; Drexler, H.; Limbourg, F.P. Notch Ligand Delta-Like 1 Is Essential for Postnatal Arteriogenesis. *Circ. Res.* **2007**, *100*, 363–371. [CrossRef] [PubMed]
60. Guan, Y.; Cai, B.; Wu, X.; Peng, S.; Gan, L.; Huang, D.; Liu, G.; Dong, L.; Xiao, L.; Liu, J.; et al. microRNA-352 regulates collateral vessel growth induced by elevated fluid shear stress in the rat hind limb. *Sci. Rep.* **2017**, *7*, 6643. [CrossRef] [PubMed]
61. Lei, Z.; van Mil, A.; Brandt, M.M.; Grundmann, S.; Hoefer, I.; Smits, M.; El Azzouzi, H.; Fukao, T.; Cheng, C.; Doevendans, P.A.; et al. MicroRNA-132/212 family enhances arteriogenesis after hindlimb ischaemia through modulation of the Ras-MAPK pathway. *J. Cell Mol. Med.* **2015**, *19*, 1994–2005. [CrossRef] [PubMed]
62. Landskroner-Eiger, S.; Qiu, C.; Perrotta, P.; Siragusa, M.; Lee, M.Y.; Ulrich, V.; Luciano, A.K.; Zhuang, Z.W.; Corti, F.; Simons, M.; et al. Endothelial miR-17~92 cluster negatively regulates arteriogenesis via miRNA-19 repression of WNT signaling. *Proc. Natl. Acad. Sci. USA* **2015**, *112*, 12812–12817. [CrossRef] [PubMed]

Article

MicroRNA-30b Is Both Necessary and Sufficient for Interleukin-21 Receptor-Mediated Angiogenesis in Experimental Peripheral Arterial Disease

Tao Wang [1,2], Liang Yang [2,3], Mingjie Yuan [2,4], Charles R. Farber [5], Rosanne Spolski [6], Warren J. Leonard [6], Vijay C. Ganta [2,7] and Brian H. Annex [2,7,*]

1. State Key Laboratory of Respiratory Diseases, Guangzhou Institute of Respiratory Health, Guangzhou Medical University, Guangzhou 510120, China; taowang@gzhmu.edu.cn
2. Robert M Berne Cardiovascular Research Center, University of Virginia, Charlottesville, VA 22908, USA; yangliang@nankai.edu.cn (L.Y.); yuanmj8341@163.com (M.Y.); vganta@augusta.edu (V.C.G.)
3. Department of Pharmacology, Nankai University, Tianjing 300071, China
4. Department of Cardiology, Renmin Hospital of Wuhan University, Wuhan 430070, China
5. Center for Public Health Genomics, University of Virginia, Charlottesville, VA 22908, USA; crf2s@virginia.edu
6. Laboratory of Molecular Immunology and the Immunology Center, National Heart Lung and Blood Institute, National Institutes of Health, Bethesda, MD 20892, USA; spolskir@nhlbi.nih.gov (R.S.); wjl@helix.nih.gov (W.J.L.)
7. Vascular Biology Center and Department of Medicine, Augusta University, Augusta, GA 30912, USA
* Correspondence: bannex@augusta.edu

Abstract: The interleukin-21 receptor (IL-21R) can be upregulated in endothelial cells (EC) from ischemic muscles in mice following hind-limb ischemia (HLI), an experimental peripheral arterial disease (PAD) model, blocking this ligand-receptor pathway impaired STAT3 activation, angiogenesis, and perfusion recovery. We sought to identify mRNA and microRNA transcripts that were differentially regulated following HLI, based on the ischemic muscle having intact, or reduced, IL-21/IL21R signaling. In this comparison, 200 mRNAs were differentially expressed but only six microRNA (miR)/miR clusters (and among these only miR-30b) were upregulated in EC isolated from ischemic muscle. Next, myoglobin-overexpressing transgenic (MgTG) C57BL/6 mice examined following HLI and IL-21 overexpression displayed greater angiogenesis, better perfusion recovery, and less tissue necrosis, with increased miR-30b expression. In EC cultured under hypoxia serum starvation, knock-down of miR-30b reduced, while overexpression of miR-30b increased IL-21-mediated EC survival and angiogenesis. In Il21r$^{-/-}$ mice following HLI, miR-30b overexpression vs. control improved perfusion recovery, with a reduction of suppressor of cytokine signaling 3, a miR-30b target and negative regulator of STAT3. Together, miR-30b appears both necessary and sufficient for IL21/IL-21R-mediated angiogenesis and may present a new therapeutic option to treat PAD if the IL21R is not available for activation.

Keywords: vascular disease; gene therapy; innate immunity; pre-clinical models

1. Introduction

Peripheral arterial disease (PAD) is caused by atherosclerosis, leading to occlusions of the arteries that supply the lower extremities; PAD affects more than 12 million people in the U.S. and millions more worldwide [1–4]. The primary problem in PAD is reduced blood flow to the leg(s), and in most patients with symptomatic, advanced PAD, there is a total blockage along the sole major arterial pathway that supplies blood to the leg, and the amount of blood flow to the lower extremity becomes entirely dependent on the extent of ischemia-induced angiogenesis to allow ambulation and avoid limb amputation [5,6].

Surgically induced hind-limb ischemia (HLI) is a widely used pre-clinical model of PAD in mice [7–9]. We utilized an unbiased discovery strategy that was based on known differences in the extent of perfusion recovery across inbred mouse strains following HLI, and showed that the IL-21R was significantly upregulated on ischemic endothelium in-vivo from a mouse strain that showed good perfusion recovery after HLI but not in a strain that showed poor perfusion recovery [10]. In addition, in IL21R-deficient ($Il21r^{-/-}$) mice on a C57Bl6 background vs. C57Bl6 wild-type (WT), and in WT mice treated with IL-21R-Fc chimera (absorbs and eliminates available IL21) vs. control IgG, there was less STAT3 activation, less angiogenesis, and poorer perfusion recovery in the ischemic limb after HLI [10]. Further, studies in cultured endothelial cells (ECs) demonstrated that IL-21 enhanced EC survival and tube formation under the PAD-relevant hypoxia serum starvation (HSS) condition, however, in EC, under normoxia IL-21 addition, it did not cause an angiogenic effect or STAT3 activation [10].

When considered as a potential therapeutic for PAD, two problems regarding the IL21/IL21R pathway became readily apparent. First, we had shown that the angiogenic effects of IL21 were receptor dependent [10]. Although we reported that, as a group, muscle biopsies from the ischemic limb of patients with one form of PAD had higher levels of the IL21R on EC when compared to controls, the data showed overlap in values in individuals in each group and the study was specific for those with intermittent claudication [11]. Therefore, we could not be certain the IL21R would always be upregulated and for activation in all patients with PAD. Second, whether administration of the ligand could be achieved without toxicity was unknown [12]. Studies on IL-21/IL-21R-signaling in vitro and in vivo have identified a number of downstream gene-expression changes based on study conditions [12–16]. With little to no data available on the IL21/IL21R pathway in PAD, we performed RNA sequencing (RNA-Seq)-based analysis of whole transcriptomes from ischemic muscle with or without IL-21/IL-21R-pathway interruption, using a soluble IL-21R-Fc chimera as described [10]. We then examined an informative time point 7 days after HLI, because the extent of perfusion recovery was comparable between the two groups; this is a strategy we used in several other publications [10,17–19]. Though 200 mRNA were differentially expressed, only a few microRNAs (miRs) clusters were changed by IL-21/IL-21R-pathway interruption. We then proceeded to validate the data from the RNA-seq and examined the expression of the identified miRs in the endothelial cells isolated from ischemic muscle and found that miR-30b was the only miR of the group modulated by IL-21. We went on to show that miR-30b is both necessary and sufficient for IL-21/IL-21R hypoxic-induced angiogenesis and identified a potential gene target for miR-30b.

2. Results

2.1. Identification of IL-21R Pathway Transcripts following Experimental PAD

Of the 32,062 transcripts quantified by RNA-seq from whole muscle, 200 mRNA were significantly different ($q < 0.1$) in animals treated with IL-21R-Fc vs. control IgG at 7 days after HLI (Supplemental Table S1). Of these transcripts, 131 mRNAs were significantly downregulated, and 69 mRNAs were significantly upregulated. Though a large number of mRNAs were differentially regulated, only six microRNA (miR)/miR clusters were differentially regulated (Table 1). Of these six miR clusters, miR-343 and miR-5103 were not detectable by qPCR in the ischemic muscle from either IL-21R-Fc or control group. From the four remaining miR clusters, when transcript levels from IL-21R-Fc group were compared to muscle from mice treated with control IgG, the miR-let7a-2 and miR-100 cluster, miR-30b within the miR-30b/miR-30d cluster, and miR-503 within the miR-503/miR-322/miR-351 showed significantly lower expression levels (0.38 ± 0.18, 0.27 ± 0.03, 0.27 ± 0.05 and 0.53 ± 0.08 fold relative to control group, n = 4/group, $p < 0.05$, Figure 1); miR-3572 showed significantly higher expression levels (1.43 ± 0.06 fold relative to control group, n = 4/group, $p = 0.003$) (Figure 1A); miR-322, miR-351 and miR-30d levels showed no significant difference (Figure 1A).

Table 1. MicroRNA (miR) and miR clusters that are differentially regulated by IL-21R-Fc in the ischemic muscle based on RNA-seq analysis. q-value indicates FDR adjusted p-value of RNA-seq data.

miRs	Fold	q-Value
miR-343	0.03	0.01
miR-5103	0.06	0.01
miR-100, miR-let7a-2	0.17	0.01
miR-30b, miR-30d	0.20	0.01
miR-322, miR-351, miR-503	0.31	0.01
miR-3572	10.08	0.01

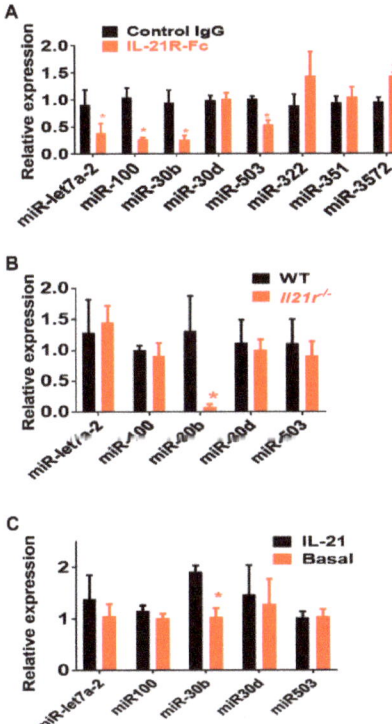

Figure 1. (**A**) Validation of RNA-sequencing analysis by real-time qPCR for microRNAs (miR) regulated by IL-21R-Fc treatment in ischemic muscle 7 days after hind-limb ischemia (HLI). When transcript levels were compared to muscle from mice treated with control IgG, miR100, miR-let7a-2, miR30b in miR30b/miR30d cluster, and miR503 in miR503/miR322/miR351 cluster showed significantly lower expression levels; 3572 showed significantly higher expression levels; miR322, miR351 and miR30d level showed no significant difference. (**B**) Expression of miR-let7a-2, miR-100, miR-30b, miR-30d and miR-503 in CD31 (endothelial cell marker)-enriched fraction of cells isolated from ischemic hindlimbs of wild-type (*WT*) and $Il21r^{-/-}$ mice were quantified by qPCR, miR30b showed significant lower expression in endothelial cells from $Il21r^{-/-}$ mice when compared to endothelial cells from WT mice (n = 4/group). However, the other four microRNAs did not show any significant difference between the two groups. (**C**) In HUVECs cultured under hypoxia serum starvation (HSS) (used to mimic ischemic muscle in vivo), 24 h treatment of IL-21 increased miR30b expression (n = 4/group, * $p < 0.05$), but did not regulate the expression of the other four microRNAs. * $p < 0.05$. Fold changes of each miR expression were calculated based on the average ΔCt value of the control group (for details, see material and methods). Data represent mean ± SEM; n = 4 per group. Control IgG indicates ischemic gastrocnemius muscle from the mice treated with control IgG, IL-21R-Fc ischemic gastrocnemius muscle from the mice treated with IL-21R-Fc.

Since we previously showed that IL-21R was upregulated in the endothelial cells under ischemia in both mouse and human PAD muscle [10,11], we isolated CD31-expressing cells from ischemic hind-limb muscles of $Il21r^{-/-}$ or wild-type (WT) mice, as described, [10] and measured the expression level of miR-let7a-2, miR-300, miR-30b, miR-503 and miR-3572. MiR-30b was the only of these miRs which showed significant difference in those from $Il21r^{-/-}$ mice (0.08 ± 0.04-fold relative to EC from WT mice, p = 0.04, Figure 1B). miR-3572 was not detectable in endothelial cells. Next, we compared the expression of miR-let7a-2, miR-300, miR-30b, miR-503 and miR3572 in HUVECs cultured under HSS conditions, with or without IL-21 treatment for 24h, and miR-30b was the only one that showed a differential expression level (1.91 ± 0.12-fold to basal, p = 0.01, Figure 1B). MiR-3572 was not detectable by qPCR. Interestingly, even though miR-30b is co-localized with miR-30d in the same microRNA cluster, miR-30d was not differentially regulated by qPCR from any of the following: (a) ischemic muscle (1.01 ± 0.12-fold to control, p = 0.14, Figure 1B), endothelial cells isolated from ischemic muscle (1.03 ± 0.11-fold to control, p = 0.46, Figure 1B) or cultured endothelial cells (1.00 ± 0.16-fold to control, p = 0.90, Figure 1C).

2.2. IL-21 Overexpression Improves Perfusion Recovery, Increases STAT3 Phosphorylation, and Upregulates miR-30b in the Ischemic Muscle after Hind Limb Ischemia (HLI) in Myoglobin Transgenic Mice

Myoglobin transgenic (MgTG) mice were used to test the effects of IL-21 overexpression following HLI for several reasons. First, compared to wild type (WT) C57BL/6 mice, these mice have impaired angiogenesis and poorer perfusion recovery following HLI [30,31]. Second, this strain was generated on a C57BL/6 background, where we previously showed findings that the IL-21R is upregulated in the C57BL/6 strain [10]. Third, the transgene includes the entire myoglobin gene as well as its striated muscle-specific promoter, so that overexpression is limited to striated muscle myocyte, and the endothelium from these mice is the same as wild-type [30,31].

As predicted, the MgTG mice on the C57BL/6 background showed a 10.6 ± 1.8-fold increase of $Il21r$ expression in the ischemic muscle when compared to non-ischemic muscle tissue, 7 days after HLI (Figure 2A). To augment IL-21 expression and IL21R pathway activation, we injected an expression plasmid containing IL-21, or its control plasmid, into the hind-limb muscles. Compared with those receiving control (scrambled sequence) plasmid, MgTG mice receiving IL-21 plasmid showed improved perfusion recovery at day-21 after HLI (67.9% ± 3.3% vs. 57.6% ± 3.2%, p = 0.02, Figure 2B) and less tissue necrosis (0 out of 6 vs. 4 out of 7, p < 0.01). Confirming the specificity of the findings, $Il21r^{-/-}$ mice receiving IL-21 cDNA did not show any changes in tissue necrosis (4 out of 12 vs. 3 out of 10, p = 0.86) or perfusion recovery at any of the same time points (54.5% ± 6.4% vs. 58.5% ± 7.1%, p = 0.68, Figure 2C). Next, we determined the expression of IL-21 at 10 days after plasmid injection and IL-21 protein expression was significantly higher in the ischemic hind-limb muscle of the mice that received IL-21 cDNA compared with mice that received scramble plasmid in both MgTG mice (IL-21/ERK2, 0.27 ± 0.08 vs. 0.06 ± 0.01, n = 5/group, p = 0.02) and $Il21r^{-/-}$ mice (IL-21/ERK2, 0.62 ± 0.08 vs. 0.22 ± 0.02, n = 6–7/group, p = 0.02) (Figure 2D). In Figure 2E, we showed that the ischemic hind-limb muscle tissue from MgTG mice receiving IL-21 plasmid had a higher capillary density compared with those receiving control plasmid (1.9 ± 0.18 vs. 1.2 ± 0.17, capillaries/fiber, n = 6–7/group, p = 0.03) 21 days after HLI.

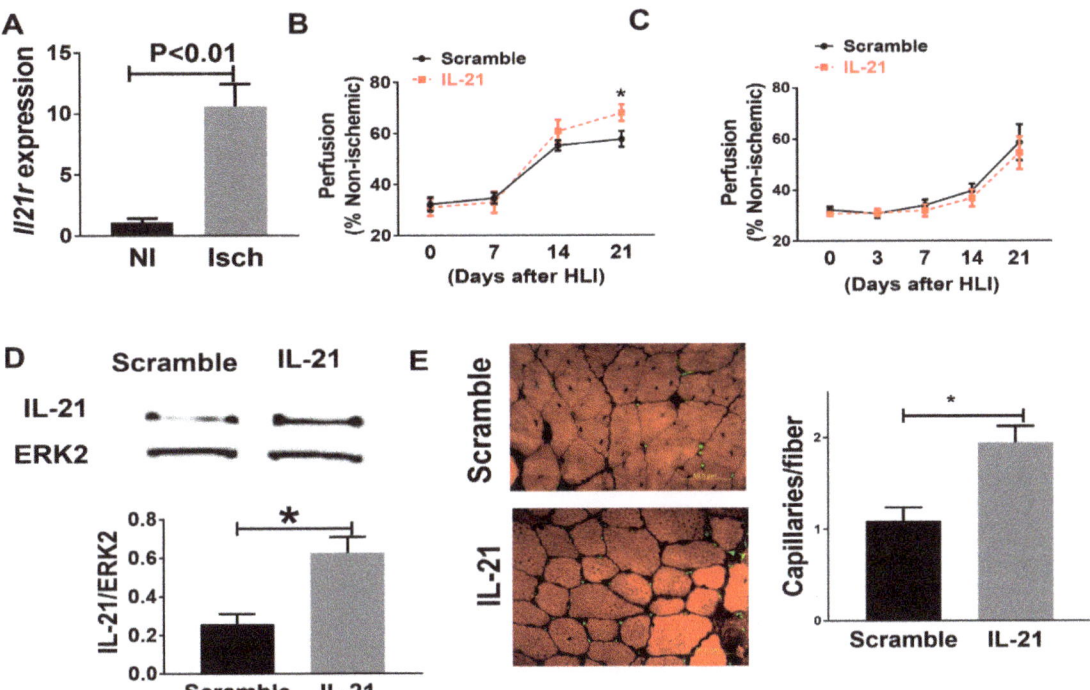

Figure 2. (**A**) Interleukin-21 receptor (IL-21R) expression is upregulated in the ischemic limb from MgTG mice when compared to the non-ischemic limb. (**B**) Data from Laser Doppler Perfusion imaging (LDPI) showed that IL-21 overexpression improved perfusion recovery in MgTG mice at day 21 (n = 6~7/group). (**C**) IL-21 overexpression did not change perfusion recovery in IL-21R knockout (KO) mice at any of the selected time point. (**D**) IL-21 plasmid transfection showed higher IL-21 protein level in the ischemic limb muscle from $Il21r^{-/-}$ mice muscle. (**E**) At day 21 after HLI, ischemic gastrocnemius muscle from MgTG mice that received IL-21 overexpression plasmid showed significant higher capillary density than mice that received control plasmid (n = 6~7/group). Average capillaries per muscle fiber: 1.9 ± 0.18 vs. 1.1 ± 0.17, p = 0.03. Data = mean ± SEM. * p < 0.05.

The IL-21R can signal via STAT1, STAT3, Akt, and ERK1/2 in different conditions [12,15] and we tested for potential differences in activation along with these pathways. Mice that received IL-21-expressing plasmid also had greater STAT3 phosphorylation (p-STAT3/STAT3, 0.72 ± 0.15 vs. 0.35 ± 0.10, n = 5/group, p = 0.02) but no change in STAT1 (p-STAT1/STAT1, 0.42 ± 0.04 vs. 0.43 ± 0.05, n = 5/group, p = 0.74), Akt (p-Akt/Akt, 0.24 ± 0.01 vs. 0.28 ± 0.01, n = 5/group, p = 0.17) or ERK1/2 (p-ERK/ERK, 0.42 ± 0.04 vs. 0.36 ± 0.05, n = 5/group, p = 0.39) phosphorylation in ischemic muscle 1 day after HLI when compared with mice receiving control plasmid (Figure 3A). Next, we measured miR-30b expression in the ischemic muscle specimen by qPCR and showed a significant higher expression from MgTG mice receiving IL-21 cDNA (2.73 ± 0.41-fold to control, n = 5, p = 0.03) (Figure 3B) when compared with those receiving control plasmid 7 days after HLI.

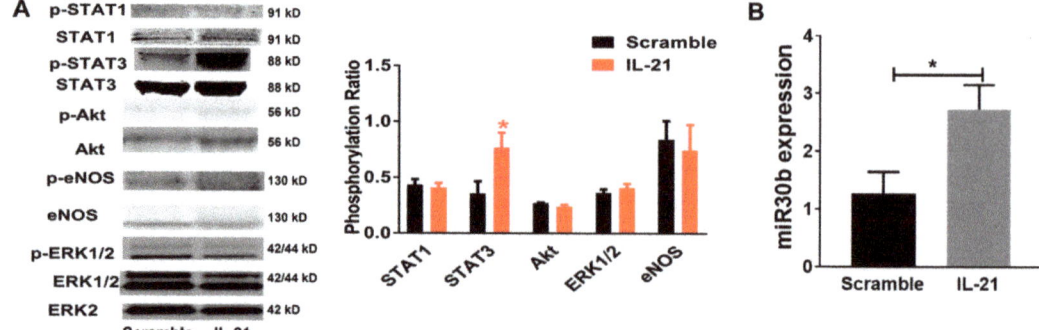

Figure 3. STAT3 phosphorylation and miR-30b expression were upregulated by IL-21 in the ischemic muscle in HLI model. (**A**) Western blots were probed for STAT1, p-STAT1, STAT3, p-STAT3, Akt, p-Akt, ERK1/2 and Erk1/2 with only ERK2 shown as loading control, IL-21 overexpression increases STAT3 phosphorylation in ischemic muscle 1 day after HLI (p-STAT3/STAT3) but shows no change in STAT1, Akt, eNOS or ERK1/2 phosphorylation. (**B**) miR-30b expression is upregulated by IL-21 overexpression in ischemic muscle 7 days after HLI, quantitative normalization of microRNA in each sample was performed using expression of small nucleolar RNA MBII-202 (sno202) as an internal control. Data = mean ± SEM. N = 4–5/group, * $p < 0.05$, IL-21 indicates IL-21 overexpression plasmid; Scramble indicates control plasmid with scramble sequence.

2.3. Overexpression of miR-30b Improves Perfusion Recovery in Il21r$^{-/-}$ Mice

To test whether miR-30b modulates the response to HLI, we overexpressed miR-30b in Il21r$^{-/-}$ mice using local intramuscular injections of pre-miR-30b overexpression plasmid 3 days before HLI with empty vector (EV) was used as control. Expression of miR-30b was assessed 10 days post plasmid injection and was increased >100 fold in ischemic hind limbs of mice vs. empty vector (Figure 2.5A). Next, we determined the effect of this gene transfer on perfusion recovery following experimental PAD, as shown in Figure 2.5B, miR-30b overexpression significantly improved perfusion recovery at day 14 (63.3% ± 3.0% vs. 49.5% ± 3.5%, n = 11~12/group, $p < 0.01$) and day 21 (74.4% ± 4.1% vs. 58.4% ± 4.6%, n = 11~12/group, $p = 0.016$) post HLI. There was a numeric but not statistically significant reduction in hind-limb necrosis rate (2/12 vs. 4/11, $p = 0.14$).

2.4. MiR-30b Is Required for IL-21-Mediated Angiogenesis In Vitro

We next altered miR-30b expression in HUVECs and assessed the effects of modulating miR-30b on cellular survival and tube formation. HUVECs transfected with miR-30b mimic had 909 ± 286-fold overexpression of miR-30b (Figure 5A, left), increased cell viability (OD450, 0.35 ± 0.01 vs. 0.27 ± 0.001, n = 8/group, $p < 0.001$, Figure 5A, right) tube formation (tube length, 3351 ± 235 vs. 2395 ± 207 μm/mm^2, n = 8/group, $p = 0.01$, Figure 5B) under HSS conditions. To investigate whether miR-30b is required for the survival effects of IL-21 treatment in hypoxic endothelial cells, HUVECs were transfected with miR-30b (hsa-miR-30b-5p) inhibitor, resulting in effective knockdown of miR-30b (Figure 5C, left). Compared with control-miR inhibitor, mir-30b inhibitor abrogated the ability of IL-21 to enhance cell viability (OD450: 0.21 ± 0.07 vs. 0.21 ± 0.005, n = 8/group, $p > 0.05$) (Figure 5C, right) and tube formation (Tube length: 2505 ± 111 vs. 2551 ± 101 μm/mm^2, n = 8/group, $p > 0.05$) (Figure 5D).

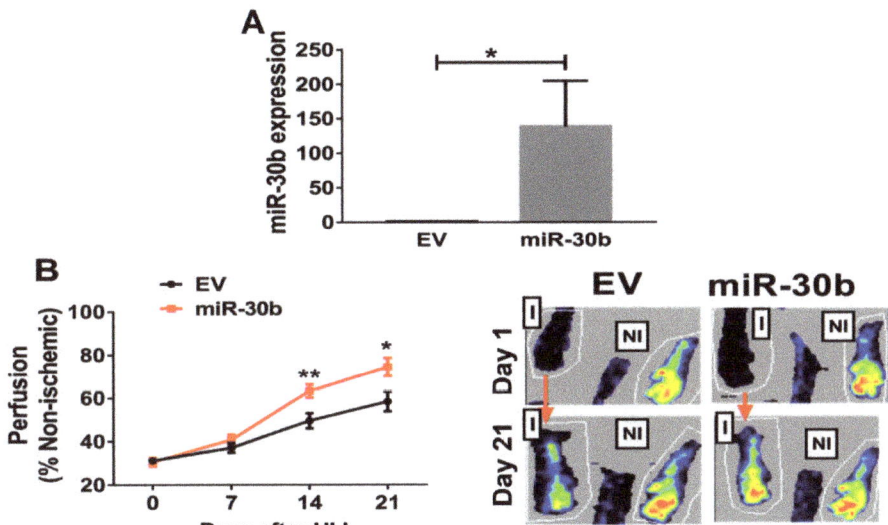

Figure 4. Augmentation of miR-30b expression in $Il21r^{-/-}$ mice improved perfusion recovery following HLI. (**A**) Higher miR-30b is expressed in hind limbs that received pCMV-miR-30b compared with those that received control plasmid (empty vector, EV), 10 days post treatment (n = 5/group, $p = 0.01$). (**B**) $Il21r^{-/-}$ mice in C57BL/6 background with miR-30b overexpression in the ischemic limb showed significantly better perfusion recovery at days 14 ($p < 0.01$) and 21 ($p = 0.016$) after HLI (n = 11–12 per group). EV indicates empty vector plasmid transfection; miR-30b indicates mice with miR-30b overexpression in the ischemic limb. Data represent mean ± SEM. I = ischemic limb, NI = non-ischemic limb, * $p < 0.05$, ** $p < 0.01$.

2.5. MiR-30b Increased STAT3 Phosphorylation and Reduced SOCS3 Expression

Suppressor of cytokine signaling 3 (SOCS3) was an important molecule that regulates STAT3 phosphorylation [32,33], and using Targetscan SOCS3 was a potential target of miR-30b in both human and mice. In mice following HLI, as shown in Figure 6A, we found that SOCS3 protein was lower in miR-30b overexpression vs. empty vector-treated mice at day 3 post HLI (SOCS3/ERK2, 0.49 ± 0.09 vs. 0.04 ± 0.02, n = 4, $p < 0.01$). Consistent with this finding, in HUVECs cultured under HSS conditions, miR-30b augmentation reduced SOCS3 protein expression (SOCS3/ERK2, 0.70 ± 0.05 vs. 0.45 ± 0.03, n = 4/group, $p = 0.01$, Figure 6B). As SOCS3 is a suppressor of STAT3, we also measured the level of STAT3 phosphorylation that is increased with miR-30b overexpression (p-STAT3/STAT3, 0.86 ± 0.08 vs. 0.28 ± 0.07, n = 4/group, $p = 0.001$, Figure 6B). These may suggest that in ischemic endothelial cells, miR-30b increases STAT3 activation by targeting SOCS3.

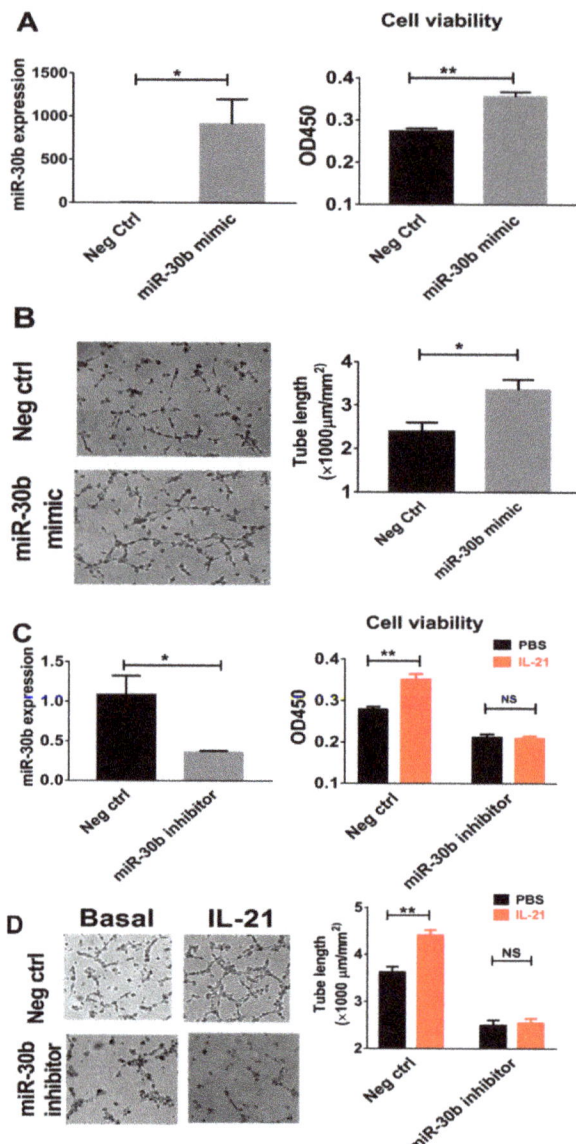

Figure 5. Effects of miR30b modulation in IL-21-mediated endothelial cell survival and angiogenesis under HSS conditions. (**A**) When miR-30b was overexpressed by miR30b mimic transfection in cultured HUVECs (left), more viable cells were detected 48 h after cultured under HSS conditions (right). (**B**) 48 h after transfection with miR-mimic negative control (neg ctrl) or miR30b mimic, HUVECs were plated in Matrigel with reduced growth factor and incubated for 6 h in basal medium. MiR-30b mimic–treated HUVECs showed enhanced tube formation, which was quantified as the tube length per area (bar graph). (**C,D**) IL-21 treatment (50 ng/mL) increased HUVEC cell viability and tube formation under HSS conditions, the IL-21 effects on HUVECs viability and tube formation were inhibited when miR30b was knocked down using miR30b inhibitor (C left). Cell viability assay is based on the cleavage of the tetrazolium salt to formazan by cellular mitochondrial dehydrogenase, OD450 indicates optical density at 450 nm. Neg ctrl = negative control RNA for microRNA inhibitor/mimic. Data represent mean ± SEM. All the above data are representative of two to three separate batches of HUVECs, n = 8 to 12 samples per group. * $p < 0.05$, ** $p < 0.01$, NS indicates non-significance.

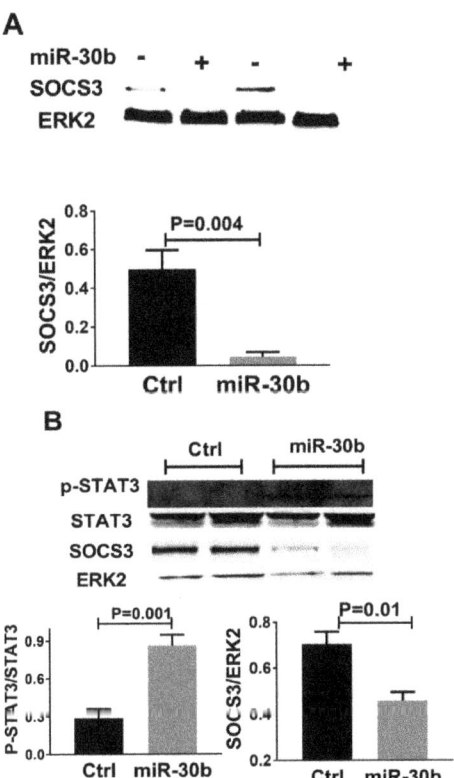

Figure 6. Protein changes in ischemic muscles with modulation of miR-30b in vivo in *Il21r$^{-/-}$* mice and in vitro human umbilical vein endothelial cells (HUVECs) cultured under hypoxia serum starvation (HSS) conditions. (**A**) Protein isolated from muscle harvested from *Il21r$^{-/-}$* mice with empty vector or the miR-30b overexpressing plasmid day-3 following HLI treated. Western blots were probed for SOCS3 and Erk1/2 with only ERK2 shown, SOCS3 expression level was significantly downregulated with miR-30b overexpression ($p < 0.01$). (**B**) Overexpression of miR-30b by miR-30b mimic in HUVECs cultured under HSS for 24 h resulted in increased STAT3 phosphorylation (p-STAT3/STAT3, $p = 0.001$) and reduced SOCS3 expression (SOCS3/ERK2, $p = 0.01$). Data represent mean ± SEM., miR-30b indicates miR-30 overexpression by pCMV-miR-30b plasmid in vivo or miR-30b mimic in vitro, Ctrl indicates negative control empty vector for miR-30b overexpression in vivo and negative control RNA for miR-30b mimic in vitro. n = 5/group.

3. Discussion

Having previously shown an unexpected role for the IL21/IL-21R pathway in mediating hypoxia-dependent angiogenesis [10], we now add several new findings regarding this unexpected pathway. First, we show that miR-30b is the only miR differentially regulated based on the presence or absence of IL-21/IL-21R-pathway activation in the hypoxia-dependent angiogenesis that occurs in experimental PAD. Second, the hypoxia dependent IL-21/IL-21R angiogenesis pathway utilizes and requires miR-30b for its context-dependent angiogenic effects. Third, the IL-21/IL-21R pathway following HLI and EC under HSS, involves activation of the STAT3 pathway, and miR-30b reduces SOCS3, which would be predicted to increase STAT3 phosphorylation. Finally, in the presence of IL-21R, IL-21-ligand overexpression improves perfusion recovery, induces therapeutic angiogenesis, and reduces tissue loss in experimental PAD. Together these finding suggest that the IL-21/IL-

21R pathway utilizes and appears dependent on the regulation on a single micro-RNA (Figure 7).

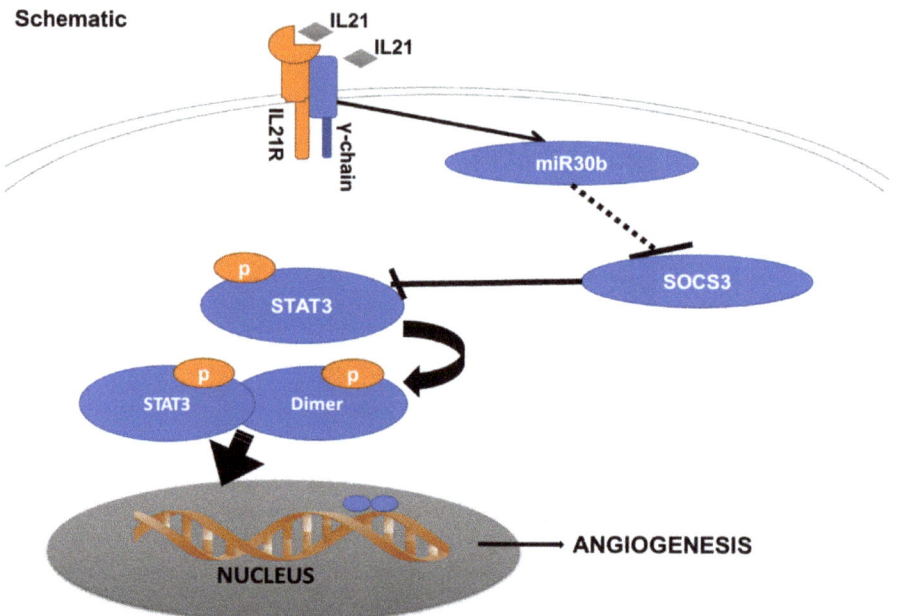

Figure 7. A schematic of proposed IL-21/IL-21R/SOCS3/STAT3 signal in ischemic endothelial cell in PAD conditions such as hypoxia serum starvation or in ischemic leg muscle with ligand (IL21) overexpression. Ligand (small green box) binds to the IL21R (red), which interacts with the common gamma chain (blue) on the cell surface. This complex activation upregulates miR-30b, leading to reduced SOCS3 expression, and results in increased STAT3 phosphorylation and greater amounts of angiogenesis.

To our knowledge, this is the first report connecting miR-30b to the IL-21/IL-21R pathway. Under experimental conditions relevant to PAD, we found that miR-30b was the only miR differentially regulated by the IL-21/IL-21R pathway in hypoxic angiogenesis. Our in vitro studies indicated that miR-30b was required for IL-21-induced angiogenesis in cultured endothelial cells and our in vivo studies indicate that miR-30b overexpression is sufficient to rescue impaired perfusion recovery after HLI in $Il21r^{-/-}$ mice. Some information is available on other miRs that are modulated by IL-21/IL-21R. Adoro et al. reported that miR-29 is required for IL-21-induced anti-viral effects in CD4+ T cells [33]; Rasmussen et al. found that miR-155a is required for IL-21-mediated STAT3 phosphorylation in CD4+ T cells from systemic lupus erythematosus (SLE) patients [34]; using microarray analysis, De Cecco et al. found IL-21 regulates CCL17, CD40, DDR1, and PIK3CD expression through miR-663b in chronic lymphocytic leukemia [35]. As we did not find differential expression of miR-29, 155a, or 663b, our data suggests an EC-specific effect of miR-30b on IL-21-induced angiogenesis. While one previous study indicated that miR-30b overexpression in endothelial cells increases vessel number and length in an in vitro sprouting angiogenesis model [36], another study demonstrated that depletion of miR-30b increased capillary-like cord formation in HUVECs under normoxia conditions [37]. These may suggest that the effects of miR-30b on angiogenesis are condition-specific.

It is interesting that our data suggest that a single miR may regulate hypoxia-dependent angiogenesis both in vitro and in vivo environments, in an unexpected biologic pathway that we previously showed involved the phosphorylation/activation of STAT3. A limited number of such examples could be found. For example, secreted from tumors, miR-9 has been shown to increase endothelial cell migration and angiogenesis via targeting SOCS5 and increasing STAT3 phosphorylation [38]. In addition, miR-337, miR-17, miR-20, miR-24 and miR-629 have been reported to regulate STAT3 activation [39,40].

A miR cluster is a set of two or more miRNAs, which are usually co-transcribed, yet we found that only miR-30b is upregulated by IL-21 in hypoxic endothelial cells, but miR-30d was not regulated. In mice, miR-30b is located within a non-coding region on chromosome 15 and is co-localized with miR-30d in the Mirc26 cluster [41], while in humans, miR-30b and miR-30d co-localized within non-coding RNA LOC102723694 on chromosome 8. Recent studies by others also showed that miRs from the same cluster could be regulated at different levels and even in divergent patterns. For example, Knudsen et al. reported that among four miRs in miR-17–92 cluster, miR-17 is the most upregulated in early colon cancer [42]; and another study using global analysis of miR clusters expression between breast tumor and adjacent tissue found that among miR-221/222 clusters, miR-221 is higher but miR-222 is lower in the tumor tissue [43]. The precise mechanisms by which this level of miR expression is controlled will require additional study.

Our data presented in this report is the first to show the therapeutic potential of modulating this IL21/IL21R hypoxia-dependent angiogenesis pathway in PAD. The specificity of the therapeutic effects of IL-21 overexpression on hypoxic angiogenesis and its receptor dependence was demonstrated by the absence of an effect of IL-21 on $Il21^{-/-}$ mice. We showed that blocking IL-21 signaling in hypoxic EC resulted in reduced miR-30b expression and IL-21R activation by ligand in ischemic muscle, resulting in increased miR-30b expression. Finally, miR-30b overexpression improves perfusion recovery in $Il21r^{-/-}$ mice. MiRs have great potential as therapeutic agents due to their ease of synthesis, stability, and ability of single miR to regulate several genes within a pathway. Modulation of a single miR has been reported to be therapeutic to PAD [44–47]. Taken together with our findings, these suggest that miR-30b is a potential therapeutic target for PAD subjects and could be an alternative to ligand administration.

Enhanced STAT3 activation is reported in ischemic tissue from a spectrum of ischemic diseases, including stroke and myocardial infarction, and functions as a protective factor to improve the recovery of these diseases [48,49]. In our previous study, IL-21R loss of function in a pre-clinical PAD model resulted in reduced angiogenesis and perfusion recovery through decreased STAT3 activation [10]. Conversely, this study demonstrated that after HLI, IL-21 overexpression increased perfusion recovery as well as STAT3 activation. Furthermore, we found that miR-30b, a molecule which is required for IL-21-induced angiogenesis under ischemia, also increases STAT3 activation with reduced SOCS3 expression. These data may suggest that miR-30b is in part or fully involved in IL-21-mediated STAT3 phosphorylation by targeting SOCS3 in hypoxia-dependent angiogenesis. In summary, we have demonstrated that a novel angiogenic IL-21/IL-21R/miR-30b/STAT3 promotes endothelial cell survival under ischemia in PAD.

4. Material and Methods

4.1. Hind-Limb Ischemia (HLI) Model, Perfusion Recovery, and In Vitro Transfection

Methods for the HLI model were performed as previously described [10,20,21]. Briefly, unilateral femoral-artery ligation and excision were performed on the left side of mice. Perfusion flow in the ischemic and contralateral non-ischemic limbs was measured on day 0, 7, 14, and 21 post-HLI with laser Doppler perfusion imaging system (Perimed, Inc., Ardmore, PA, USA), as described previously [20,21]. Perfusion was expressed as the ratio of the left (ischemic) to right (non-ischemic) hind limb. Limb necrosis was determined with ordinal values, as described [22].

For IL-21 overexpression, a full-length murine IL-21 cDNA was purchased from Open Biosystems (Huntsville, AL, USA). The cDNA was directionally cloned in the p-cytomegalovirus (CMV)-TnT (Promega Corp., Madison, WI, USA) plasmid that directs high expression from the CMV promoter upon transfection into mammalian cells. Resulting clones were sequenced to verify that there were no errors introduced during the PCR cloning reactions. Mice were grouped to receive pCMV-TnT plasmid vectors delivering IL-21 cDNA or a scrambled DNA sequence as a control. For mmu-miR-30b transfection, expression plasmid of mmu-miR-30b-5p (pCMV-miR-30b) and empty vector (EV) control were purchased from Origene (Rockville, MD, USA). Mice were grouped to receive pCMV-miR-30b plasmid vector or EV. Plasmids were transfected into mouse hind-limbs via electric, pulse-mediated gene transfer as described previously [21]. Mice were allowed to recover for 3 days before use in experimental PAD studies. Animal studies were approved by the Institutional Animal Care and Use Committee and conformed to the Guide for the Care and Use of Laboratory Animals published by the US National Institutes of Health.

4.2. Identification of Transcripts Regulated by IL-21R following HLI

Male C57BL/6J mice (14 to 18 weeks of age) were purchased from the Jackson Laboratory (Bar Harbor, ME) for IL-21R-Fc treatment. A mouse IL-21R Fc chimera (IL-21R-Fc) fusion protein was prepared in the Protein Expression Laboratory, National Cancer Institute and was used to neutralize IL-21 [23]. IL-21R-Fc were injected at a dose of 0.2 mg/mouse intra-peritoneally to C57BL/6J mice immediately (0) and 1, 3 and 5 days after surgical HLI (n = 4), as previously described [10], and an equivalent dose of mouse IgG (Sigma-Aldrich, St. Louis, MO, USA) was used in the control group (n = 4). Seven days after experimental PAD, mice were euthanized, and gastrocnemius muscles from the mice with IL-21R-Fc or control antibody treatment were harvested for RNA isolation and sequencing.

4.3. RNA Isolation and Sequencing

Total RNA was isolated from the ischemic gastrocnemius muscle (n = 4/group) using PureLink® RNA Kit (Life Technology, Grand Island, NY, USA), as previously described [10,19]. The quality of RNA was monitored by RNA gel electrophoresis following the manufacturer's instructions. RNA sequencing (RNA-Seq) was performed as described previously [24]. Briefly, RNA-seq libraries were constructed from 1 µg of total RNA using the TruSeq Stranded Total RNA prep kit (Illumina, San Diego, CA, USA). The resulting barcoded libraries were sequenced on an Illumina NextSeq 500 using 150-cycle High Output Kits (Illumina). Quality control of the resulting sequence data was performed using FastQC (http://www.bioinformatics.babraham.ac.uk/projects/fastqc (accessed on 28 November 2021)). All data were analyzed using Illumina BaseSpace applications. Sequencing reads were aligned to the mouse mm10 reference transcriptome and using the TopHat Alignment v1.0.0 application [25]. Expression values (FPKM; Fragments Per Kilobase of transcript per Million mapped reads) were then generated for each transcript and differentially expressed transcripts (at a false discovery rate (FDR, q-value) of <0.1) were identified using Cufflinks Assembly and Differential Expression v1.1.0 application.

4.4. Quantitative RT-PCR (qPCR)

To validate data from the RNA-seq, qPCR was performed using primer/probes from Applied Biosystems (Foster City, CA, USA), similar to methods described previously [10,26]. Small nucleolar RNA MBII-202 (sno202) transcript was used for normalization of miRs loading, and hypoxanthine-guanine phosphoribosyltransferase-1 (*Hprt-1*) was used for normalization of mRNA loading. The $2^{-\Delta\Delta Ct}$ method was used to calculate fold changes as described previously [27,28].

4.5. Protein Analysis

Levels of selected target proteins were analyzed by Western blotting using antibodies to total and phosphorylated (p)-signal transducer and activator of transcription (STAT) 1 (p-Y701), p-STAT3 (p-Y705), STAT3 (Cell Signaling, Danvers, MA, USA) and Suppressor of cytokine signaling 3 (SOCS3, Cat# 626601, Biolegend, San Diego, CA, USA), as previously described [29]. Western blots were analyzed by Odyssey Infrared Imaging System (LI-COR Biosciences, Lincoln, NE, USA) and quantified by ImageJ software (National Institute of Health, Bethesda, MD, USA).

4.6. In Vitro Transfection, Cellular Viability and Angiogenesis Assay

Pooled human umbilical vein endothelial cells (HUVEC) were purchased (Cell Applications Inc., San Diego, CA, USA). To mimic the endothelial cells under ischemic condition in HLI models, HUVECs were exposed to HSS conditions, as previously described. In vitro transfection was performed to overexpress/knockdown using miR-30b (hsa-miR-30b-5p) mimic or inhibitor, as previously described [18].

For cellular viability studies, HUVECs were plated in a 96-well plate at a density of 1×10^4 cells/well (n = 8/group), and then cultured under HSS conditions for 48 h. At the end of the incubation, cell viability was assessed using tetrazolium dye incorporation (BioVision, Milpitas, CA, USA). In vitro angiogenesis assay was performed as previously described [10,17–19]. Briefly, after exposure to HSS conditions for 24 h, transfected HUVECs were plated at a density of 1×10^4 cells/well on 96-well dishes which were coated with growth factor-reduced Matrigel (BD Biosciences, San Jose, CA, USA), and then exposed to HSS conditions for 6 h with rhIL-21 (50 ng/mL) or with vehicle alone to assess tube formation. Each condition was performed in 6 wells. The degree of tube formation was determined by measuring the length of the tubes and the number of loops from each well under 40× magnification using the online WimTube application module (Wimasis GmbH, Munich, Germany). Each experiment was repeated on at least 2 different batches of HUVECs.

4.7. Data Analysis and Statistics

Statistical analysis was performed with GraphPad Prism software. An unpaired *t* test was used for comparison between 2 groups, and comparisons in experiments with ≥3 groups were performed with one-way ANOVA and the Tukey post hoc test. Differences in necrosis rate after HLI were analyzed by chi-square test. Statistical significance was set at $p < 0.05$. Statistical methods of RNA-seq data are described in above in Section 4.3.

Supplementary Materials: The following supporting information can be downloaded at: https://www.mdpi.com/article/10.3390/ijms23010271/s1.

Author Contributions: T.W. designed the study, performed the experiments, analyzed the data, and wrote the manuscript; L.Y. and M.Y. participated in conducting the experiments; C.R.F. performed RNA-Sequencing and data analysis: R.S. and W.J.L. edited the manuscript; V.C.G. participated in designing the study and editing the manuscript; B.H.A. supervised the study and edited the manuscript. All authors have read and agreed to the published version of the manuscript.

Funding: This research was supported in part by the Division of Intramural Research, National Heart, Lung, and Blood Institute, NIH. This work was supported by R01 HL121635, R01HL141325, R01HL148590, R01HL101200, R01GM129074 and R01 (to B.H.A.); 7R01HL146673-02 (to V.C.G.); Natural Science Foundation of China No. 81700426 and No. 81970046 to T.W.

Institutional Review Board Statement: All animal studies were approved by the University of Virginia Institutional Review Board.

Informed Consent Statement: Not applicable.

Data Availability Statement: Data will be made available upon reasonable request to the corresponding author.

Acknowledgments: All authors acknowledge and consent to the study.

Conflicts of Interest: The authors declare no conflict of interest.

Disclosure: W.J.L. and R.S. are inventors on NIH patents related to IL-21.

References

1. Annex, B.H. Therapeutic angiogenesis for critical limb ischaemia. *Nat. Rev. Cardiol.* **2013**, *10*, 387–396. [CrossRef]
2. Fowkes, F.G.; Rudan, D.; Rudan, I.; Aboyans, V.; Denenberg, J.O.; McDermott, M.M.; Norman, P.E.; Sampson, U.K.; Williams, L.J.; Mensah, G.A.; et al. Comparison of global estimates of prevalence and risk factors for peripheral artery disease in 2000 and 2010: A systematic review and analysis. *Lancet* **2013**, *382*, 1329–1340. [CrossRef]

3. Criqui, M.H.; Aboyans, V. Epidemiology of peripheral artery disease. *Circ. Res.* **2015**, *116*, 1509–1526. [CrossRef] [PubMed]
4. Fowkes, F.G.; Aboyans, V.; Fowkes, F.J.; McDermott, M.M.; Sampson, U.K.; Criqui, M.H. Peripheral artery disease: Epidemiology and global perspectives. *Nat. Rev. Cardiol.* **2017**, *14*, 156–170. [CrossRef]
5. Tanaka, M.; Taketomi, K.; Yonemitsu, Y. Therapeutic angiogenesis: Recent and future prospects of gene therapy in peripheral artery disease. *Curr. Gene. Ther.* **2014**, *14*, 300–308. [CrossRef] [PubMed]
6. Troidl, K.; Schaper, W. Arteriogenesis versus angiogenesis in peripheral artery disease. *Diabetes Metab. Res. Rev.* **2012**, *28* (Suppl. S1), 27–29. [CrossRef]
7. Niiyama, H.; Huang, N.F.; Rollins, M.D.; Cooke, J.P. Murine model of hindlimb ischemia. *J. Vis. Exp.* **2009**, *23*, 1035. [CrossRef]
8. Dokun, A.O.; Keum, S.; Hazarika, S.; Li, Y.; Lamonte, G.M.; Wheeler, F.; Marchuk, D.A.; Annex, B.H. A quantitative trait locus (LSq-1) on mouse chromosome 7 is linked to the absence of tissue loss after surgical hindlimb ischemia. *Circulation* **2008**, *117*, 1207–1215. [CrossRef]
9. Limbourg, A.; Korff, T.; Napp, L.C.; Schaper, W.; Drexler, H.; Limbourg, F.P. Evaluation of postnatal arteriogenesis and angiogenesis in a mouse model of hind-limb ischemia. *Nat. Protoc.* **2009**, *4*, 1737–1746. [CrossRef]
10. Wang, T.; Cunningham, A.; Dokun, A.O.; Hazarika, S.; Houston, K.; Chen, L.; Lye, R.J.; Spolski, R.; Leonard, W.J.; Annex, B.H. Loss of interleukin-21 receptor activation in hypoxic endothelial cells impairs perfusion recovery after hindlimb ischemia. *Arter. Thromb. Vasc. Biol.* **2015**, *35*, 1218–1225. [CrossRef]
11. Wang, T.; Cunningham, A.; Houston, K.; Sharma, A.M.; Chen, L.; Dokun, A.O.; Lye, R.J.; Spolski, R.; Leonard, W.J.; Annex, B.H. Endothelial interleukin-21 receptor up-regulation in peripheral artery disease. *Vasc. Med.* **2016**, *21*, 99–104. [CrossRef] [PubMed]
12. Spolski, R.; Leonard, W.J. Interleukin-21: A double-edged sword with therapeutic potential. *Nat. Rev. Drug Discov.* **2014**, *13*, 379–395. [CrossRef]
13. Spolski, R.; Leonard, W.J. Interleukin-21: Basic biology and implications for cancer and autoimmunity. *Annu. Rev. Immunol.* **2008**, *26*, 57–79. [CrossRef] [PubMed]
14. Booty, M.G.; Barreira-Silva, P.; Carpenter, S.M.; Nunes-Alves, C.; Jacques, M.K.; Stowell, B.L.; Jayaraman, P.; Beamer, G.; Behar, S.M. IL-21 signaling is essential for optimal host resistance against Mycobacterium tuberculosis infection. *Sci. Rep.* **2016**, *6*, 36720. [CrossRef] [PubMed]
15. Leonard, W.J.; Wan, C.K. IL-21 Signaling in Immunity. *F1000Research* **2016**, *5*. [CrossRef] [PubMed]
16. Shi, X.; Que, R.; Liu, B.; Li, M.; Cai, J.; Shou, D.; Wen, L.; Liu, D.; Chen, L.; Liang, T.; et al. Role of IL-21 signaling pathway in transplant-related biology. *Transpl. Rev.* **2016**, *30*, 27–30. [CrossRef]
17. Ganta, V.C.; Choi, M.; Kutateladze, A.; Annex, B.H. VEGF165b Modulates Endothelial VEGFR1-STAT3 Signaling Pathway and Angiogenesis in Human and Experimental Peripheral Arterial Disease. *Circ. Res.* **2017**, *120*, 282–295. [CrossRef]
18. Hazarika, S.; Farber, C.R.; Dokun, A.O.; Pitsillides, A.N.; Wang, T.; Lye, R.J.; Annex, B.H. MicroRNA-93 controls perfusion recovery after hindlimb ischemia by modulating expression of multiple genes in the cell cycle pathway. *Circulation* **2013**, *127*, 1818–1828. [CrossRef]
19. Ganta, V.C.; Choi, M.H.; Kutateladze, A.; Fox, T.E.; Farber, C.R.; Annex, B.H. A MicroRNA93-IRF9-IRG1-Itaconic Acid Pathway Modulates M2-like-Macrophage Polarization to Revascularize Ischemic Muscle. *Circulation* **2017**, *135*, 2403. [CrossRef] [PubMed]
20. Dokun, A.O.; Chen, L.; Lanjewar, S.S.; Lye, R.J.; Annex, B.H. Glycaemic control improves perfusion recovery and VEGFR2 protein expression in diabetic mice following experimental PAD. *Cardiovasc. Res.* **2014**, *101*, 364–372. [CrossRef]
21. Dokun, A.O.; Chen, L.; Okutsu, M.; Farber, C.R.; Hazarika, S.; Jones, W.S.; Craig, D.; Marchuk, D.A.; Lye, R.J.; Shah, S.H.; et al. ADAM12: A genetic modifier of preclinical peripheral arterial disease. *Am. J. Physiol. Heart Circ. Physiol.* **2015**, *309*, H790–H803. [CrossRef]
22. Li, Y.J.; Hazarika, S.; Xie, D.H.; Pippen, A.M.; Kontos, C.D.; Annex, B.H. In mice with type 2 diabetes, a vascular endothelial growth factor (VEGF)-activating transcription factor modulates VEGF signaling and induces therapeutic angiogenesis after hindlimb ischemia. *Diabetes* **2007**, *56*, 656–665. [CrossRef]
23. Spolski, R.; Wang, L.; Wang, C.K.; Bonville, C.; Domachowske, J.; Kim, H.P.; Yu, Z.X.; Leonard, W. IL-21 promotes the pathologic immune response to Pneumovirus infection. *J. Immunol.* **2012**, *188*, 1924–1932. [CrossRef]
24. Calabrese, G.; Bennett, B.J.; Orozco, L.; Kang, H.M.; Eskin, E.; Dombret, C.; De Backer, O.; Lusis, A.J.; Farber, C.R. Systems Genetic Analysis of Osteoblast-Lineage Cells. *PLoS Genet.* **2012**, *8*, e1003150. [CrossRef]
25. Kim, D.; Pertea, G.; Trapnell, C.; Pimentel, H.; Kelley, R.; Salzberg, S.L. TopHat2: Accurate alignment of transcriptomes in the presence of insertions, deletions and gene fusions. *Genome Biol.* **2013**, *14*, R36. [CrossRef]
26. Wang, T.; Cunningham, A.; Dokun, A.; Hazarika, S.; Chen, L.; Lye, R.; Leonard, W.; Annex, B. Angiogenic Properties of Interleukin-21 under Hypoxic Conditions. *Blood* **2014**. e-Letter.
27. Wang, T.; Satoh, F.; Morimoto, R.; Nakamura, Y.; Sasano, H.; Auchus, R.J.; Edwards, M.A.; Rainey, W.E. Gene expression profiles in aldosterone-producing adenomas and adjacent adrenal glands. *Eur. J. Endocrinol.* **2011**, *164*, 613–619. [CrossRef]
28. Wang, T.; Rowland, J.G.; Parmar, J.; Nesterova, M.; Seki, T.; Rainey, W.E. Comparison of aldosterone production among human adrenocortical cell lines. *Horm. Metab. Res.* **2012**, *44*, 245–250. [CrossRef]
29. Hazarika, S.; Dokun, A.O.; Li, Y.; Popel, A.S.; Kontos, C.D.; Annex, B.H. Impaired angiogenesis after Hindlimb ischemia in type 2 diabetes Mellitus—Differential regulation of vascular endothelial growth factor receptor 1 and soluble vascular endothelial growth factor receptor 1. *Circ. Res.* **2007**, *101*, 948–956. [CrossRef]

30. Hazarika, S.; Angelo, M.; Li, Y.J.; Aldrich, A.J.; Odronic, S.I.; Yan, Z.; Stamler, J.S.; Annex, B.H. Myocyte Specific Overexpression of Myoglobin Impairs Angiogenesis After Hind-Limb Ischemia. *Arterioscl. Throm. Vas.* **2008**, *28*, 2144–2175. [CrossRef]
31. Meisner, J.K.; Song, J.; Annex, B.H.; Price, R.J. Myoglobin Overexpression Inhibits Reperfusion in the Ischemic Mouse Hindlimb through Impaired Angiogenesis but Not Arteriogenesis. *Am. J. Pathol.* **2013**, *183*, 1710–1718. [CrossRef]
32. Yuan, J.; Zhang, F.; Niu, R. Multiple regulation pathways and pivotal biological functions of STAT3 in cancer. *Sci. Rep.* **2015**, *5*, 17663. [CrossRef] [PubMed]
33. Adoro, S.; Cubillos-Ruiz, J.R.; Chen, X.; Deruaz, M.; Vrbanac, V.D.; Song, M.; Park, S.; Murooka, T.T.; Dudek, T.E.; Luster, A.D.; et al. IL-21 induces antiviral microRNA-29 in CD4 T cells to limit HIV-1 infection. *Nat. Commun.* **2015**, *6*, 7562. [CrossRef]
34. Rasmussen, T.K.; Andersen, T.; Bak, R.O.; Yiu, G.; Sorensen, C.M.; Stengaard-Pedersen, K.; Mikkelsen, J.G.; Utz, P.J.; Holm, C.K.; Deleuran, B. Overexpression of microRNA-155 increases IL-21 mediated STAT3 signaling and IL-21 production in systemic lupus erythematosus. *Arthritis. Res. Ther.* **2015**, *17*, 154. [CrossRef]
35. De Cecco, L.; Capaia, M.; Zupo, S.; Cutrona, G.; Matis, S.; Brizzolara, A.; Orengo, A.M.; Croce, M.; Marchesi, E.; Ferrarini, M.; et al. Interleukin 21 Controls mRNA and MicroRNA Expression in CD40-Activated Chronic Lymphocytic Leukemia Cells. *PLoS ONE* **2015**, *10*, e0134706. [CrossRef]
36. Bridge, G.; Monteiro, R.; Henderson, S.; Emuss, V.; Lagos, D.; Georgopoulou, D.; Patient, R.; Boshoff, C. The microRNA-30 family targets DLL4 to modulate endothelial cell behavior during angiogenesis. *Blood* **2012**, *120*, 5063–5072. [CrossRef]
37. Howe, G.A.; Kazda, K.; Addison, C.L. MicroRNA-30b controls endothelial cell capillary morphogenesis through regulation of transforming growth factor beta 2. *PLoS ONE* **2017**, *12*, e0185619. [CrossRef]
38. Zhuang, G.; Wu, X.; Jiang, Z.; Kasman, I.; Yao, J.; Guan, Y.; Oeh, J.; Modrusan, Z.; Bais, C.; Sampath, D.; et al. Tumour-secreted miR-9 promotes endothelial cell migration and angiogenesis by activating the JAK-STAT pathway. *EMBO J.* **2012**, *31*, 3513–3523. [CrossRef]
39. Yu, H.; Lee, H.; Herrmann, A.; Buettner, R.; Jove, R. Revisiting STAT3 signalling in cancer: New and unexpected biological functions. *Nat. Rev. Cancer* **2014**, *14*, 736–746. [CrossRef]
40. Zhang, M.; Liu, Q.; Mi, S.; Liang, X.; Zhang, Z.; Su, X.; Liu, J.; Chen, Y.; Wang, M.; Zhang, Y.; et al. Both miR-17-5p and miR-20a alleviate suppressive potential of myeloid-derived suppressor cells by modulating STAT3 expression. *J. Immunol.* **2011**, *186*, 4716–4724. [CrossRef]
41. Gaziel-Sovran, A.; Segura, M.F.; Di Micco, R.; Collins, M.K.; Hanniford, D.; Vega-Saenz de Miera, E.; Rakus, J.F.; Dankert, J.F.; Shang, S.; Kerbel, R.S.; et al. miR-30b/30d regulation of GalNAc transferases enhances invasion and immunosuppression during metastasis. *Cancer Cell* **2011**, *20*, 104–118. [CrossRef] [PubMed]
42. Knudsen, K.N.; Nielsen, B.S.; Lindebjerg, J.; Hansen, T.F.; Holst, R.; Sorensen, F.B. microRNA-17 Is the Most Up-Regulated Member of the miR-17-92 Cluster during Early Colon Cancer Evolution. *PLoS ONE* **2015**, *10*, e0140503. [CrossRef] [PubMed]
43. Guo, L.; Yang, S.; Zhao, Y.; Zhang, H.; Wu, Q.; Chen, F. Global analysis of miRNA gene clusters and gene families reveals dynamic and coordinated expression. *BioMed Res. Int.* **2014**, *2014*, 782490. [CrossRef] [PubMed]
44. Christopher, A.F.; Kaur, R.P.; Kaur, G.; Kaur, A.; Gupta, V.; Bansal, P. MicroRNA therapeutics: Discovering novel targets and developing specific therapy. *Perspect Clin. Res.* **2016**, *7*, 68–74. [CrossRef] [PubMed]
45. Gibson, N.W. Engineered microRNA therapeutics. *J. R. Coll. Physicians Edinb.* **2014**, *44*, 196–200. [CrossRef]
46. Barger, J.F.; Nana-Sinkam, S.P. MicroRNA as tools and therapeutics in lung cancer. *Respir. Med.* **2015**, *109*, 803–812. [CrossRef]
47. Lei, Z.; Sluijter, J.P.; van Mil, A. MicroRNA Therapeutics for Cardiac Regeneration. *Mini Rev. Med. Chem.* **2015**, *15*, 441–451. [CrossRef]
48. Krishnamurthy, P.; Rajasingh, J.; Lambers, E.; Qin, G.; Losordo, D.W.; Kishore, R. IL-10 inhibits inflammation and attenuates left ventricular remodeling after myocardial infarction via activation of STAT3 and suppression of HuR. *Circ. Res.* **2009**, *104*, e9–e18. [CrossRef] [PubMed]
49. Hoffmann, C.J.; Harms, U.; Rex, A.; Szulzewsky, F.; Wolf, S.A.; Grittner, U.; Lattig-Tunnemann, G.; Sendtner, M.; Kettenmann, H.; Dirnagl, U.; et al. Vascular Signal Transducer and Activator of Transcription-3 Promotes Angiogenesis and Neuroplasticity Long-Term after Stroke. *Circulation* **2015**, *131*, 1772–1782. [CrossRef]

MDPI AG
Grosspeteranlage 5
4052 Basel
Switzerland
Tel.: +41 61 683 77 34

International Journal of Molecular Sciences Editorial Office
E-mail: ijms@mdpi.com
www.mdpi.com/journal/ijms

Disclaimer/Publisher's Note: The title and front matter of this reprint are at the discretion of the . The publisher is not responsible for their content or any associated concerns. The statements, opinions and data contained in all individual articles are solely those of the individual Editors and contributors and not of MDPI. MDPI disclaims responsibility for any injury to people or property resulting from any ideas, methods, instructions or products referred to in the content.

www.ingramcontent.com/pod-product-compliance
Lightning Source LLC
LaVergne TN
LVHW070637100526
838202LV00012B/824